Gen Z and the Future of Food

Gen Z and the Future of Food

Diana Bogueva | Dora Marinova

Curtin University, Australia

World Scientific

NEW JERSEY • LONDON • SINGAPORE • BEIJING • SHANGHAI • HONG KONG • TAIPEI • CHENNAI • TOKYO

Published by

World Scientific Publishing Co. Pte. Ltd.
5 Toh Tuck Link, Singapore 596224
USA office: 27 Warren Street, Suite 401-402, Hackensack, NJ 07601
UK office: 57 Shelton Street, Covent Garden, London WC2H 9HE

British Library Cataloguing-in-Publication Data
A catalogue record for this book is available from the British Library.

GEN Z AND THE FUTURE OF FOOD

ISBN 978-981-98-2261-4 (hardcover)
ISBN 978-981-98-2262-1 (ebook for institutions)
ISBN 978-981-98-2263-8 (ebook for individuals)

For any available supplementary material, please visit
https://www.worldscientific.com/worldscibooks/10.1142/14564#t=suppl

Desk Editor: Yipeng Hu

Typeset by Stallion Press
Email: enquiries@stallionpress.com

To all Gen Z and other generations who help regenerating life on Earth.

Foreword: Food and the Sorcerer's Apprentice

The past 75 years have seen the biggest shift for food in human history for 10,000 years. We have moved from an agriculture-based culture to a city civilisation, losing touch with the source of food and nature. Most of our families are migrants, from the land to the cities.

For generations before 1950, food was a treasure, much respected. It was never given for granted. It was scarce, much work, and a daily chore.

For baby-boomers like me, food was a pleasure. Food became a given, it was ubiquitous, easy. It was the fuel that enabled unlimited growth visions. Chemicals were necessary and undisputed, fertilisers and pesticides for production, additives for processing, transportation and distribution

For Generation X, food became a serious health issue, a danger rather than a pleasure, scary and difficult to control. Food became global, erasing local, whenever and wherever possible. It was still an individual trip for everyone.

For Generation Y, peer pressure started. Around 11 or 12, most were confronted with anorexia or other eating disorders. Most parents were unaware, teachers were unprepared, it grew and grew. Misinformation ruled.

For Generation Z, social media are far more powerful. Generation Z is much more aware and informed about food issues. They have no doubts about its key role, and in many ways are more comfortable than previous generations in confronting the issues.

The vegan movement became mainstream, though remaining a small minority of the total population. Diversity is accepted, and even championed.

Food is now a key component of personal, regional and national identity. It is a daily commitment for Generation Z. They are more comfortable and knowledgeable than previous generations, giving an optimistic view for all other generations. Maybe they have mastered social media, with a positive outcome.

In a world that seems to become more and more complex, this book gives much needed hope and courage for the future, I am deeply grateful to the authors.

Very big thanks to my children, who taught me so much, Maya from Generation X, Edouard and Florence from Generation Y, and especially Jade from Generation Z.

Edouard Cointreau
President and founder
Gourmand Awards
July 2, 2025

Author Bios

Diana Bogueva is a social scientist and Research Fellow at the Curtin University Sustainability Policy Institute in Australia, where she also teaches in the Master of Environment and Climate Emergency and Built Environment programs. She holds a PhD from Curtin University and specialises in sustainable food consumption, generational consumer behaviour, food and masculinity, alternative proteins, novel food technologies and food system harmonisation.

Her research aligns with the global transition toward sustainable, ethical and resilient food systems, emphasising consumer-driven innovation and interdisciplinary collaboration. Dr Bogueva has received several international awards, including the 24th and 28th Gourmand Awards for her books Environmental, Health and Business Opportunities in the New Meat Alternatives Market (2019) and Food in a Planetary Emergency (2022). Her recent co-edited volume Nutrition Science, Marketing Nutrition, Health Claims and Public Policy (2023) was a finalist in the prestigious PROSE Awards. Eight of her papers have been ranked in the top 5% in the world for media attention.

Diana is the President of the Global Harmonisation Initiative (GHI), an international organisation with over 1,500 scientific members across 72 countries. In this role, she leads global efforts to align food standards, strengthen science communication and elevate consumer perspectives in food policy. Her work combines academic excellence with a commitment to social impact and planetary health.

Dora Marinova is Professor of Sustainability at the Curtin University Sustainability Policy (CUSP) Institute, Australia which she established in 2007 and where she was Director between 2015 and 2018 and Deputy Director between 2007 and 2015. She is also a Professor at the Institute for Economics and Politics at the University of National and World Economy, Sofia, Bulgaria where she commenced her academic career. Her areas of research interest cover sustainability, innovation and sustainometrics.

Dora has more than 400 refereed publications and has supervised 85 PhD students to successful completion. In 2023, the book *Food in a Planetary Emergency* which she co-authored with Diana Bogueva received "Best in the World" in the Future of Food category of the prestigious 28th Gourmand Award. Ten of her papers are rated by Altmetric in the top 5% in the world for the social media attention they have received. Since 2013, Dora regularly serves as a member of different peer review panels of Australia's National Health and Medical Research Council (NHMRC). She is an Elected Member of the Australian Institute of Aboriginal and Torres Strait Islander Studies (AIATSIS) and Elected Fellow of the Modelling and Simulation Society of Australia and New Zealand (MSSANZ). For the period 1975–2023, Scopus lists her as Australia's top author and 21st in the world in the area of sustainability. According to ScholarGPS, Dora is in the top 0.5% scholars worldwide for 2024.

Contents

List of Tables

List of Figures

Chapter 1

Gen Z: The Generation Shaping Food's Future

Generation Z or Gen Z, a generation born between 1995 and 2010, is not merely going to witness the future of food, it is actively creating it. As it comes of age, completes its education, enters the workforce and some even start raising families of their own, this generation is shaping how food is produced, marketed, consumed and understood. Gen Z is not defined just by its food preferences, what the young people eat and what they like or dislike about food — although these are important, but by how its members engage with the world around them. They are also defined not solely by their traits, e.g. being tech-savvy, globally connected, sustainability oriented and sensitive to social injustice, but by how these socially conscious young people are deeply aware of their impact on people and the planet. These traits however only scratch the surface. What truly sets Gen Z apart is their way of being — how they navigate contradictions, question norms and reimagine possibilities. They are rewriting the rules of engagement — not just with food, but with identity, culture, technology and society at large, and the systems that govern our lives. This realisation emerged for us after numerous deep and meaningful, candid and insightful conversations with these young people, where they spoke openly about their experiences, challenges and perspectives. Their opinions painted a picture of a generation both idealistic and pragmatic, anxious yet ambitious, sceptical yet hopeful and deeply connected, informed and curious.

This book is a way to share Gen Z's voices with a larger audience. We were privileged to spend time with Gen Zers from across the world, from different cultures and geographies, although the majority of the conversations were held in Australia. Armed with a human research ethics approval from Australia's Curtin University and without being able to secure formal research funding, we embarked on a two-year long journey to listen — really listen — to what Gen Z thinks, feels, envisions and has to say about food. These conversations happened in a variety of places: across camping sites, in public transport on buses and trains, around dinner tables, coffee shops and in park areas as well as in more formal settings in university campuses. Talking with Gen Z about food was effortless. These young people were passionate, enthusiastic, informed, knowledgeable, with firm opinions and eager to share — as if they instinctively knew that food is not just a necessity or a trend, but a canvas for change. They knew that not only the future belongs to them, but the future of food too. We love Gen Zers, we are very thankful for their insights and allowing us to share them in this book. Our faith in the future of food is in good hands.

It is important not to make the mistake of seeing Gen Z as monolith. It is incorrect to assume that all Gen Zers around the world are the same or that they have similar aspirations. Their experiences and ambitions differ across contexts, cultures and continents. There are however some powerful global trends that have shaped this generation uniquely irrespective as to which place they call home. The smart phones revolution, the explosion of social media, climate change and rising climate anxiety, the COVID-19 pandemic, economic insecurity and the mental health crisis — these are some uniting experiences for Gen Z for better or for worse, and Gen Z seems to know a lot about them. These are not just background conditions — they are formative forces. Gen Z is the first generation to grow up with instantaneous access to information and algorithm-curated realities. If they do not know something, they know how to find out faster and often more critically than their parents and any previous generations.

While Gen Z was maturing, many changes were happening in the food arena. This book covers and explores some of the shifts Gen Z is engaging with — from personalised nutrition to lab-grown meat to zero-waste packaging, precision fermentation and food transparency powered by blockchain. These innovations do not merely reflect technological advances, they echo Gen Z's values.

All of this, however, is unfolding against a sobering background. While we conducted our research interviews with Gen Z, the planet's climate experienced its hottest year on record in 2024, with temperature soaring of 1.55^{o}C above pre-industrial levels and 2015–2024 being the warmest decade ever recorded (WMO, 2025). Our planet's biodiversity continued its rapid collapse and being seriously threatened with a catastrophic decline of 73% of wildlife populations over the last 50 years (WWF, 2024). Issues of mental health, especially of young people, worsened and socio-economic inequality raised higher than ever before (Haidt, 2024). These are not just statistics. All of these issues are related to our food choices, interconnected with how we produce, distribute and consume food (Marinova and Bogueva, 2022), but this book is not about them. We do not want the reader to feel overwhelmed or powerless as this is not how Gen Z feels. Gen Z themselves reject despair. Our conversations portrayed these young people as cautiously optimistic, and this is how we want you to also see them. What we found was a generation navigating complexity with cautious optimism, seeking solutions, not just pointing out problems. They are aware of the stakes, but they are also building communities of care, creativity and resistance.

Let us start this journey of listening to Gen Z's words supported by academic literature from around the world, global case studies and cross-disciplinary insights. What matters to Gen Z, what are their food styles, are they embracing green eating? How are they redefining what it means to eat well, eat ethically or eat together? Why is transparency so important to them, why are they mainly sober at parties and why they opt for plant-based milks? Are they leaning into sustainability or rethinking it altogether? These are all topics covered in the book. Gen Z will redefine and shape the food landscape of 2050 when they will be in their 40s and 50s and would have passed the baton to Generation Alpha. For now, however, let us stay in the moment with this remarkable generation of Gen Z and listen to what they have to say. Their vision might just be the key to a more just, joyful and resilient food future.

Who Is Gen Z

Raised in an environment of overprotection and caution with parents trying to shield them from the world's many dangers, Gen Z has spent too much time on their electronic devices. *It's the computer all the time... We used to kick a ball around the neighbourhood at their age* (Female, Gen X, born 1974) — a Gen Z mum shared. That same mother worryingly described that parents are afraid to let their children play out because the neighbourhood is not the same anymore, and the world is a scary place.

Gen Z is a cohort shaped by globalised Internet networks, visual and interactive communication, and the relentless rise of technology, with social media becoming their playgrounds — something like vibrant ecosystems that replace the backyards and neighbourhood streets of earlier generations. These are the spaces where they connect, express themselves and mobilise, fostering relationships that transcend geographic boundaries. Through our conversations with Gen Z, it became clear that their immersion in a hyper-connected and complex world has fostered an early maturity, one marked by a profound sense of responsibility but also weighed down by the burden of high expectations.

Raised mostly by Generation X, Gen Z are realists. They understand how scary the world can be, having grown up in times of terrorism, suffering the consequences of a global recession and increasing school violence as well as experiencing a global pandemic. What truly stands out is the transformative power of their collective influence and early maturity, born from constant exposure to a world struggling with pressing issues like climate change and sustainability, inequality and mental health (Bogueva *et al.*, 2024). This is also reflected in the way they view food.

As one Gen Z shared with us, everything is:

> *"...so new and different, and it feels so much stranger than what our parents, their parents, and even Millennials before us experienced. For us Gen Zs, it seems like the world has gone mad.*

> *Everything happening around us affects our mental health, and this chaos is mirrored in the food we are expected to consume — crickets, bugs, artificial meat, cultured meat, algae. It is as if the food industry and others with vested interests are pushing these alternatives on us, without fully considering how they impact our sense of normality or well-being."* (Male, born 1999)

Gen Z is practical but incredibly social, communicating with more people than any generation before them. Where their Generation X parents might divide social circles into just two groups — close friends and casual contacts, Gen Z's connections are far more layered and varied. Their relationships span so many levels of closeness and interaction that they could rival the number of words for "snow" in the Eskimo languages. Just as those languages describe snow in numerous ways, Gen Z manages a wide range of relationships, each with its own meaning and role. Relationships start in direct messages in Instagram, Twitter/X, Snapchat, discord servers, TikTok comments, gaming communities, identity exploration, situationships and other. Gen Z does not assume one "default" relationship path. This ability to navigate such a complex web of connections shows how well they have adapted to living in a world that is more connected than ever.

Generation Z's intrinsic connection to the global heartbeat allows them to steer and reshape cultural norms, often in ways that feel both innovative and profound. Their impact, we came to realise, is not a question of size but of their ability to inspire change and redefine how we interact with one another and the world. They are not just shaping their personal choices but driving a broader transformation in how food is produced, marketed and consumed. They are not just participants in this change. It will not be too ambitious to say that they are its architects. Through their voices and actions, they embody the promise of a generation poised to leave an indelible mark on the future.

Why Gen Z Holds the Key to the Future of Food

Imagine a generation born into a world saturated with digital noise, where navigating social norms feels like decoding an invisible rulebook. Gen Z — those born loosely between 1995 and 2010 as other timeframes (e.g. 1997–2012) are also suggested — comprise nearly one-third of the global population. They are characterised by their interconnectedness, tech-savviness and deeply ingrained awareness of global issues. As the most ethnically diverse and digitally immersed cohort in human history, their influence extends far beyond TikTok trends and Instagram aesthetics. Gen Z actively shares its views widely and openly, leveraging digital platforms to express opinions and influence conversations (Marinova and Bogueva, 2022).

This is the generation rewriting the rules of engagement with food, often in ways that are as intriguing as they are transformative. For Gen Z, food is not merely sustenance. It is a language, a statement, embracement of diversity and ethnicity and sometimes a coping mechanism. Gen Z's food requirements, expectations and choices reflect its values, whether it is the push for plant-based options, ethical sourcing or reduction of food waste. However, from our conversations with young representatives of this population cohort, we have come to understand that their relationship with food is deeply ambiguous — marked by a mix of comfort, rebellion, trust and necessity.

Why does it matter what Gen Z thinks about food and how it experiences the new developments in this area? The answer is simple and straightforward: because Gen Z is driving some of the most significant shifts in the global food landscape. This generation's demands for sustainability, transparency and innovation are disrupting established systems, compelling industries to evolve or risk becoming irrelevant. Gen Z's profound influence extends far beyond its population numbers, as this youth cohort challenges and pushes industries and businesses to innovate, adapt and rethink the future. However, these demands stem from a context that remains largely misunderstood by outsiders. To better understand Gen Z, we tried to penetrate their world by engaging directly and talking to them and putting together insights from studies across the globe that have specifically examined this population cohort. There has been a lot of interest in Gen Z and we are not alone in trying to understand what motivates this youth, what the driving force behind their actions is, why they behave in a certain way, whether their choices are conscious or unconscious, in other words, what makes them tick.

We do this through the prism of food. As George Bernard Show once wrote: "There is no love sincerer than the love of food" (Oxford Reference, 2025). We all share this love but we do this very differently. In a constantly changing biophysical environment, there is ample evidence that many of us in the West need to shift our food preferences and we have discussed this in our earlier research (e.g. Marinova and Bogueva, 2022). Food production and the global food systems are also changing offering some exciting opportunities but also options which make some frown or just be uncertain. What are Gen Z's attitudes and reactions?

In this chapter, we explore the unspoken rules shaping Gen Z's food choices, the digital pressures that make their lives uniquely complex, and the ways they band together to navigate this complex maze. We argue that understanding this generation is not just important. It is essential for anyone, hoping to predict or influence the future of food.

Starting with Breakfast

The future of food is intricately linked to small daily rituals that carry big implications. Breakfast, often an overlooked meal in a busy daily routine, emerges as one of the most intentional and personal moments in the everyday life of Gen Z. Unlike previous generations, Gen Z views breakfast as a gateway to health, personalisation and convenience (Young, 2022). It is no longer just about fuelling the body, breakfast has become a cultural touchstone, a space where questions of well-being, identity and trust in innovation converge. Centred on prevention, personalisation and purpose, breakfast provides a unique lens through which to explore broader food system transformations. It is a mirror of newness, personality, trust and experiences.

Research evidence from the US shows that Gen Z tends to prefer lighter healthy breakfast options which are easy to prepare, portable or ready to eat (Fernandez, 2025). Smoothies and plant-based milks are often part of Gen Z's breakfast choices. Almost half of the UK Gen Z cohort considers breakfast their healthiest meal, opting for nutrient-dense choices like porridge and overnight oats, rich in calcium and protein (Young, 2022). The ways Gen Z reimagines the first meal of the day are similar around the globe as these young people see it as the cornerstone of their health-conscious lifestyles. Unlike older generations, many Gen Z individuals eat breakfast shortly after waking, a habit rooted in their commitment to energy, mindfulness and wellness. Convenience is a cornerstone in breakfast choices (Goldschmidt, 2024) which also reflect innovation and identity.

Innovation: Where Health Meets Technology

Food innovations are rapidly evolving to meet Gen Z's expectations and preferences. Advances in nutrigenomics and artificial intelligence (AI) now allow for personalised nutrition, including age-specific, protein-enhanced yogurts, powders blended into smoothies, nutrient-optimised cereals, AI-generated plant-based smoothies or cultured dairy alternatives tailored to individual cognitive and physiological needs. These innovations mark a shift. Food is no longer just habitual but increasingly therapeutic, supporting not only gut health and mobility but also emotional well-being and mental clarity. Understanding the benefits from healthy breakfast food choices also drives another trend among Gen Z. They are happy to enjoy breakfast types of foods at any time of the day, including as snacks, lunch, dinner or late-night meal (Technomic, 2017). It is interesting to note that two-thirds of Gen Z in the USA snack at least once per day and one-third say they do this twice per day (Food Insights, 2022). Having the healthy breakfast types food options seems a good way to respond to hunger cues, particularly given Gen Z's lifestyle highly dominated by computers, iPhones and social media.

In a world challenged by climate change and disrupted supply chains, food systems' resilience becomes essential. Gen Z is open to incorporating climate-smart ingredients; so, in addition to the oats and traditional wheat-based cereal options they are happy to see available and use locally produced millet and sorghum-based cereals, seaweed and vertically farmed greens. These innovative healthy options are considered appropriate by Gen Zers for their breakfasts as they could support both sustainability and regional sovereignty, offering decentralised, adaptive solutions to global instability.

Identity: Food as a Reflection of Self

For Gen Z, food is not just fuel. It is a declaration of identity. Their dietary choices are deeply connected with personal values and cultural roots. The breakfast menu becomes a canvas for self-expression, where what is eaten speaks volumes about who they are and what they believe in. They embrace plant-based food, have preference for transparent labelling and share interest in traditional recipes reimagined with a modern twist which reveals a generation searching for continuity and connection. Many draw upon cultural heritage while simultaneously seeking novel tastes that reflect their globalised experiences. This dual desire to honour tradition while embracing change is reshaping the food landscape.

Breakfast is also a time for emotional grounding. It connects Gen Z to comforting childhood rituals, family traditions and a sense of belonging. Whether it is a warm bowl of oats or a culturally inspired dish reinterpreted through a sustainable lens, these meals help the young people maintain emotional and cultural continuity in a rapidly changing world.

This generation is redefining eating as an intentional and value-driven act. Starting with breakfast, its priorities — health, sustainability and authenticity — are shaping how they consume and interact with food.

One participant summarised their perspective succinctly:

> *"It's all about alignment — health, sustainability and authenticity. I don't just eat for myself. I think about the bigger picture. If my choices can reduce my carbon footprint, avoid harming animals, and still taste amazing, why wouldn't I go for traditional plant-based or locally sourced foods? It's like voting for the world I want to live in, one meal at a time."* (Female, born 1997)

This opinion encapsulates the generational shift toward mindful consumption. Food is no longer just about individual preference but about collective impact, supporting sustainability initiatives, reducing carbon footprints and making ethical considerations part of daily life. Food reflects the broader trends in Gen Z's attitudes. With its food choices, this young generation makes "a statement about who they are and the world they want to inhabit" (Lempert, 2024, para. 3).

Trust beyond Innovation to Transparency

Innovation alone is not enough. Introducing innovative food options and alternatives to traditional foodstuff, such as lab-grown eggs, precision-fermented milk or AI-designed meals will fail without the crucial ingredient of trust. Gen Z is notably open to innovation and keen to trying new things, but only when this is matched with transparency, authenticity and ethical framing (Marinova and Bogueva, 2022). Our research, particularly among Gen Z consumers, reveals that while they are curious and accept new technologies in food, they demand that these innovations reflect their values. Foods must still feel "real". They want to know the origins of the ingredients used, the conditions under which they were produced, and the broader impact on the environment and society (Marinova and Bogueva, 2022).

To succeed, food innovations must meet a high standard — not just nutritional efficacy but they need to also have emotional and ethical resonance. They must earn trust through facts and data as well as through storytelling, shared values and a clear commitment to the planet and community well-being.

Experiences: Food as a Social and Sensory Journey

Gen Z is shifting consumer culture by valuing experiences over possessions (Jolly, 2024). In Australia, 94% of Gen Z young people prioritise things that they find enjoyable and 90% prefer spending money on experiences rather than on buying things (Jolly, 2024). They live in the moment, like things that make them happy and enjoy spending time with friends. An earlier survey (OC&C Strategy Consultants, 2019) across nine countries — Brazil, China, France, Germany, Italy, Poland, Türkiye, UK and USA, showed that attitudes and behaviours are most common among Gen Z compared to the previous generations of Millennials, Generation X and Baby

Boomers. That study also indicated that preference for experiences rather than buing products is more prevalent among Gen Z (OC&C Strategy Consultants, 2019). This mindset transcends attitudes from travel to entertainment and deeply influences food behaviours. For Gen Z, food is an immersive experience, a means of exploration, connection and creativity. As exemplified in the case of breakfast, food becomes an event and expression, not a chore.

Gen Z favours food environments that combine convenience with quality and value-aligned design with multisensory appeal. They seek foods that push boundaries, surprise the palate and reflect social consciousness. Whether it will be the comfort of fermented food and the porridge from their grandmother's recipe or perhaps a novel personalised, lab-grown breakfast bowl served via an AI-curated app, Gen Z's food journey is one of constant evolution. As the food system evolves, understanding these intersections will be key to crafting sustainable, ethical and meaningful futures. By watching how Gen Z reshapes breakfast and any other meal of the day, we glimpse the broader trajectory of food itself: intentional, personalised, resilient, ethical, sustainable and deeply human.

Food as Activism and Collective Responsibility

This generation's approach to food is highly intentional. Irrespective as to whether they adhere to any particular dietary style (more about this in Chapters 3 and 4), these young people see food also as expression of individual activism and a collective stance for the things they value.

What we have learned from the numerous conversations and interactions with Gen Z individuals as well as from formal surveys of university students, is that many of them are deeply concerned about the planet's future, particularly about the issues related to climate change, food security and sustainability (Bardales Salguero *et al.*, 2024; Bogueva and Marinova, 2020). For this generation, food is a way to take action and contribute to a better world. They view their food choices as meaningful acts of resistance against harmful industries, such as factory farming and unsustainable agriculture, while striving to create a positive impact on both society and the environment. Their approach often reflects a desire to harmonise personal health with global sustainability, bridging the gap between individual actions and collective responsibility.

When asked how Gen Z stays motivated to uphold these values, social media emerged as a significant source of support and inspiration. As one of the Gen Zers we talked to explained:

> *"Social media, honestly... is very important. TikTok and Instagram are full of influencers who make sustainable eating look effortless and creative. It's not just about trends — it's about discovering brands or recipes that align with my values and realising that making a difference can actually be exciting and delicious."* (Female, born 2004)

This generation's engagement with food is powered by digital platforms, where they explore innovative recipes, discover ethical brands and find community in shared values. These young people's ability to merge creativity with conscientious consumption demonstrates why Gen Z is not just a consumer group but a transformative force in reshaping the future of food.

Although it is impossible to claim that they are all the same and numerous differences exist, many in this generation embrace a more mindful, conscious approach to what they eat, opting for locally sourced, organic and ethically produced foods. Being the most diverse and largest global generation, Gen Z is fostering an openness to worldwide cuisines, fusion foods, ethnic and cultural food practices. Its global outlook fuels interest in foods that reflect inclusivity and celebrate cultural heritage. With a curiosity for global ethnic cuisines and new food trends, such as fermented foods, kefirs and yogurts, Gen Z is open to many options and their choices are sending bold messages.

Gen Z is also reshaping the meaning of activism. A study of European Gen Z's attitudes towards changing the food system (Seisto and Tamlander, 2025) identified four categories of social engagement, represented roughly equally but with very different outlooks. The first category are those who are proactive and technology-oriented. They are deeply concerned about the global environment and the impacts of food production which results in proactive engagement, including dietary changes, such as reduction in meat consumption and acceptance of technological solutions, such as plant-based alternatives. These consumers drive change through their sustainable behaviours. The second category are socially proactive and engaged. Although their belief in technology is moderate, they use social influence, information sharing and collaboration to engage others around them to motivate change. Both of these categories represent the face of Gen Z's food activism with young people being the change they also want to see in others. The other two categories, namely practical and change-resistant and disengaged and sceptical, are resistant to dietary shifts and resent the power of individual actions. There were also gender differences with more men represented in the last two categories (Seisto and Tamlander, 2025).

With unparalleled access to information, change-driven Gen Z leverages social media, apps and online reviews to inform food choices. Platforms like TikTok, Instagram and YouTube have become key influencers in discovering food trends and recipes. They are the ones that will define the future of food. The change-resistant groups are likely to follow as is always the case with the diffusion of innovations (Roger, 2003) and technological transitions.

Without generalising about Gen Z, in this book we are allowing voices from all categories of engagement to be heard. This does not take away the fact that Gen Z's consumption is values driven. It only positions young people differently in regards to innovation and changes in the food system. Gen Z continues to prioritise sustainability, ethical sourcing and social responsibility, influencing brands to adopt environmentally friendly and transparent practices. Compared to older generations, Gen Z consumers continue to be more health-conscious and intentional, often consuming less alcohol and avoiding dairy, gluten, eggs and animal proteins for ethical, environmental and wellness reasons (Williams, 2023). The chapters to follow are about all of these changes that Gen Z brings to the fore.

Conclusion: The Legacy of Generation Z

This book is an example of an area identified as generational research (e.g. Seemiller and Grace, 2024). Although some may oppose this concept (e.g. Costanza *et al.*, 2023), it has a very practical value in areas, such as marketing (Parment, 2015), workforce dynamics (Yardney, 2024) and consumer behaviour (Angmo and Mahajan, 2024). Also, alternative explanations, such as lifespan theory and social constructionist frameworks, cannot explain the impact and scale sustainability and environmental concerns have on people's perceptions, values and actions. A generational approach helps understand the changes in the world shaped by geopolitical events, economic and technological developments, health and well-being trends as well as social and environmental concerns. All of these circumstances have never had such a strong influence on a particular global population cohort as it has been the case with Gen Z. Contributing factors are the scale of these events and developments as well as the globalised and tech-savvy nature of Gen Z. Another important consideration is that it is the first time in history that the changes caused by human activities on climate change in particular are giving only a short window of opportunity to protect life on this planet from existential threats. Gen Z is fully aware of these circumstances and feels the responsibility about the future.

By analysing the future of food through the eyes of Gen Z we are not trying to simplify the issues, stereotype consumer behaviour or discriminate against other perspectives. It is about capturing voices and opinions of those who will be seriously affected by the changes happening in the world and offer their responses. Despite considered individual action and choice, food is part of the global landscape and Gen Z's cohort makes a mark on how we think about nourishment.

Generation Z's influence on the food system is undeniable, driven by their values, technological fluency and global perspective. As they continue to advocate for sustainability, innovation and inclusivity, their legacy will be a transformed food landscape that prioritises health, ethics and environmental stewardship. This generation's commitment to shaping the future of food promises a world where food systems are more equitable, sustainable and resilient.

As we began writing these lines, we found ourselves circling back to a fundamental question: why does it matter what Gen Z (or any other generation, for that matter) thinks about food? Why should we care about their tastes, their habits and their relationship with what they eat? Do they live on convenient, fast food or more sophisticated foodstuff? Are they drawn to growing their own produce? Are they experimental eaters, eagerly embracing novelty or are they more cautious scrutinising every detail from how food is sourced and packaged to how it is consumed and what happens to it afterward?

The more we talked, the more we realised that this is a generation that approaches food with genuine curiosity and excitement. They care about what is on their plates, what is in their food, how it is produced and how much it costs economically, socially and environmentally. Gen Z is plugged into the latest trends and movements shaping the food industry, while also grappling with larger issues, including climate change, waste management and sustainability. As we reflected on these questions, we wondered whether we could even define a general taste for a generation so diverse and dynamic. This is why we took the approach of letting Gen Z speak for itself.

In the following chapters, we delve into the above aspects, exploring them in some depth and uncovering the nuances that make Gen Z's relationship with food so fascinating. However, as much as we were captivated by these young people's potential to redefine the food landscape, we were also careful not to idealise them. Instead, we aimed to present an honest and balanced portrait of this generation through their own words, their own stories and their own perspectives. Let us uncover what food really means to them and why their choices are shaping the future in ways we never anticipated.

References

Angmo, P. & Mahajan, R. (2024). Virtual influencer marketing: A study of millennials and gen Z consumer behaviour. *Qualitative Market Research*, 27(2), 280–300. https://doi.org/10.1108/QMR-01-2023-0009

Bardales Salguero, R., Bogueva, D., & Marinova, D. (2024). Australia's university Generation Z and its concerns about climate change. *Sustainable Earth Reviews*, 7, 8. https://doi.org/10.1186/s42055-024-00075-w

Bogueva, D. & Marinova, D. (2020). Autonomous sensory meridian response (ASMR) for responding to climate change. *Sustainability*, 12, 6947. https://doi.org/10.3390/su12176947

Bogueva, D., Marinova, D., Wäechter, N., & Tekiner, I. H. (2024). Societal concerns. In Seemiller, C. & Grace, M. (eds) *Gen Z Around the World: Understanding the Global Cohort Culture of Generation Z* (pp. 119–127). Emerald Publishing, Leeds, UK.

Costanza, D. P., Rudolph, C. W., Hannes Zacher, H. (2023). Are generations a useful concept? *Acta Psychologica*, 241, 104059. https://doi.org/10.1016/j.actpsy.2023.104059

Fernandez, J. (2025). Gen Z's breakfast habits: How the youngest generation approaches the first meal of the day. YouGov. https://business.yougov.com/content/51614-gen-zs-breakfast-habits-how-the-youngest-generation-approaches-the-first-meal-of-the-day

Food Insights. (2022). 2022 Food and Health Survey Spotlight: Generation Z. https://foodinsight.org/spotlight-generation-z/

Goldschmidt, B. (2024). Convenience Key to Breakfast for Consumers. https://progressivegrocer.com/convenience-key-breakfast-consumers

Haidt, J. (2024). *The Anxious Generation: How the Great Rewiring of Childhood is Causing an Epidemic of Mental Illness.* Alen Lane, London.

Jolly, N. (2024). 'Experiences over possessions': Gen Z study finds a generation living for today. https://mumbrella.com.au/experiences-over-possessions-gen-z-study-finds-a-generation-living-for-today-817711

Lempert, P. (2024). The Importance of Understanding Gen Z's Food Philosophy and How to Satisfy Their Needs. https://retailconsumerprofessionals.com/content-library/article/5oRngG/the-importance-of-understanding-gen-z-s-food-philosophy-and-how-to-satisfy-their-needs

Marinova, D. & Bogueva, D. (2022). *Food in a Planetary Emergency.* Springer, Singapore.

OC&C Strategy Consultants. (2019). A generation without borders: Embracing Generation Z. https://www.occstrategy.com/wp-content/uploads/2024/03/a-generation-without-borders_us.pdf

Oxford Reference. (2025). George Bernard Shaw 1856–1950: Irish dramatist. https://www.oxfordreference.com/display/10.1093/acref/9780191843730.001.0001/q-oro-ed5-00009969S

Parment, A. (2015). *Marketing to the 90s Generation: Global Data on Society, Consumption, and Identity*. Palgrave Macmillan, London.

Rogers, E. M. (2003). *Diffusion of Innovations*, 5th ed. Free Press, New York, NY.

Seemiller, C. & Grace, M. (eds) (2024). *Gen Z Around the World: Understanding the Global Cohort Culture of Generation Z*. Emerald Publishing, Leeds, UK.

Seisto, A. & Tamlander, T. (2025). From activists to sceptics: The many faces of Gen Z in the changing food system. https://www.vttresearch.com/en/news-and-ideas/activists-sceptics-many-faces-gen-z-changing-food-system

Technomic. (2017). Breakfast brief: Generation Z. https://s3.us-east-2.amazonaws.com/winsight-cdn/technomic/files/public/documents/Smuckers-Breakfast-Brief/SmuckersBrief_GenZ_20170124.pdf

Williams, L. A. (2023). Gen Z Consumers Influence Everything from Taste to Expansive Food and Beverage Trends. Prepared Foods. https://www.preparedfoods.com/articles/128129-gen-z-consumers-influence-everything-from-taste-to-expansive-food-and-beverage-trends

Yardney, M. (2024). Beyond the Buzzwords: Decoding Generational Research. https://propertyupdate.com.au/beyond-the-buzzwords-decoding-generational-research/

Young, A. (2022). Gen Z believe breakfast is their healthiest meal — with over a third opting for porridge. https://www.express.co.uk/life-style/food/1696279/generational-eating-habits-breakfast-healthy-food-porridge

Chapter 2

What Matters to Gen Z: Values, Societal Concerns and the Planet

In January 2022, before the military actions in Ukraine and the Palestinian conflict in Israel and surrounding countries, we conducted a global survey of university students, all from Gen Z, to understand what matters to them. The survey covered 33 countries (Australia, Austria, Belgium, Botswana, Brazil, Bulgaria, Canada, China, Croatia, Cyprus, Ecuador, Germany, Hong Kong Special Administrative Region, India, Indonesia, Israel, Italy, Jordan, Lebanon, Montenegro, the Netherlands, Philippines, Portugal, Romania, Russia, Serbia, Singapore, South Africa, Spain, Türkiye, United Arab Emirates, United Kingdom and United States of America) with more than 21 thousand participants in total, 71% of whom lived in urban areas. University Gen Z is a special cohort of young people who are on the pathway to becoming well-educated and take up professional positions within society. By the time of the survey, the COVID-19 pandemic caused by the severe acute respiratory syndrome coronavirus 2 (SARS-CoV-2), had already subsided and most young people were starting to enjoy normal life and opportunities without major health-related impositions, such as lockdowns and mandatory vaccinations. While on the health front things were getting quieter, there were other global challenges looming. Although there have always been conflicts across the globe, the military escalations that commenced in Ukraine in February 2022 and in the Gaza Strip in October 2023, exacerbated geopolitical tensions, put the world in direct confrontation, interrupted supply chains and imposed high risks on the entire global population, including Gen Z. In February 2025, at the time we were writing this book, the peace prospects for the world were looking better with talks and negotiations happening between the important players in the conflicts. In a way, things were transitioning back to the normal state that our survey captured. This later changed with new military conflicts in the Middle East. What mattered to Gen Z at the start of 2022? If we understand what these young people's values and concerns were then, we may expect that these same insights will hold in in a relatively peaceful and stable world.

In addition to any geopolitical challenges, there are other warrying trends that have deteriorated during these last few years. The Earth is getting hotter; plastics pollution is now spreading everywhere at an unprecedented scale and we are continuing to lose species which once inhabited this planet. According to the World Meteorological Organisation (WMO, 2025), 2024 was the hottest year on record and the past ten years have been the warmest Top Ten with record-breaking temperatures. For the first time, the global mean temperature in 2024 was 1.55°C above the 1850–1900 average and above the Paris Agreement's desirable target of 1.5°C (WMO, 2025). The high temperatures affected the land and ocean surfaces, impacting on the planet's ecosystems but also on people's livelihoods, economies and ability to provide food. In addition, the rate of land, soil, freshwater and marine pollution with plastics continues to put at risks all ecosystems of the planet. Microplastics and nanoplastics (broken-down plastics smaller than 5 mm and 100 nm, respectively) are found in human and animal bodies, placentas, food and drinks (IUCN, 2024). According to a 2024 study, 52 million tonnes of plastics leak in the natural environment every year (Cottom *et al.*, 2024), with 88% of them being macroplastics generated predominantly from urban environments (IUCN, 2024) but threatening human health, wildlife, food production and all life on Earth. The world now produces 450 million tonnes of plastics per year, a 225-fold increase from the just two million tonnes made in 1950 at the onset of the plastics industry (Ritchie *et al.*, 2023). When these plastics are not managed properly, they become serious environmental hazards and source of pollution that has been spreading beyond human control.

In fact, only 9% of all plastics in the world are recycled and 49% end up in landfill (Ritchie *et al.*, 2023). It is hard to find a place on the planet not affected by plastics. Climate change combined with pollution and destruction of natural habitat is leading to fast species extinction. Estimates show that up to 6 million animal and plant species are facing extinction in the next 50 years (Wiens and Zelinka, 2024); this is close to 330 species per day. Although we are yet to properly understand biodiversity and how it matters to humans, we know that food production is a major contributor to land use changes and decimation of the natural habitats of many non-human species (Marinova and Bogueva, 2022).

Is Gen Z across the globe concerned about these issues? What are Gen Z's interests? Are there causes that they rally behind and others that are low on their priority lists? Let's approach this from top-down and bottom-up perspective, that is what matters and what does not matter to them. We base a lot of this discussion on the global Gen Z survey of university students but also complement it with other literature sources and direct quotes from these young people.

Who Is Gen Z

"Will you share this photo with us?" — asked one of the Gen Z youngsters who were riding the train with us (see Figure 2.1). Despite our desire to do so, they were fast to leave the train station without giving us a chance to explain what we were doing and why. As happy and confident as they look, Gen Z are navigating a world that is very different to ours. Their world is the future and the better we understand them, the clearer it would be what opportunities and challenges lay ahead of them and for all of us.

A lot has already been written about Gen Z — the people born between 1995 and 2010, after Generation Y and before Generation Alpha (e.g. Seemiller and Grace, 2024). Generational analysis is common in many social science areas, including sociology, management, psychology, marketing or leadership. Although the boundaries around who belongs to which generational cohort, and who does not, can be quite arbitrary and blurred, it is widely accepted that a generation starts from when people are born and extends to the

Figure 2.1. An encounter with Generation Z on the train in Perth, Western Australia.

age when they can biologically have children. This demographic definition is now accepted to span around 15 years (McCrindle Research, 2025). In addition to the purely biological factors, there are many other circumstances that make generational analysis justified sociologically, technologically, geopolitically and even environmentally. It is common in social studies to disaggregate according to gender, education, income, religion or race (Seemiller and Grace, 2024); what generational analysis does is look at population through a time perspective. In fact, it is very similar to disaggregation according to age, but the age brackets are set at 15 years to capture common elements and factors affecting childhood and adolescence as a demographic cohort progresses through its lifespan exposed to similar societal, geopolitical and natural events leading towards adulthood and older stages of life.

The problems Baby Boomers (born between 1950 and 1964) had, were related to a post-war recovery with the aim of building a strong economy, housing and infrastructure in a global order where industrialisation would define development. Although most female Baby Boomers have been engaged in the economy, they had relatively high fertility rates contributing to global population increase as well as disproportionate use of resources and generation of pollution by the developed parts of the world. Technology use was not a defining feature of the Baby Boomers who applied themselves to becoming useful in the workforce guided by remuneration. Demographers like to think of the Baby Boomers as the grandparents of Gen Z who laid the foundations of the progress but also the controversial issues these young people need to master. Gen Z's parents come from Generation X (born between 1965 and 1979) — people with financial freedom, business and entrepreneurial expertise who lived in a relatively peaceful time with economic opportunities, and Generation Y (born between 1980 and 1994) who embraced technology as well as flexibility in working hours with opportunities for travel and nice food.

While Baby Boomers created wealth at all costs, Generation X and Generation Y were exploiting the benefits from this development gradually realising its limiting factors, such as house market prices, climate change and work-life balance. The global fertility rates also drastically decreased from close to 5 children per woman in 1950 to 2.2 children in 2021 (IHME, 2024). With the Second World War and the Iron Curtain's Cold War era being the political background for the Baby Boomers, the Vietnam War was shaping the political views of Generation X while September 11, 2001 and the confrontation it instigated significantly impacted Generation Y.

The differences between generations go beyond the geopolitical events that they experience in their formative ages and extend to the technological changes that shape their lives (Haidt, 2024). Starting with radio, then television, computers, personal computers, the Internet, mobile phones, high-speed broadband Internet, smart phones and the social media, consecutive generations were shaped differently by technology (Twenge, 2023). Food has also been a defining generational characteristic — from sustenance to local markets and shops, supermarkets, global supermarket chains, fast food outlets and ready-made meals, eating out and gourmet choices. The COVID-19 pandemic affected all people across the planet but the younger generations embraced the use of online technologies faster aspiring for the convenience to work remotely with the flexibility of time and place. They also changed their outlook at the world at present and for the future.

"Loyal, Compassionate, and Open-Minded. These are just some of the self-described characteristics of Generation Z" (Seemiller and Grace, 2024, p. 1). They also point out to the following factors as greatly motivating Gen Z in their actions: seeing the fruits of their efforts, having an opportunity for personal advancement as well as advocating for something they believe in, learning and being better at something and not letting others down. What Seemiller and Grace (2024) also argue is that Gen Z in this day and age creates a global peer personality which links these young people across the globe generating greater world-wide interdependence and connectivity than what we have seen in previous generations.

Our global sample of university Gen Z from the 2022 survey is very particular. In 2019, Gen Z became the world's largest population cohort comprising 2.47 billion people (Spitznagel, 2020) and representing 32% of the global population estimated at 2.62 billion in 2025 (Noor, 2025). Statistically, with 21,371 responses, our sample represents the global Gen Z cohort with a standard error of 0.00342, 0.68 relative standard error and confidence interval of 0.00670. Notwithstanding this, our sample is skewed in the following ways:

- 75% of the respondents identified as white while Gen Z is the most racially diverse generation that has lived in the world;
- 65% of our respondents were women, 31% men, 2% non-binary/gender-fluid, 1% preferred not to state and the rest were other, while the global population is very close to 50%:50% (women:men);
- Being all university students, our Gen Z falls in a very high education bracket, while only 40% of the global population has attained some kind of tertiary education (Galan, 2025a). The majority of participants also come from families with established educational traditions with only 13% being the first to attend university. By comparison, the share of first-in-family students

in Australia varies between 13% and 50% depending on the educational institution where they are enrolled (Patfield, 2022).

It is also interesting to note that the majority of our Gen Z people come from middle-income families. As far as religious beliefs systems are concerned, 39% of them follow an organised religion while 30% are spiritual but do not follow an organised religion and 30% are not at all spiritual. Only 14% of the global population are not affiliated with any religious group (Galan, 2025b) which shows that our Gen Z is much more independent in its worldviews.

Ideologically, our Gen Z is left-leaning (see Table 2.1) — 59%, followed by 22% reporting no particular ideology. The countries with the highest share of left-leaning young people were Germany (92%), Austria (86%) and Montenegro (79%). At the other end of the spectrum with the highest right-leaning share of young people were India (73%), Israel (60%) and Portugal (56%). There is, however, another characteristic of Gen Z; they often do not openly declare their political affiliation and are prone to not telling the truth. Almost half (namely 48%) of Gen Z voters in the USA lied to people close to them about whom they voted for (Talev, 2024). This may indicate Gen Z wanting to have their views private but also that they change opinion, adjust or try to avoid conflict.

Another specific but worrying characteristic of our Gen Z group is the psychological discomfort that these young people feel. More than half of them, namely 54% reported some psychological condition, such as anxiety and depression. This is highly concerning for a young generation that is yet to engage professionally, have children and

Table 2.1. Political ideology of Generation Z.

	Share of Gen Z (%)	Country with the highest share
Left	59	Germany (92%)
Centre	18	Bulgaria, Cyprus, Türkiye (36%)
Right	11	India (73%)
No ideology	22	Australia (56%)
No answer	8	Australia (19%)

make its mark on the world. What are some of the issues that impact on Gen Z and what are the causes they care about?

How Gen Z Sees Itself

When we asked Gen Z to share who they are, the answers were never just bullet points in a resumé or curriculum vitae (CV). This generation does not define itself through titles or accolades. Gen Z defines itself through purpose. Through conversations about everything from technology to food, what stood out most was how powerfully these young people connect their personal choices to global impact. For Gen Z, who they are is inseparable from what they stand for.

Food, especially, becomes a way of telling that story. It is not just about taste; it is about traceability, tradition and trust; it is activism on a plate. Whether talking about climate change, mental health or cultural identity, Gen Z speaks through its meals. These young people see what is grown, how it is made, who benefits and what values it carries. Their language is one of care and curiosity, not certainty, and they do not shy away from complexity.

As one young Australian Gen Zer put it, it is not about having all the answers. It is about asking better questions and staying real while doing it. The "who Gen Z is" is not about individualism, it is collective, connected and evolving. It starts not with a title, but with a truth.

> *"Look, I'm still figuring myself out, but I know this much... I care. I care about people, the planet, and making sure we leave things better than we found 'em. I grew up online, yeah, but that doesn't mean I'm disconnected. If anything, I've learned to spot the real from the fake faster than most. I don't pretend to have all the answers, but I'm not afraid to ask the hard questions either. I just want to live honestly, support my mate, and do something that matters."* (Male, born 2005)

This is not a generation obsessed with certainty or fixed identity. For many Gen Zers, who they are is not a static definition, but an evolving story shaped by the world they are inheriting and questioning. They embrace complexity, often using humour and humility to speak truths that older generations might avoid. Their sense of self is less about where they have come from and more about what kind of future they want to help create. Another Gen Z voice, reflecting on their role in the wider world, said:

> *"I'm part of a generation that doesn't just sit back... we speak up. Whether it's climate change, mental health or calling out unfair stuff, I feel like it's on us to have a crack at fixing things. I'm proud of being open-minded, and I reckon listening is just as powerful as speaking up. I'm not perfect, and I don't try to be... but I do try to stay kind, informed and ready to stand for what I believe in. That's who I am, at least for now."* (Female, born 2002)

This statement clearly speaks about Gen Z's relationship with activism and authenticity. There is no romanticism here, just a grounded, practical idealism. These young people are not looking for perfection in themselves or the world. What they want is meaning and in the context of food, as we discuss later, this shows up in how they seek transparency, ethical sourcing and a sense of social responsibility in what they consume.

However, Gen Z does not exist in a vacuum. When asked about their identity in relation to older generations, many reflected with respect and clarity:

> *"My parents are Gen X — tough, practical, did their best with what they had. I've learnt heaps from them, like working hard, not whinging and keeping your head down when things get rough. Honestly, they are like a rock... my big supporters I can always rely on for anything... I love them much.... My grandparents? They're old-school — Boomers. Big on tradition, loyalty, and they always reckon a meat pie and a cuppa can fix just about anything. I respect that. I really do. But I reckon our generation's just wired a bit differently. We're more likely to speak up, even if our voice shakes. We care more about the planet, about people who don't look like us or live like us. We talk openly about mental health, identity and stuff that used to be swept under the rug. [It] doesn't mean we've got it all sorted — far from it — but we're trying to break the cycle, not just survive it. I guess what makes us different is we're not afraid to question everything, even if that makes things*

uncomfortable. And yeah, we still love a yarn with Nan and Pop — but we also want to leave behind a world they'd be proud of too." (Female, born 2003)

This generational distinction is delicate but powerful. While Generation X might have prioritised stability, resilience and silent endurance, their children — Gen Z, are choosing visibility, emotional honesty and collective responsibility. Their grandparents — the Baby Boomers, represent tradition and perseverance, often symbolised by simple pleasures and stoic attitudes. Gen Z respects this, even draws strength from it, but the young people also feel the need to challenge inherited habits that no longer serve the planet or people.

What emerges is a generation deeply relational. It defines itself not only by contrast, but by connection — what should be kept, what should shift, and what it hopes to transform. Gen Z is fluent in navigating uncertainty and while it often feels the weight of global crises from climate change to inequality, it does not carry it with despair. Instead, this young cohort carries it with cautious hope.

This generation is not just shaping the future of food. These young people are reshaping what it means to belong, to care and to take responsibility across generations. Their honesty may be confronting, but it is also refreshing. They are not afraid to say: "We don't have all the answers, but we are ready to ask the right questions" and that, perhaps, is their most defining strength.

What Matters to Gen Z

Gen Z has surely taken up its place in academic interest. A Google Scholar search for publications with the keyword "Generation Z" generates millions of entries, which span in a multitude of research areas — from demography, sociology and cross-cultural studies to management and marketing to labour market research, tourism trends, teaching and future studies. There is hardly any social sciences and humanities field that has not turned its attention to one of the youngest generations alive. All these publications employ different methodologies and theoretical approaches as well as differ in their geographic coverage and economic settings. The peak of

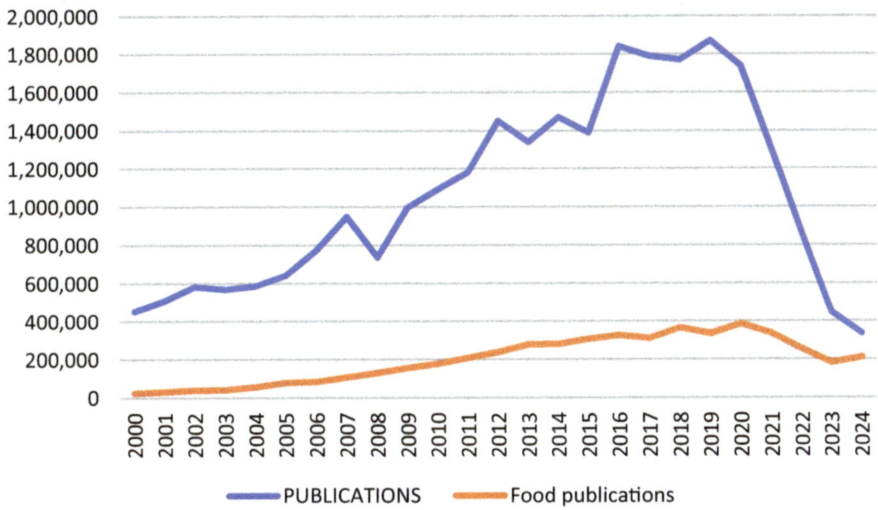

Figure 2.2. Number of Google Scholar publications per year with the keywords "Generation Z" (blue line) and "Generation Z" and "Food" (orange line), 2000–2024 (data extracted 20 February 2025).

academic interest appears to have been in the late 2010s and it seems that it has sharply decreased since the 2020s (see Figure 2.2). This may mean that we have already developed a sound understanding of Gen Z through the accumulation of these efforts. However, what we did in 2022 was to take a global snapshot of university Gen Z using the same questions at the same time across 33 countries.

Included in Figure 2.2 is also the trend of publications which were relating Gen Z to food. Their share started at 5.5% in 2000 but reached 63% in 2024 indicating the increasing importance of the issues related to food when investigating Gen Z. Although the food publications also dipped since the early 2020s, their relative share increased making them a defining feature of more recent Gen Z research.

A big motivator for us to conduct the global survey was to understand how Gen Z views the world and what matters to them. Some of the findings were expected, given the vast body of already existing literature, while others came to us as a surprise. As the list of societal issues was prescribed by us, the surprises were in the levels of concerns assigned by the participants.

The main surprise from our global survey was the importance Gen Z puts on sustainability. We both have been working in the area of sustainability for 20 years, but this in no way impacted the conduct of the survey (which had an official ethics approval by Curtin University Human Research Ethics Committee). More than 70 colleagues from a range of disciplines contributed to the data collection (more information about them can be found in Seemiller and Grace, 2024). Sustainability was the societal issue with the highest level of concern by Gen Z across the globe (see Table 2.2).

Overall, 70% of the participants stated that they were concerned or very concerned about sustainability. In fact, previous commentators have called Gen Z the "sustainability generation" (Petro, 2021) or the "eco conscious generation" (Shaw, 2024). There is already evidence that these young people are prepared to pay higher prices for more sustainable products (Ramsing *et al.*, 2023). The question is also whether Gen Z is changing its behaviour to be more sustainable and this is where the information is conflicting. As most young

Table 2.2. Societal issues of high concern to Generation Z.

#	Issue	Not concerned (%)	Slightly and somewhat concerned (%)	Concerned and very concerned (%)	Weighted level of concern (%)
1	Sustainability	3	27	70	60.0
2	Racism	8	26	65	57.2
3	Education	6	30	64	56.4
4	Access to healthcare	8	29	63	56.2
5	Poverty	6	32	62	55.8
6	Sexism	11	28	61	55.4
7	Access to viable and affordable housing	7	32	61	55.0
8	Climate change	9	31	60	54.6
9	Limitations on personal freedoms	8	33	59	54.4
10	Cost of higher education	8	33	59	54.2

people, Gen Z follows fashion trends as well as adheres to other prevailing social norms, but also sets up its own expectations and requirements. For example, having plant-based milks has become a defining characteristic of Gen Z and this has environmental as well as health benefits (for more see Chapter 8). On the other hand, Gen Z is addicted to fast fashion and even when they look for garments made from more sustainably produced materials, online shopping and the low prices of new fashionable items take precedence over environmental and other ethical considerations (Makkar and Spry, 2024). The intention-behaviour gap, that is, between ethically minded consumers and their actual purchasing behaviour (Carrington *et al.*, 2010), can be quite wide not only in relation to clothes, but also in the need to have the latest technological gadgets as well as health and wellness products. Often Gen Z finds itself in paradoxical situations where its good sustainability intentions contradict its actual actions. In the case of food, this is manifested, for example, with relatively high awareness of the environmental impacts of livestock-based products but still excessive consumption of meat (Bogueva and Marinova, 2022). Notwithstanding this, Gen Z's formal education achievements –the highest compared to previous generations in USA (Dennington, 2021) and Australia (McCrindle, 2025), are helping it to live more sustainably (EY and JA Worldwide, 2023). On a positive note, a survey in Australia found that Gen Z are shopping more sustainably in 2024 compared to what they did two years earlier (McNamara, 2024).

Climate change is closely related to sustainability and is another societal issue that global Gen Z is seriously concerned about. There is ample evidence that Gen Z sees climate change as anthropogenic — 96% of Australian Gen Zs believe this to be the case (YouthSence, 2019), and the share of those who are concerned or very concerned in our global survey, namely 60%, is very similar to the results from the 2024 global survey by Deloitte covering 44 countries, which show that 62% of Gen Z have felt worried about climate change in the last month (Segal, 2024). This has also been an area of massive activism and actions through movements, such as School Strike for Climate or Fridays for Future (Marinova and Bogueva, 2022), which had to adjust its tactics during the COVID-19 pandemic and are now calling for climate justice (Mayes and Talukdar, 2023). The grown-up Gen Z are also prepared to put pressure on their employers to respond to climate change and even change jobs when there is lack of action (Segal, 2024). When it comes to lobbying or campaigning for climate change-related issues, Gen Z does not necessarily engage with the traditional forms of volunteering in-person for a community organisation of a cause, signing petitions or raising money, but instead use technology to voice their concern (Bardales Salguero *et al.*, 2024).

Social justice is another strong characteristic of Gen Z. Similar to climate change, the issue is perceived with global dimensions. In the words of an American Gen Zer: "This is the first time, in terms of our generation, that we've had such a connection with people, not just among our communities, but across the world because of the internet" (Graeter, 2024, para. 9). The three main aspects of social justice that emerged among the concerns in our survey are racism, sexism and limitations on personal freedoms (see Table 2.2). While racism and sexism are based on prejudices and discrimination, personal freedoms are strongly linked to the right and power of expression. Gen Z are bold in their opinions, want to be heard and are against all barriers that prohibit people's advancement in society. Because they can take different expressions, it is difficult to give definitions as to what racism, sexism and limitations of freedoms are (European Institute for Gender Equality, 2025). Negative attitudes and experiences still exist in society, often expressed through existing stereotypes, and this is exactly where Gen Z concerns lay. Limitations on personal freedoms can also relate to political views and opinions and they are part of a human rights agenda. However, if Gen Z is disillusioned with the lack of action and hope, it can shift towards populist extremes at the left or right (Lissowsky *et al.*, 2024). We have not seen such extremes in our global Gen Z sample; however, one in five young people (22%) did not associate with any ideology indicating lack of interest, distrust or resentment towards the current political ideology.

Opportunities to properly engage in the economy are another set of concerns for the global university Gen Z. They start with education and the cost of higher education, extend to access to viable and affordable housing, particularly when they leave the parental dwellings, access to healthcare and ultimately, avoiding poverty (see Table 2.2). These conditions can also be perceived as economic justice for Gen Z (Bogueva *et al.*, 2024). As the global survey is of university Gen Z, they value education highly but are also concerned as to what this would cost. There are vast differences between countries where high education is free, e.g. Germany, and others where students take a loan or have to pay upfront fees (e.g. Australia and USA). At the other end of the spectrum are poverty concerns which apply not only to the individuals themselves but also as a societal consideration. Particularly concerning is economic injustice manifested with the increasing gap between the rich and the poor (Perna, 2022). Access to health care as well as to good housing is similarly related to economic justice. With half of our sample reporting mental health conditions, including stress, anxiety and depressions (Annie E. Casey Foundation, 2021), it is not surprising that health care is ranked very high at #3 as a concern (see Table 2.2).

Gen Z is the first generation globally to grow up with the Internet, smart phones and social media. They are educationally transformed not only in terms of years of study and levels of achievement but also by using technology. McCrindle (2024, para. 1) describes them as digital integrators: "they have integrated technology seamlessly into their lives, and having used it from the youngest age, it is almost like the air that they breathe, permeating almost all areas of their lifestyle and relationships". Notwithstanding this, the preferred (for 77% of the people surveyed) and actual (for 79% of the people surveyed) method of communication for our university Gen Z is in-person conversations. At the time of the survey, Instagram was the preferred social media platform (only 5% not using it), followed by WhatsApp (17% not using it). Personal websites and blogs are no longer attractive communication ways with respectively 86% and 84% of the university Gen Z not using them.

Interestingly, in 2022 TikTok was not used by 37% of the participants, but this has drastically changed more recently. In 2024, the popularity of Instagram continued (with 91% of Gen Z using it) while TikTok rose in prominence with 86% of Gen Z seeing it as a favourite channel for entertainment, news and staying in contact with family and friends (Smith, 2024). Although YouTube is behind in popularity as the third most frequently used social media platform, it allows Gen Z to engage with longer video material (Smith, 2024) and a young male Gen Zer admits: "YouTube is an ocean of entertainment" (Ryan Jenkins, 2024, quote 9).

LinkedIn, the professional platform for finding jobs, internships and building networks, is not used by 62% of the university Gen Z. Even the popularity of the search engine Google has now fallen behind Instagram and TikTok for Gen Z (Koetsier, 2024).

What Matters Less to Gen Z

There seems to be more knowledge about what triggers reaction and responses in Gen Z than about the issues that they are less concerned about. Policy makers and politicians need to have a reality check when it comes to Gen Z as a lot of well-established behaviours and assumptions no longer apply. In the workplace, conventional theories that explain how to attract and retain talent, such as employer branding and talent recruiting, have limited applicability when it comes to Gen Z (Vieira *et al.*, 2024).

For example, the least common motivating factors for Gen Z in our study are competing with others to win and knowing that someone would return the favour. Only around a quarter (27%) of the respondents were strongly motivated by such factors, a similar percentage found them not motivating at all and the remainder were neutral. This shows that Gen Z are not interested in competition and are happy to be altruistic in their actions as long as they believe in and are fully committed to a cause. A study in India analysed the level of psychological altruism of Gen Z which implies acting for the well-being of others without paying attention to self-interest (Maheshwari *et al.*, 2020). It found that 88% of Gen Z was reporting considerable (average or above) levels of altruism with females being more altruistic. As far as competition is concerned, Gen Z has accepted that the job market is tight, particularly after

COVID-19, and instead opt for "job hopping" and travel to get to know the world (Australian Human Rights Commission, 2023). There is some realistic pragmatism in acknowledging that they have different interests and opportunities: "Older Australians... see us as less likely to want to work hard like they did to climb the corporate ladder and believe us to be lazy and want everything handed to us... whereas most of us are working harder than ever just to try and get ahead" (Australian Human Rights Commission, 2023, p. 19). There is further evidence that Gen Z is more intelligent as a result from favourable economic prosperity, social conditions and life history speed (Pietschnig *et al.*, 2015) which also leads to increasing empathy among young Americans (Konrath *et al.*, 2025). Hence, Gen Z should not be underestimated and if they are not happy with the culture and requirements at the workplace, they will find a new job on LinkedIn straightaway (Chan, n.d.).

In our survey, we gave Gen Z a long list of social issues asking about their level of concern. All of these issues are expected to be relevant to young people but we wanted to find out the degree of uneasiness. Table 2.3 shows the bottom two issues from the list; they are border security and national debt. A 2024 study of American Gen Z similarly shows that only 26% of them are concerned with border security (A Decisive Block, 2024) compared to 29% in our 2022 survey. Being digital savvy and globally connected, these young people realise that borders are not essential for today's interconnected world. Their use of technology and social media extends beyond political borders and as an Australian Gen Zer explains: "We can spread our thoughts and opinions faster and

Table 2.3. Societal issues of low concern to Generation Z.

	Not concerned (%)	Slightly and somewhat concerned (%)	Concerned and very concerned (%)	Weighted level of concern (%)
Border security	25	46	29	38.6
National debt	21	44	35	42.2
War	12	40	48	48.4

find like-minded people who want to make a difference" (Australian Human Rights Commission, 2023, p. 21).

Among all societal issues, border security has the highest share of people who are not concerned. Other issues about which a relatively high proportion of Gen Z show no concern are gun rights, access to voting, transphobia and homophobia, although there may be national differences in the ways they are being interpreted.

In Australia, Gen Z is concerned about finances more than any other generation (ASIC, 2024). However, this seems to be more in relation to personal finances rather than at a national level. Gen Z believes in intergenerational equity and that they should not be suffering financially for decisions and actions taken by other generations (Strickland, 2024). As Gen Z does not see itself yet in a position of power, in many ways it does not feel responsible for government decisions that are impacting their future but they are still cautious as to what this financial future would be like with 79% of the Gen Z people in our sample being concerned. This level of concern however is much lower than the other issues that they feel passionate about, such as sustainability. As Gray (2020, para. 4), observes about youth in the USA: "Rather than starting with a flawed assumption that younger Americans are ignorant to the federal debt's threat, policymakers and observers need to grapple with the challenge that younger Americans are informed and engaged — they just prioritise other issues over the national debt".

In our survey, we did not explicitly ask about political events, historical conflicts or present-day confrontations, such as the Cold War, the wars in Vietnam, Afghanistan, Iraq, Syria, Sudan, the Democratic Republic of Congo or Myanmar, but we had "war" as one of the possible societal issues that may raise concern (see Table 2.3). Only a tenth (12%) of the young people were not concerned about war with the remaining 88% showing some level of concern. As it happened, the February 2022 war in Europe between Russia and Ukraine started a month after we completed the survey. A year and a half later, the conflict between Palestine and Israel escalated. In 2025 and 2026, wars broke up between Israel and USA, and Iran. These events would have certainly pushed the levels of concern about war higher. In fact, a 2024 survey shows that the war in Gaza activated Gen Z more than any other conflict (Davis, 2024). Student demonstrations at university campuses across the world ignited solidarity and requested answers with social media making information easily accessible to Gen Z. According to an Australian study, "Gen Z are broadly driven by a deep sense of social awareness, and the horrors of the Israel-Palestinian conflict is this 'moment' for the next generation when they can't look away" (Meade, 2024, para. 8). One of our Gen Z interviewees explains:

> *"We're not numb, we're just not buying into the same systems. We care, deeply, but we're not going to shout about every issue just because we're told to. It's not that we don't worry about borders or voting or even war, it's just we see the world differently. Our fight isn't about flags... it's about fairness. We want a planet that's liveable, leaders that actually listen, and a future we don't have to fight for. Yeah, we're stressed about money,... but not 'cause we blew it. We're carrying the debt, the climate mess and the fallout from choices we didn't make. We're not powerless, but we are bloody tired of being ignored."* (Female, born 2003)

On the one hand, Gen Z is technologically smart and actively engaged with social media that shape their social awareness, interest in international affairs and news, but on the other, it is the generation less likely to actively get involved in conflicts. For example, a UK study found that Gen Z is the less likely generation to join the military professionals or perceive a military career as a life-long choice (Dascalu *et al.*, 2024). Under-recruitment for military service is also a major challenge for a country like the USA (Cox, 2024; Crowley, 2025). In short, for Gen Z it is not a question of war and peace, but of a fair sustainable development that gives a chance to all on the planet.

Gen Z and Food

This book is about Generation Z and food. We did include several issues explicitly related to food amongst the list of potential societal concerns (see Table 2.4). It is interesting to note that the levels of concern about health and nutrition and sustainable food systems

Table 2.4. Food-related societal issues of concern to Generation Z.

	Not concerned [%]	Slightly and somewhat concerned [%]	Concerned and very concerned [%]	Weighted concern [%]
Health and nutrition	4	40	56	26.1
Sustainable food systems	4	38	57	26.2
Factory farming/unhealthy food production	11	41	48	24.4

were virtually the same with only 4% of the survey participants being not concerned and 56–57% being concerned or very concerned. Untrustworthy food production methods, such as factory farming and other unhealthy systems attracted slightly lower levels of concern — 11% not concerned and 48% concerned or very concerned (see Table 2.4). It appears that Gen Z across the 33 countries of the survey has quite a good knowledge of the challenges posed by food security.

Part of the explanation about this food awareness is Gen Z having higher food insecurity compared to other generations, with 1 in 3 young Americans not being able to afford food in the first half of 2022 (Polzin *et al.*, 2022). These youth lacked money and other resources to provide food for themselves. Education (e.g. without a college degree), income (e.g. below the poverty line) and race (e.g. Black and Hispanic) were the three driving factors for food insecurity. By comparison, 94% of the Baby Boomers were in a position to provide food for themselves (Polzin *et al.*, 2022). Similarly, in the UK, food insecurity was the highest among Gen Z at 32% compared to 6% for the Baby Boomers, 14% for Generation X and 18% for the Millennials (Galler, 2024). It is not surprising then that Gen Z is concerned about nutrition and consequently its health. University students are particularly affected as they need to borrow money, draw from savings or rely on their parents for support. In 2020, during the COVID-19 pandemic, 3.8 million college students in the USA lacked access to sufficient food resulting in skipping meals or eating less (GAO, 2024).

A solution to stable supply of nutritious and healthy food are sustainable food systems which deliver nutrition to all current and future generations without compromising the ability to produce food in the future (FAO, 2018). Such systems do not just produce food; they also provide synergistically economic opportunities, care for the natural environment and broad-based social benefits (Sustainable Development Goals, n.d.). The share of Gen Z who are seriously concerned about the sustainability of the global food systems is very high (see Table 2.4). In the absence of sustainable food systems, young generations are exposed to a situation of overconsumption by the wealthiest countries and sections of the society while leaving others without the necessary nutrition. Over-reliance of livestock-based foods, mainly in the Western types of diet, is contributing to the overexploitation of the planet's natural resources, including land, freshwater and fertilisers, and generation of air, soil and water pollution (Marinova and Bogueva, 2022). Gen Z is placing a lot of importance on the quality and sustainability of the food they eat. In the USA, 70% of Gen Z support climate-smart agricultural practices and more than half of these young consumers are prepared to pay higher prices for sustainably produced foods (Mitchell, 2024). A Canadian study revealed that the majority of Gen Z is in a stage of changing its food choices towards sustainable dietary behaviours (Ruzgys and Pickering, 2024). However, in situations where Gen Z are financially constrained, sustainability concerns give way to considerations about price, health and taste (Halicka *et al.*, 2025).

Gen Z's relationship with food is quite complex. They are often described as "foodies" who prioritise taste and food experiences, but are also rallying against social and environmental injustice. Factory farming and unhealthy food production practices is an important aspect of how they perceive food. There have been many influential books and documentaries exposing the cruelty of the current livestock practices — "Animal Liberation" (Singer, 1975, 2023), "The Omnivore's Dilemma" (Pollen, 2006), "Eating Animals" (Safran Foer, 2009); "Earthlings" (2005), "Cowspiracy" (2014), "Dominion" (2018), to name a few. The majority of Gen Z has not read or seen all of them; however, they have a good knowledge about the environmental impacts of livestock animals. Although younger people are more likely to abstain from animal-based foods and transition to plant-rich dietary choices, they seem to be doing this mainly for health benefits. Many of them however have started to question the origin of the foods they eat, what are the production practices and what impact they have. In our survey, close to 90% of Gen Z had a level of concern about the way food is produced. Two-thirds of Australian Gen Z reported that animal cruelty affected their spending when it comes also to cosmetic products; these young people are gradually shifting to sustainably produced meat, they are against live exports, caged eggs and factory farms (year13, 2022). Gen Z is not necessarily driving a vegan trend – the number of vegans among American Gen Zers is only 3% (Zulkosky, 2023), but Gen Z in principle objects against the unethical and inhumane treatment of livestock animals and other unhealthy food production practices. An Australian Gen Zer explains his food decisions: "Knowing that products are sourced in a manner that does not involve avoidable harm to animals and the environment can influence me to spend a few extra bucks knowing I'm supporting businesses that source products ethically" (year13, 2022, para. 16).

Conclusion

In our global survey, more than half of the university Gen Z participants described themselves as: loyal (77%), responsible (64%), thoughtful (63%), compassionate (62%), open-minded (61%), curious (58%), realistic (55%), sensible (54%), authentic (53%), cooperative (53%) and determined (51%). These are the qualities that will define their own future as well as the future of food. In the following chapters of the book, we offer some insights as to what the future of food is likely to be from the perspective of Gen Z. As with any forecast, we only see some indicative signs and there will be may other factors that will ultimately shape our dietary habits. With sustainability being a defining feature of Gen Z, it is likely that the interest in nutritious, healthy and sustainably produced foods will translate into consumer choices and improved production practices.

Exactly 50% of the university Gen Z believe that people are inherently good, meaning that the other 50% disagree with such a statement. This leaves all options open and the future will tell whether the glass is half full or half empty and whether we will see a transition to a more compassionate, healthier and environmentally better world. With Gen Z being at the cusp of all changes, they will not be acting alone. The older Baby Boomers, Generation X and the Millennials together with the younger Generations Alpha and Beta will also shape the future of food and the planet. For now, it is extremely reassuring to also see that the Gen Z young people have a positive outlook about this future — 74% feel optimistic about their own future and 72% believe that good things will happen to them. They are embracing the challenges created by previous generations on this planet, including in relation to food, and we can only look at the signs as to what their responses will be.

References

Annie E. Casey Foundation. (2021). Generation Z and mental health. https://www.aecf.org/blog/generation-z-and-mental-health

Australian Human Rights Commission. (2023). Talking about my generation. https://humanrights.gov.au/sites/default/files/document/publication/talking_about_my_generation.pdf

Australian Securities and Investments Commission (ASIC). (2023). Gen Z more concerned about finances than any other generation in Australia. https://asic.gov.au/about-asic/news-centre/find-a-media-release/2023-releases/23-302mr-gen-z-more-concerned-about-finances-than-any-generation-in-australia/

Bardales Salguero, R., Bogueva, D., & Marinova, D. (2024) Australia's university Generation Z and its concerns about climate change. *Sustainable Earth Reviews*, 7, 8. https://doi.org/10.1186/s42055-024-00075-w

Bogueva, D. & Marinova, D. (2022). Australian Generation Z and the nexus between climate change and alternative proteins. *Animals*, 12, 2512. https://doi.org/10.3390/ani12192512

Bogueva, D., Marinova, D., Wäechter, N., & Tekiner, I. H. (2024). Societal concerns. In Seemiller, C. & Grace, M. (eds) *Gen Z Around the World: Understanding the Global Cohort Culture of Generation Z*, pp. 119–127. Emerald Publishing, Leeds, UK.

Carrington, M. J., Neville, B. A., & Whitwell, G. J. (2010). Why ethical consumers don't walk their talk: Towards a framework for understanding the gap between the ethical purchase intentions and actual buying behaviour of ethically minded consumers. *Journal of Business Ethics*, 97, 139–158. https://doi.org/10.1007/s10551-010-0501-6

Chan, F. (n.d.). How to retain and nurture Gen Z employees. https://recruiters.prosple.com/resources-all/post/how-to-retain-and-nurture-gen-z-employees

Cottom, J. W., Cook, E., & Velis, C. A. (2024). A local-to-global emissions inventory of macroplastic pollution. *Nature*, 633, 101–108. https://doi.org/10.1038/s41586-024-07758-6

Cox, L. (2024). Gen Z's Military Malaise. Berkeley Political Review. https://bpr.studentorg.berkeley.edu/2024/02/27/gen-zs-military-malaise/

Crowley, T. (2025). 5 Reasons Why Military Recruitment is Struggling in the US in 2025. https://www.nsin.us/why-military-recruitment-is-struggling

Dascalu, D., Ogden, T., Retter, L., & Utting, K. (2024). Generation Z and joint professional military education. Final Synthesis Report. https://www.rand.org/pubs/research_reports/RRA3249-1.html

Davis, R. (2024). Why does the war in Gaza activate Gen Z more than any other current conflict? https://www.laloyolan.com/e2024/why-does-the-war-in-gaza-activate-gen-z-more-than-any-other-current-conflict/article_c60526d2-226b-5f65-97d3-5687bf7ce326.html

Dennington, A. (2021). Gen Ztressed: A new generation of college students. https://timelycare.com/blog/generation-z-college-students

European Institute for Gender Equality. (2025). Sexism at work: how can we stop it? https://eige.europa.eu/publications-resources/toolkits-guides/sexism-at-work-handbook/part-1-understand/what-sexism?language_content_entity=en

EY and JA Worldwide. (2023). How can we empower the next generations to build a more sustainable future. Generational Sustainability Survey. https://www.ey.com/content/dam/ey-unified-site/ey-com/en-gl/about-us/corporate-responsibility/documents/ey-ja-2023-sustainability-report-27-july-2023.pdf

Food and Agriculture Organisation of the United Nations (FAO). (2018). Sustainable food systems: Concept and framework. https://openknowledge.fao.org/server/api/core/bitstreams/b620989c-407b-4caf-a152-f790f55fec71/content

Galan, S. (2025a). Education worldwide — statistics & facts. Statista. https://www.statista.com/topics/7785/education-worldwide/#topicOverview

Galan, S. (2025b). Share of global population affiliated with major religious groups in 2022, by religion. https://www.statista.com/statistics/374704/share-of-global-population-by-religion/

Galler, G. (2024). Decoding generational preferences in food values, taste and security. New Food Magazine. https://www.newfoodmagazine.com/news/197404/decoding-generational-preferences-in-food-values-taste-and-security

Government Accountability Office (GAO). (2024). Supplemental Nutrition Assistance Program: Estimated Eligibility and Receipt among Food Insecure College Students. https://www.gao.gov/products/gao-24-107074

Graeter, D. (2024). Daphne's Dispatch: Gen-Z a mouthpiece for social justice. The Post. https://www.thepostathens.com/article/2024/10/gen-z-mouthpiece-social-justice-change-activism

Gray, G. (2020). The politically active generations: Millennials, Gen Z care about the debt — and more. George W. Bush Institute. https://www.bushcenter.org/catalyst/federal-debt/gray-politically-active-generation-cares-about-national-debt

Haidt, J. (2024). *The Anxious Generation: How the Great Rewiring of Childhood is Causing an Epidemic of Mental illness.* Alen Lane, London.

Halicka, E., Kaczorowska, J., Rejman, K., & Plichta, M. (2025). Investigating the consumer choices of Gen Z: A sustainable food system perspective. *Nutrients*, 17(3), 591. https://doi.org/10.3390/nu17030591

Institute for Health Metrics and Evaluation (IHME). (2024). The Lancet: Dramatic declines in global fertility rates set to transform global population patterns by 2100. https://www.healthdata.org/news-events/newsroom/news-releases/lancet-dramatic-declines-global-fertility-rates-set-transform

Koetsier, J. (2024). GenZ Dumping Google for TikTok, Instagram As Social Search Wins. Forbes. https://www.forbes.com/sites/johnkoetsier/2024/03/11/genz-dumping-google-for-tiktok-instagram-as-social-search-wins/

Konrath, S., Martingano, A. J., Davis, M., & Breithaupt, F. (2025). Empathy trends in American youth between 1979 and 2018: An update. *Social Psychological and Personality Science*, 16 (3), 252–265. https://doi.org/10.1177/19485506231218

Lissowsky, M., El Khoury, M., & Reiland, P. (2024). Gen Z Survey: Committed to Human Rights, Worried about Democracy. Friedrich Naumann Foundation. https://www.freiheit.org/human-rights-hub-geneva/gen-z-survey-committed-human-rights-worried-about-democracy-0#

Maheshwari, A., Vaghani, J., & Mistry, N. (2020). Analyzing levels of altruism in generation Z. *The International Journal of Indian Psychology*, 8(1), https://doi.org/10.25215/0801.025

Makkar, M., & Spry, A. (2024). Gen Z supports sustainability — and fuels ultra-fast fashion. How does that work? The Conversation. https://theconversation.com/gen-z-supports-sustainability-and-fuels-ultra-fast-fashion-how-does-that-work-238874

Marinova, D. & Bogueva, D. (2022). *Food in a Planetary Emergency.* Springer, Singapore.

Mayes, E. & Talukdar, R. (2023). As school students strike for climate once more, here's how the movement and its tactics have changed. The Conversation. https://theconversation.com/as-school-students-strike-for-climate-once-more-heres-how-the-movement-and-its-tactics-have-changed-217663

McCrindle. (2024). Digital. https://generationz.com.au/articles/digital/

McCrindle. (2025). What comes after Generation Z? Introducing Generation Alpha. https://mccrindle.com.au/article/what-comes-after-generation-z-introducing-generation-alpha/

McCrindle Research. (2025). The generations defined. https://mccrindle.com.au/article/topic/demographics/the-generations-defined

McNamara, L. (2024). Generation Z are committed to sustainability, but are their shopping habits? https://mumbrella.com.au/generation-z-are-committed-to-sustainability-but-are-their-shopping-habits-830522

Meade, A. (2024). 'Can't look away': Israel-Gaza war driving rise in news consumption among gen Z Australians, report suggests. The Guardian. https://www.theguardian.com/media/article/2024/jun/17/israel-gaza-war-increase-gen-z-media-consumption-digital-news-report-research

Mitchell, J. (2024). Gen Z Hungry for Transparency in Food Sustainability, Quality Amid Climate Concerns, According to New Survey. Quality Assurance & Food Safety. https://www.qualityassurancemag.com/news/gen-z-hungry-for-transparency-in-food-sustainability-quality-amid-climate-concerns-according-to-new-survey

Noor, A. (2025). What is Gen Z? and What is their age range? Qureos. https://www.qureos.com/hiring-guide/gen-z-statistics

Patfield, S. (2022). 'They don't expect a lot of me, they just want me to go to uni': first-in-family students show how we need a broader definition of 'success' in year 12. The Conversation. https://theconversation.com/they-dont-expect-a-lot-of-me-they-just-want-me-to-go-to-uni-first-in-family-students-show-how-we-need-a-broader-definition-of-success-in-year-12-196284

Perna, M. C. (2022). Deloitte: Almost half of Gen Z workers live with financial anxiety every day. https://www.forbes.com/sites/markcpern

a/2022/05/23/deloitte-almost-half-of-gen-z-workers-live-with-financial-anxiety-every-day/?sh5366738f07073

Petro, G. (2021). Gen Z Is Emerging as the Sustainability Generation. Forbes. https://www.forbes.com/sites/gregpetro/2021/04/30/gen-z-is-emerging-as-the-sustainability-generation/

Pietschnig, J. & Voracek, M. (2015). One century of global IQ gains: A formal meta-analysis of the Flynn effect (1909–2013). *Perspectives on Psychological Science*, 10(3). https://doi.org/10.1177/1745691615577701

Pollen, M. (2006). *The Omnivore's Dilemma: A Natural History of Four Meals.* The Penguin Press, New York, NY.

Polzin, S., Wahdat, A. Z., & Lusk, J. (2022). Why food insecurity among Gen Z is so much higher than for other age groups. The Conversation. https://theconversation.com/why-food-insecurity-among-gen-z-is-so-much-higher-than-for-other-age-groups-187736

Ramsing, R., Santo, R., Kim, B. F., Altema-Johnson, D., Wooden, A., Chang, K. B., Semba, R. D., & Love, D. C. (2023). Dairy and plant-based milks: Implications for nutrition and planetary health. *Current Environmental Health Reports*, 10, 291–302. https://doi.org/10.1007/s40572-023-00400-z

Ritchie, H., Samborska, V., & Roser, M. (2023). Plastic Pollution. Our World in Data. https://ourworldindata.org/plastic-pollution

Ruzgys, S. & Pickering, G. J. (2024). Gen Z and sustainable diets: Application of The Transtheoretical Model and the theory of planned behaviour. *Journal of Cleaner Production*, 434, 140300. https://doi.org/10.1016/j.jclepro.2023.140300

Ryan Jenkins. (2024). 14 Generation Z quotes that expose their unique and high tech lives. https://www.ryanjenkins.com/blog/14-generation-z-quotes-that-expose-their-unique-and-high-tech-lives

Safran Foer, J. (2009). *Eating Animals.* Little, Brown and Company, New York, NY.

Seemiller, C. & Grace, M. (eds) (2024). *Gen Z Around the World: Understanding the Global Cohort Culture of Generation Z.* Emerald Publishing, Leeds, UK.

Segal, M. (2024). Over 40% of Gen Z & Millennials Have or Plan to Change Jobs Over Climate Concerns: Deloitte Survey. ESG Today. https://www.esgtoday.com/more-than-40-of-gen-z-millennials-would-change-jobs-over-climate-concerns-deloitte-survey-finds/

Shaw, J. (2024). Gen Z: the Eco Conscious Generation. https://alfred24.com/gen-z-the-eco-conscious-generation/

Singer, P. (1975). *Animal Liberation: A New Ethics for our Treatment of Animals.* HarperCollins, New York, NY.

Singer, P. (2023). *Animal Liberation Now: The Definitive Classic Renewed.* Revised ed. HarperCollins, New York, NY.

Smith, A. (2024). How Gen Z uses social media and what that means for brands. https://sproutsocial.com/insights/gen-z-social-media

Spitznagel, E. (2020). Generation Z is bigger than millennials — and they're out to change the world. https://nypost.com/2020/01/25/generation-z-is-bigger-than-millennials-and-theyre-out-to-change-the-world/

Strickland, E. (2024). Australia's fiscal debt will cost Gen Z's future. The Ethics Centre. https://ethics.org.au/australias-fiscal-debt-will-cost-gen-zs-future/

Sustainable Development Goals. (n.d.). Fast Facts — What are sustainable food systems? United Nations. https://www.un.org/sustainabledevelopment/fast-facts-what-are-sustainable-food-systems/

Talev, M. (2024). America's youngest voters become major election liars. Axios. https://www.axios.com/2024/10/30/election-gen-z-voting-lies

Twenge, J. M. (2023). *Generations: The Real Differences Between Gen Z, Millennials, Gen X, Boomers, and Silents — and What they Mean for the Future.* Atria Books, New York, NY.

Union for Conservation of Nature (IUCN). (2024). Plastic pollution. https://iucn.org/resources/issues-brief/plastic-pollution

Vieira, J., Gomes da Costa, C., & Santos, V. (2024). Talent management and Generation Z: A systematic literature review through the lens of employer branding. *Administrative Sciences*, 14(3), 49. https://doi.org/10.3390/admsci14030049

Wiens, J. J. & Zelinka, J. (2024). How many species will Earth lose to climate change? *Global Change Biology*, 30, e17125. https://doi.org/10.1111/gcb.17125

World Meteorological Organisation (WMO). (2025). WMO confirms 2024 as warmest year on record at about 1.55°C above pre-industrial level. https://wmo.int/news/media-centre/wmo-confirms-2024-warmest-year-record-about-155degc-above-pre-industrial-level

year13. (2022). Gen Z's Animal Cruelty Awakening. https://www.youthsense.com.au/youth-insights/gen-zs-animal-cruelty-awakening/

YouthSence (2019). 96% of Australian Gen Zs Believe in Human-made Climate Change. https://youthsense.com.au/employers/gen-z-believe-human-made-climate-change/

Zulkosky, C. (2023). Gen Z, Veganism, and the Future. The Food Institute. https://foodinstitute.com/focus/gen-z-vegans-and-the-future/

Chapter 3

The Food Style of Today and Tomorrow: Gen Z's Food Preferences

Out of a total global population of 8.2 billion, a remarkable 1.9 billion belong to Generation Z (DataReportal, 2025, p. 19). This is not only a large demographic, it is the most educated generation in history. The majority of them are interested or already pursuing university education. In the USA, 50% of the adult Gen Z are enrolled in universities (GilPress, 2024). Globally, the shares of girls and boys attending university were respectively 43% and 39% in 2023 (Our World in Data, 2023). The attitudes and values of university Gen Z are very progressive and quite informative about life today and in the future. Gen Z is becoming increasingly influential in the food area with their preferences and direct spending. It is estimated that in 2021, these young people spent 23% of their income on food products (Lepicier, 2023). Globally, spending on food and non-alcoholic beverages is the second top item in Gen Z's budget at US$2,036 after housing, water, electricity and gas at US$3,447 (NielsenIQ, 2024).

In Chapter 2, we outlined the main societal issues of concern to global university Generation Z with the top five being sustainability, racism, education, access to healthcare and poverty. In a very unusual and probably unexpected way, we find that they are reflected in Gen Z's food related preferences and habits. The links are suggestive and open for interpretation but to us, there are clear identifying characteristics of the food style of Gen Z that define their attitudes and preferences. They are not only linked to the taste and appearance of dishes, but also to their hidden characteristics, such

Table 3.1. Characteristics of the food style of Generation Z.

#	Top societal issues for Gen Z (see Chapter 1)	Food style characteristics	Comments
1	Sustainability	Sustainability	Environmental footprint, including carbon emissions, land use and protection of biodiversity; preference for local food; animal welfare; biodegradable packaging
2	Racism	Global cuisine	Interest in a variety of cuisines, including traditional and local foods, new experiences, novel tastes
3	Education	Reliable information	Labelling, authenticity, online searches for confirmation, social media feedbacks, researching foods, discovering new ideas
4	Access to healthcare	Health consciousness	Healthy eating for physical and mental health, continuous energy, organic ingredients, freshly prepared options
5	Poverty	Speedy preparation and delivery	Online orders, build your own food, simple recipes, salads and bowls

as sustainability, globalised world, availability of reliable information, health credentials and speed of preparation or service, which reflect the values and expectations of this young cohort. Table 3.1 shows how the societal issues are reflected in the food style of Gen Z and we expand on these characteristics below.

In the park area of a decommissioned coal power station in Perth, Western Australia, young people are enjoying the communal space sitting on benches as part of a festival event (see Figure 3.1). There is music playing and the atmosphere is quite upbeat. The Gen Z people are also relaxing and relishing in the sunset glimpses above the stunning river. What one also notices in this idyllic picture is that the bench tables where there is normally food are empty. Some of the young people are having beer but there are no signs of food. Is this a food-free event? No, it was not supposed to be a food-free event but the organisers had little understanding of the food style of Gen Z and

Figure 3.1. Generation Z at a community event in Perth, Western Australia, 2025.

got everything wrong. The organisers were expecting Gen Z to bring their own food as for a picnic or join the queue to order pizza from the nearby truck. They were expecting that Gen Z would also drink a copious amount of alcohol, mainly beer, as previous generations still love doing. None of these assumptions are valid for Gen Z. One thing the organisers got right was the setting for the event — a lovely outdoor place close to nature with a beautiful relaxing view, but they did not know that Gen Z does not like to prepare and take picnic food and also does not like queuing for pizza. The rest of this chapter describes the main characteristics of Gen Z's food style which can inform similar events as well as any organisations producing, marketing or selling food to Gen Z.

Sustainability

Sustainability is a defining characteristic of Gen Z not only in our global university student survey but also when it comes to food. The connections are clear and they manifest in all sustainability aspects — environmental, social and economic. Gen Z does not equate sustainability only with eco-consciousness, environmental footprints of foods, use of chemicals, loss of native habitats and overexploitation of the oceans. This young cohort is very concerned about social values, such as modern slavery, gender equality, child labour, animal well-being or the rights of Indigenous people. For example, a survey in the USA, shows that the alignment of restaurants in supporting causes which resonate with Gen Z's social values, is important for 70% of the young people (Zuluaga, 2024).

Some analysts comment that Gen Z's true sustainability is challenged when they make decisions based on price (e.g. Quad, 2025), but in a market economy the economic dimension is very important for sustainability. Affordability represented by the price of food, be it at home or a restaurant, is a logical component of Gen Z's sustainability values. As one Gen Z puts it:

> *"If I'm buying something... like food or anything else... I want to know who made it, how they were treated, and what the company stands for. But yeah, I still have a budget, so it's about balance. Give me the facts and I'll decide what matters most."* (Male, born 2000)

This view encapsulates transparency and trustworthiness as core demands, even when economic limits are real. It is not apathy, it is pragmatism. Gen Z consumers are widely recognised for caring deeply about sustainability, labour rights and brand values, but they also navigate financial constraints that shape how those values play out in practice. A study in the UK using the Theory of Planned Behaviour found that Gen Z has strong ethical awareness and a desire to make responsible choices, but frugality and limited purchasing power often restrict their ability to act consistently on those intentions (Djafarova and Foots, 2022). What Gen Z also wants is clear and transparent information about the sustainability aspects of the businesses that produce, manufacture, distribute and supply food products.

Ideologically, Gen Z in our survey is distributed between left (29%), centre (26%) and right (11%) with one in five (22%) not adhering to any particular political views, that is not supporting any ideology (see Figure 3.2). This shows that sustainability is not related to any political ideology and is a uniting issue that transcends different worldviews.

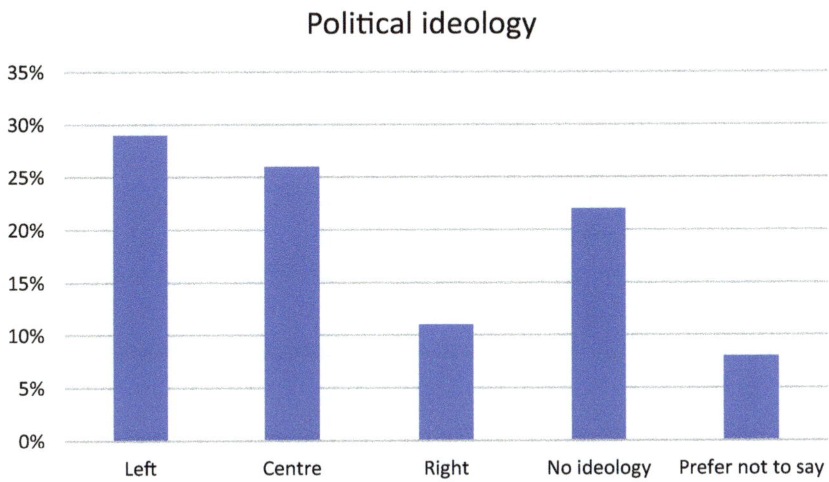

Figure 3.2. Political ideology of the global university Gen Z sample, 2022.

Gen Z however have a different attitude towards the sustainability of the products they use or consume. Many prefer and expect the foodstuffs to be sustainably produced, including local and organic options, and are prepared to pay extra for this (Martinez, 2025; PDI Technologies, 2023; Shaw, 2025). On average Gen Z is prepared to pay 10% more for sustainable products (NielsenIQ, 2024) and almost all (91%) want to buy from companies whose practices are sustainable (PDI Technologies, 2023). Also, Gen Z youth is not keen to recycle the packaging of any food or beverages, but instead expects convenience. This may mean the use of biodegradable and compostable packaging or alternatively, to be the responsibility of local government (Nuttall, 2025) or the producers/suppliers (please also refer to Chapter 13 for more details on packaging). In this way, Gen Z is pushing producers to lift their game, incorporate sustainability practices and offer better food products. In fact, by doing this Gen Z also changes the consumer behaviour of other generations (NielsenIQ, 2024). They want sustainability information to be available and easily accessible. Gen Z engages in conversations about sustainability and push for a more eco-conscious society (Noor, 2025). They encourage food products to have multiple sustainability credentials, such as being planet-positive, in compostable packaging and local. This young cohort does not want slow progress but instead as many sustainability criteria should to be addressed at once, as time is running out to stop climate change, biodiversity loss, deterioration of social and environmental conditions. Sustainably performing brands can enjoy loyalty with Gen Z and improved brand reputation about their products, including food items. It is likely that soon Gen Z will be checking and making purchase decisions based on the environmental, social and governance (ESG) reporting of food producers — economic factors that are shaping the current business environment, and the youth will also be fast in uncovering any greenwashing.

Gen Z has sustainability-oriented attitudes combined with digital literacy and desire for eco-friendly behaviour. A study in India shows that when compared to the older Generation Y, environmental knowledge and peer influence play an important role in shaping Gen Z's purchasing intentions towards sustainable products (Ghouse *et al.*, 2024). As far as food is concerned, this relationship is more complex. Gen Z in Australia has good level of knowledge about climate change but continues to consume high quantities of red meat — one of the biggest contributors to greenhouse gas emissions (Bogueva and Marinova, 2022). In Greece, university students are similarly not reducing their meat intake and sustainable food consumption is expressed only to occasionally buying fruit and vegetables that are in-season or locally grown (Kamenidou *et al.*, 2019). However, when US Gen Z is segmented according to levels of environmental involvement and ecological consciousness, there seems to be a direct correlation — the higher the environmental characteristics, the more attention young people pay to the sustainability and health attributes of food (Su *et al.*, 2019).

Packaging is very important for conscious Gen Z consumers who look for more sustainable food options. They also expect labelling and information that confirm the sustainability credentials of the food and its packaging. This is equally valuable for online orders, delivery and take-away options. Gen Z extends its sustainability values to other essential aspects of these young people's lives, such as clothing (Manley *et al.*, 2023), transport preferences (Fisu *et al.*, 2024), including e-bikes and other modes of active transport (Pickup, 2022), walkable neighbourhoods (Haynes, 2024) and communal living spaces with shared facilities (Parker, 2025).

The young people also realise that not everything can be perfectly aligned with their desires and admit that they have to be practically realistic. This however does not stop them to aspire for things to be better. The mindset of imperfect progress is echoed in the words of a young Gen Z who explained:

> *"It's not always perfect, right? I'll still grab a burger now and then, but I'll also ride my bike, thrift my clothes, and call out greenwashing when I see it. For me and most of my friends, it's about doing better where we can and not pretending to be saints."* (Female, born 2001)

This opinion shared with us reflects Gen Z's grounded approach to sustainability — realistic, socially influenced and deeply values-driven, even when behaviour does not always align neatly with intentions. These young people are also aiming to create jobs and employment careers that reflect their sustainability aspirations putting the planet and society first. They support businesses with sustainable and responsible practices and actively participate in sustainability initiatives looking to demonstrate real impact. This is the first generation to firmly position itself at the intersection of environmental, economic and social priorities (see Figure 3.3).

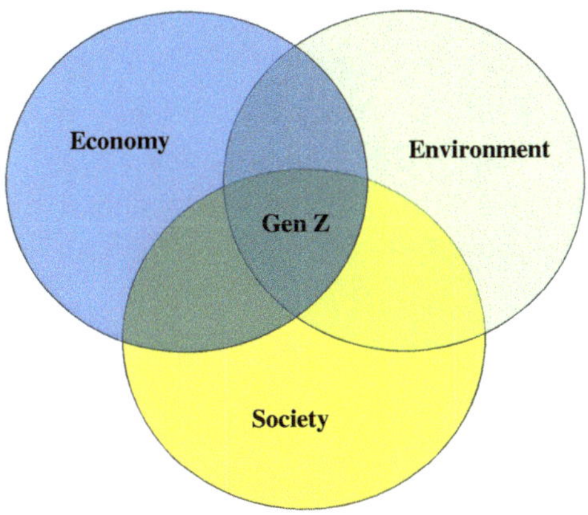

Figure 3.3. Gen Z at the centre of sustainability.

Global Cuisine

Racism as a form of discrimination or stereotyping, was listed as the second top issue of concern for our global university Gen Z sample. One reason for this is because Gen Z is often caught in racism manifestations in real life or online, either as bystanders, witnesses, victims and even perpetrators (Rismawati *et al.*, 2025). Experiencing racism is unpleasant and can be worse when perpetrated by people in power, such as police, educators, healthcare or the court system. This has deep ramifications for a person's life as it denigrates self-esteem, can cause trauma and other mental health problems. Often racism is expressed in the form of hurtful jokes, as this girl explains: "Starting in middle school, almost every day I would be asked if I ate cats or dogs..." (Abdelkader *et al.*, 2021, para. 2).

Being digitally connected across the continents, the ethnically and racially diverse global Gen Z has a progressive outlook at the world and low level of tolerance for racism. Globally, Gen Z is the least Western Generation (NielsenIQ, 2024). Diversity is particularly demonstrated in migrant countries, such as Australia, Canada, New Zealand and USA, where Gen Z is more racially, ethnically, linguistically and religiously diverse than previous generations. Many African countries, including Uganda, Liberia, Kenya and Nigeria are also highly ethnically diverse. According to the World Population Review (2025), diversity around the world is increasing due to population movements with increased global job economic opportunities and modern ways of travelling. In the USA, Gen Z believes that racial justice should be a top government priority (The Annie E. Casey Foundation, 2024) while in Australia, it sees racism as systemic, deep-rooted and widespread (Grants Assist, 2025). A UK Gen Zer describes her country having "deeply entrenched unconscious bias" (West, 2025, para. 5) and this is something young people want to change. Gen Z rallies around social causes and also joins forces in refusing to buy products from companies whose advertising campaigns are perceived as racist — a 2018 survey shows 79% of them would do this (Francis and Hoefel, 2018) but in 2022 this percentage dropped to 42% despite still being higher than the other generations

(Statista, 2024). It is inaccurate to argue that all members of the Gen Z cohort have the same outlook towards racism, but it is very revealing that in our global university sample from 33 different countries they voiced their concerns very highly. There are also those who think that racism is a cover-up for bigger and more serious issues within society: "It is a tool that has become consistently utilised and weaponised to divide us as a people and distract us from other issues" (West, 2025, para. 7).

There is a strong global connection between racism and food at a national and international level. It is intertwined with other social determinants of health and health equity, such as poverty, unemployment and disability (Odoms-Young and Marino, 2019). Racism can also be ingrained in the cultural perceptions about food, including who has historically consumed or produced certain varieties of foods. For example, foods associated with black culture in America, e.g. sweet potato pie, are considered by the general population less healthy and inferior (Fielding-Singh, 2021). In Brazil, experiences with racial discrimination were associated with higher food insecurity with the community food environment (including number, location and opening hours of stores) being an effect modifier, potentially indicating the impact of structural racism (de Vargas *et al.*, 2024).

Another aspect of the link between racism and food has been the dominance of Western or white people's cuisine in restaurants and shops with other cultural foods being seen as lower class or at the best exotic (Feller, 2021). Asian, African or Latin American dishes are being westernised to please the dominant culture in places, such as Australia, Canada, European countries and USA, with their flavours, aromas and tastes toned down. Looking down on other cultures' food has been part of racism that Gen Z no longer subscribes to.

Gen Z's curiosity about global foods goes beyond taste. For many, embracing global cuisine is also a form of solidarity, support and cultural respect, as one Gen Zer explained to us:

> *"For me and my friends, digging into another culture's dish isn't just about chasing bold flavours. You know..., it's kinda... about*

backing the people behind them. It's their way of saying, 'We see you, we respect your roots, and we're keen to stand alongside you'. It's something that the older generation will never get it... Sharing food becomes an act of solidarity, ... if you wish... a way of bridging worlds with a fork. Whether it's quietly slurping pho with your friends or showing up for a community cook-up..., I hope you get it... it's about honouring the story behind the pho or Pad thai, or tacos or whatever we are eating, not just the spices, the taste.... To me and my friends, every bite's a handshake across cultures." (Female, born 1999)

These words reflect a wider Gen Z mindset, one that embraces food as a form of cultural appreciation and everyday activism. For this generation, eating globally is not only about flavour, but about fostering inclusivity and challenging outdated hierarchies. In fact, Gen Z is intrigued by the richness and opportunities different cuisines around the world can offer. These young people believe that bringing diversity into society is beneficial for offering new experiences, especially through food.

Another contributing factor to Gen Z's interest in global cuisines is the fact that they travel a lot and are directly exposed to different cultures. They are attracted to street food common in places outside the Western world. Food trucks offering menus from different parts of the globe have now become part of the life on university campuses as well as festivals and other celebrations. Gen Z is motivated to discover new tastes, unfamiliar flavours and novel experiences. As digital natives, these young people also like taking and sharing photos of the food they eat making the experience very personal, photogenic and joyful. This user-generated content then becomes very useful for others to follow.

Gen Z likes food sharing, which again is more attuned to communal living than the Western individualistic lifestyles. Tapas, casseroles, hotpots, salads, pizzas, dips, sushi, rice paper rolls, nachos, dumplings, cheese platters and Indian snacks are all options from international cuisines that young people enjoy sharing together.

"Sharing food, for me, is how I push back against the disconnect. When the world feels divided, passing a plate is my way of saying,

> *'You're welcome here'. It's culture, care and connection... no big speeches, just real moments. A bite shared isn't just flavour... it's respect."* (Male, born 2000)

This sentiment captures the laid-back but deeply values-driven attitude many Gen Zers bring to the table. For them, food is not just about filling up, it is a way of building bridges, showing up for friends and embracing the rich mix of cultures that make up our modern world, including in nowadays Australia. Eating together also contributes to positive emotions making people feel happier and more satisfied (Albright, 2022).

Fruit juices, smoothies and carbonated soft drinks are more commonly drunk by Gen Z which some describe as "the sober curious generation" (Witherington, 2025). The level of alcohol consumption is much lower than that of previous generations although it may be replaced by energy drinks. Sober socialising is again more characteristic of global communal eating than Western gathering but it also helps create an atmosphere of well-being and joy.

In addition to hot coffee in a wide range of flavours, cold categories such as iced coffee, frozen blended coffees and cans are preferred by Gen Z who tend to start with this stimulating drink at a younger age (Fromm, 2023). Teas reminiscent of exotic cultures and traditions, including matcha, chai, fruity, herbal teas and blended varieties — hot and cold, are also popular with Gen Z who seek innovative flavours and the mood effects of these soothing liquids.

Gen Z's interests in desserts also exhibits global tastes, diverse ingredients and a wide range of origins, including from Asia, Europe and Latin America. This young cohort is familiar with flavours, such as lychee, dragon fruit, red bean, spirulina, passion fruit, papaya and macadamia. They like natural ingredients and small portions which bring up sweet memories from travelling but also simple childhood pleasures. According to consumer research from T. Hasegawa, "[s]howcasing high-protein, plant-based treats and those with more functional, nourishing ingredients can encourage [young] consumers to see them as a pleasurable part of a balanced diet, not as an indulgence" (Shoup, 2022, para. 15).

Reliable Information

The tech-savvy Gen Z is the most educated generation that has lived on this planet. In countries, such as USA and Australia, Gen Z students have low drop-out rates in high school and are more likely to pursue university studies than previous generations (Parker and Igielnik, 2020). Gen Z is also likely to have parents who are more educated than the previous generation (Dennington, 2021; McCrindle, 2025). With the global gap of educational achievement closing around the globe, in 2020 only 13% of people alive were without basic education compared to 46% in 1950 (Ritchie *et al.*, 2023). The education gap is also closing between genders with girls having higher university enrolment rates than boys (Our World in Data, 2023). Gen Z young people are the ones who most benefit from these positive changes.

Educational levels affect how people think but also how they find information about issues that matter to them. Education empowers people to lead more interesting and fulfilling lives and avoid hardship and poverty. It is not surprising then that one of the top societal concerns of Gen Z is education. These young people approach education with autonomy, equipped with digital technology and independent learning styles linking it to real-world problems (West, n.d.). They value the opportunity to find information and develop knowledge about the things that matter to them. Food is one area which they are very keen to understand.

In the search for personalised and tailored to their need answers, Gen Z is motivated in adopting generative artificial intelligence (GenAI), e.g. Chat Generative Pre-trained Transformer (ChatGPT), but concerns about reliability of information and misinformation remain (Chan and Lee, 2023). In response, Gen Z prefers combining integrating different sources of information and hybrid approaches which combine images, videos, audio and text (Seemiller and Grace, 2017). Functions, such as start, stop, fast-forward, fast-backward and replay are useful options in following a particular story or explanation (Ang *et al.*, 2022). Information from influencers is perceived as genuine, sincere and trustworthy as often they are viewed as peers and TikTok and Instagram are popular search engines (Camphouse, 2024). Polished advertising does not seem to have the same effect as for other age groups. Gen Z young people are very efficient in exploring a subject of interest and are prepared to challenge divergent opinions using technology and social media.

When it comes to food, Gen Z is likely to conduct its own investigation, be it about a particular dish, restaurant, nutritional profile, environmental footprint or origin. If Gen Z is checking up a particular food, it is likely to use multiple sources, including finding out from where the information used is and how reliable that is (Content Science, 2025).

They also expect manufacturers to be transparent and provide enough information on labels and the food packaging so that they can form their own opinion. In studies of Australian Gen Z's acceptance of alternative proteins in general and cultured meat in particular, transparency came as a major expectation and criticism towards current industry practices (Bogueva and Marinova, 2020, 2022). Transparency is needed to alleviate any suspicions as the ones shared by a participant:

> *"I heard and believe clean meat is part of big corporations' vast interests. It has nothing to do with meat scarcity or environment issues. It's part of another big scheme, like the production of drugs."* (Female, born 2002)

The same concerns are echoed by their European counterparts who also want evidence-based information and support from the food brands and governments promoting particular dietary options (EIT Food, 2019). These attitudes are explained by the pervasive lack of trust in businesses. According to the brand management platform Camphouse (2024), which organises marketing campaigns, by a 5 to 1 margin, Gen Z does not trust businesses to act in the interest of society. Gen Z is then compelled to find out the facts for itself and expose the truth for others while sending a clear message to the producer or supplier involved. According to an American Gen Zer: "Brands need to understand the problem they are causing and that their company will not exist if they don't fix the system. Their priority should be the same as the people's priority" (The Gen Z Reckoning, 2025).

Gen Z is also quick to share its experience and opinion online and through social media to contribute towards the available information on the web and inform others. Things that this group of young people values among others are: authenticity of the products with clear provenance and list of ingredients, original ideas about possible new ways of using familiar foods and confirmation of the quality of the products. They have a sense of community with responsibility that comes from it. These consumers also value ethics and moral responsibility of those producing foodstuffs as well as manufacturers following guidelines and meeting safety requirements (Czernyszewicz and Wiśniewska, 2024). Most importantly, Gen Z knows its rights and powers as consumers in the digitally connected world. With a global spending power expected to reach US$ 12 trillion by 2030 (Canphouse, 2024), Gen Z's insistence for information should not be ignored.

On a negative side however, there is the potential of information overload, the fear of missing out (FOMO) and conflicting messages which young people need to navigate on the social media as well as product information labels. This can leave them feeling overwhelmed and have mental health implications. Stress, anxiety and depression are common among Gen Z, including university educated young people. They search for responses in the same way — as a community of young people who want to do the right thing.

There is need to understand Gen Z as a new population cohort that faces a future with unprecedented challenges related to food, such as climate change and exploitation of species. They however handle these challenges collectively and head-on by expecting and searching for reliable information and then sharing their experiences (Katz *et al.*, 2021). By approaching food from an information lens and using digital and social media, Gen Z is taking ownership of its own health and destiny.

Health Consciousness

Described as "fresh foodies", Gen Z has a passion for eating and interest in learning more about food (Kılıç *et al.*, 2021). These young people are interested in food that supports and improves their physical and mental health which they see interconnected. A Gen Zer describes what being healthy means emphasising the link between physical condition and mental focus: "The state of only being able to move the body to do what has to be done cannot be said to be a healthy state. Personally, I think that doing any task with 100% concentration is a healthy state" (Heo and Jang, 2023, p. 1666)

Compared to previous generations, Gen Z is facing serious mental health challenges with one in three young people reporting depression, anxiety, stress and other common mental health problems (Tapper, 2024) while in our global university students sample that share was 54% (see Chapter 2), possibly under the pressure for academic performance, financial burden and social factors. The COVID-19 pandemic contributed to these high rates of mental health problems but so have the use of digital devices, the climate change prospects and geopolitical conflicts. Gen Z's struggle with mental health seems to be happening all across the world — in the Americas, Asia, Europe, Africa and Australia (Akbar *et al.*, 2024). Certain personality characteristics of Gen Z, such as being open, conscientious and empathic, also contribute to the decline in mental health (Akbar *et al.*, 2024). Although many strategies and interventions exist in the academic literature to improve Gen Z's mental health, ranging from counselling, emotional support and monitoring social media use to improved literacy, increased social engagement and self-management techniques, there is a distinctive lack of connections with food.

In response to the scale of the mental health problems, Gen Z is less interested to talk to the doctors and psychologists but instead is taking control of its well-being through what they are good at, namely social media. "People don't understand that anyone who's struggling can find a community online and share resources that work for them" — a female Gen Zer explains (Tapper, 2024, para. 38). Social media content and trends endorse and promote healthy, nutritious eating and are used by Gen Z as a source of information and advice. It seems that Gen Z's interest in healthy food is also a reflection of taking control over its life through this daily necessity. For Gen Z, healthy means above all fresh and natural as a Korean university student explains: "Many people exercise regularly and continuously or eat only natural food. Such self-care seems to be the first step to health" (Heo and Jang, 2023, p. 1666).

In Europe, young people see healthy eating as integral to their physical and mental health with 71% of them trying to do so while at home; although this share drops to 55% when eating out for convenience (EIT Food, 2019). Healthy eating means whole, organic and plant-based foods and they "are put off by processed foods, which 79% consider unhealthy" (EIT Food, 2019, p. 3). Counting calories to maintain a healthy weight is practised by only around a third (36%) of Gen Z with the majority thinking that this contributes to mental health issues (EIT Food, 2019). They even reject the need for caloric information on food labels and only around half (54%) consider low-calorie options as healthy (EIT Food, 2019). One Gen Zer explains:

> *"Look, I'm not counting every calorie or cutting out meat completely, but I am paying attention. I want food that's fresh, makes me feel strong, and doesn't add guilt or drama to my day. I'm definitely more aware of what goes on my plate. If it's fresh, makes me feel good, and doesn't stress me out, I'm in. It's not about extremes, it's about feeling balanced, staying true to myself, and knowing my choices matter. I'm not here for perfection. I'm here for food that respects me back."* (Female, born 2005)

This shared opinion reinforces Gen Z's intuitive and mental-health-informed approach to eating, while nodding to the idea of unconscious flexitarianism without naming it. Intuitive eating, which is not a diet but a way to respond to your body's needs, is gaining popularity as a personalised approach to eating (Jennings, 2023).

Relying on social media for information however comes with the potential pitfalls of misinformation — the spread of wrong information without the aim to deceive, and disinformation — deliberately misleading with false information (EIAT, n.d.). Young people can feel overwhelmed, confused and misled, particularly as food is a popular topic for both, content providers and users. Below are the words of a Gen Zer:

> *"There's so much info flying around online, it's hard to know what's legit. But I reckon if it's not processed and your nan would recognise it, you're probably on the right track. Like, if you can imagine her cooking it or growing it, it's probably not gonna mess you up. I've learned to trust real food over fancy marketing. The louder the health claim, the more I question it."* (Male, born 2000)

The scale of the problem can reach what has been described as infodemic — a flood of excessive and unreliable information that can threaten situation which can potentially cause public health risks (WHO, 2025). Gen Z may be tech-savvy, but that does not mean they are immune to algorithmic influence or viral misinformation. In fact, their digitally native status makes them more exposed to unverified trends, pseudoscience and influencer-driven health claims. What emerges from our conversations is a desire for clarity, honesty and simplicity — food advice grounded in culture, common sense and lived experience, not just clicks. Gen Z does not just want "clean eating", they want clean facts and a way to filter signals from noise.

There is a role for nutritionists, food scientists, researchers, educators and communicators to encourage Gen Z to develop critical thinking to be able to navigate the prolific information sources (Diekman *et al.*, 2023). Young people specifically want better advice on the connection between the food choices they make and their mental health. This advice needs to also take into consideration that Gen Z's information and social needs are difficult to separate — these young people want to know what others "like them" think and how relevant the information is to them personally (Hassoun *et al.*, 2023).

The foods that young people consider healthy generally align well with recommendations from reputable nutritional and health-related bodies, such as official dietary guidelines or advice from government health bodies. However, they also add the expectation that these foods will be fresh, all natural (that is with minimal processing), organic and a good sustainable source of nutrients. Taste and presentation are similarly important to create feel-good vibes

(Bambridge-Sutton, 2024). Functional foods which have additional health benefits, e.g. probiotics for improved gut health and adaptogens to help manage stress, are also appealing to Gen Z (Rand, 2025).

When it comes to food, there is always the possibility of discrepancies between intention and actual behaviour. For example, only 26% of older Gen Z in France and other European countries consume the recommended amount of fruit and vegetables (Bambridge-Sutton, 2024). Similar gaps are found almost everywhere where people have access to a wide choice of foods.

There is one highly important area, however, where Gen Z in wealthy countries, such as Australia, USA and UK, does not appear to be making the right choices, neither from a health, nor environmental point of view. This is in relation to the consumption of red meat. A 2024 survey of UK Gen Z shows that these young people are more likely than the total population to associate red meat (beef, pork, lamb and mutton) with healthy eating and they were less aware that such consumption should be in moderation (Forkes-Reece, 2024). Similar attitudes prevail in Australia, USA and many European countries. Such views are in conflict with existing scientific evidence (e.g. Marinova and Bogueva, 2022) as well as with national dietary guidelines. Red meat, and beef in particular, also has one of the highest environmental footprints. Despite Gen Z's reluctance to give up fully or partially red meat and move to diets, such as vegan, vegetarian or flexitarian, evidence from supermarket purchases shows that young people have reduced the amount of red meat they buy at a greater rate than other shoppers (Forkes-Reece, 2024). A possible explanation is that Gen Z is mainstreaming flexitarianism where meat is only one of many ingredients in the dish or meal. This group of young people appears to have become de-facto or unconscious (Forkes-Reece, 2024) meat reducers in the search for healthier diets.

Speedy Food Preparation and Delivery

Gen Z are progressively entering the workforce with even the youngest representatives — of the age of 14–15, allowed to be employed in some capacity. According to the Australian Bureau of Statistics (ABS, 2024a), 78% of Gen Z in Australia are working or studying full time and 11% are doing so on a part-time basis. Although use of digital devices is highly common around the world for work and leisure with the global average being 6.5 hours (Duarte, 2025), Gen Z spends longer on mobile and computer screens. The screen time is used to stream videos, TV, music and podcasts, play video games, engage with social media, browse websites and shop online. For Australian Gen Z, this screen time is 7.3 hours per day (ABS, 2024b) while in the USA, it is estimated to be more than 9 hours per day and 2 hours longer than the country's average (Duarte, 2025). Exercise, participating in sports, spending time in outdoor activities and travelling are all popular among Gen Z. Australian Gen Z spends an average of 1.7 hours per day on exercise, sport and outdoor activity (ABS, 2024b) while travel is a must for 34% of American Gen Z (Nuttall, 2025). New experiences, beautiful photographic opportunities which can be shared on social media and cultural encounters are highly valued by Gen Z.

The combination between work, study and leisure with high use of digital devices makes Gen Z time-poor with a myriad of demands on their time and attention. This means that young people are not inclined to spend a lot of time for meals or for their food to be ready. In fact, it is estimated that in the USA the average consumers have only 52 minutes per day to prepare and eat their meals and for Gen Z this time is down to 30 minutes (Nelson, 2024). They want to be able to earn money to satisfy their interests and needs with poverty or inability to do so being a major concern. Around 30% of Gen Z in the USA feel financially insecure and almost half of them (namely 46%) live from a paycheque to paycheque (GilPress, 2024). In addition to studying, many young people hold several jobs simultaneously with different rosters which puts even more pressure on their time. However, food is still important to them but their eating style is different. It is oriented towards speedy food preparation and delivery.

Gen Z are not elaborate chefs and at home usually cook for one taking shortcuts with many prepared, pre-cooked ingredients and ready-to-cook meals. As one of them explains: “I have a busy schedule, so convenience is a big thing...I’m the one who cooks in my apartment. I get garlic that is already cooked and ready to go” (Hyer, 2024, para. 20). They rarely use cookbooks but instead follow advice from TikTok videos “where you can find anything” (Hyer, 2024, para. 23), including “idiot-proof recipes” (Cost, 2025, para. 11). Youtube is similarly a popular source of information about new dishes as is Instagram. The homemade meals of Gen Z are easy to prepare with many fresh or frozen ingredients where presentation is important. With convenience and speed being top considerations, Gen Z is happy to use the microwave to prepare its meals.

Some describe the cooking style of Gen Z as a failure as the simplest techniques, such as whipping an egg, making an omelette, a simple soup or a stir-fry, appear to be too stressful and demanding (Cost, 2025). Only a third of Gen Z consider themselves skilled in the kitchen, but time is the most often-cited reason by two-thirds of them for not cooking at home (FunFacts, 2023). It is not surprising then that Gen Z often opt for eating out or ordering food delivery online.

Speed of service combined with quality food and innovative menus is a big drawcard for Gen Z when they dine out. They prefer mobile ordering, QR codes, from a kiosk or handheld devices which can speed up the process. The COVID-19 pandemic also made online ordering and home delivery popular and Gen Z are willing to explore different menu options from their favourite restaurants. Checking social media rankings and reading other customers’ feedback are a popular habit for this tech-savvy generation when it comes to exploring a new location or an unfamiliar restaurant. Food tourism is similarly gaining in popularity domestically (e.g. in China, Ding *et al.*, 2022) and internationally (e.g. creating new foodscapes, Park and Widyanta, 2022). Gen Z is also attracted to new dining concepts outside of the traditional restaurants, such as cafés, pizzerias and gourmet markets where they can get unique personalised authentic and fast experience. They also enjoy the build-your-own-meal concept where they can select the ingredients themselves and pay by weight. This allows them to select food ingredients that are healthy and fresh as well as to control their portions with small dishes being popular. In fact, this is similar to the way Gen Z "cooks", that is prepares its meals, at home. Customising your meal, including spices and condiments, is popular amongst Chinese hotpot restaurant goers. Make-your-own dessert is similarly attractive to young people where they can unleash their creative fantasy not just in taste but also with presentation.

In addition to price, Gen Z is also very sensitive to the service being fast when dining in (Zuluaga, 2024). When under pressure from the costs of living and inflation, Gen Z is less keen to start cooking and eating at home more frequently because it requires time that they do not have. Instead, Gen Z opts for saving more and tracking its spendings while eating at cheaper restaurants and ordering less expensive options.

Cooking and eating meals together have been traditionally ways to bring families, friends and communities together in a safe shared space. They also improve mental and physical health in times of hardship and other challenges. Gen Z are gradually losing this ability to enjoy shared quality meals at home which also reduces the opportunities for deep and meaningful social interactions. Researchers have identified these worrying trends and are putting forward initiatives and interventions that can restore some of the healing and binding functions of food by designing sharing-oriented food and kitchen products (Yuan *et al.*, 2023), offering online food preparation classes, including with online apps (Moh *et al.*, 2023) as well as going back to developing basic cooking skills. Young people in Brazil continue to engage with social media for cooking content but are looking for easy, speedy and simple recipes presented in a succinct way (de Camargo *et al.*, 2024). One participant explains: "A recipe that doesn't require a lot of time in the oven or in the fridge [to set], one you can use the blender and don't need to beat, mix, or grate... I do love carrot cake, but to peel and cut carrots is a lot of work, so it's easier to make a chocolate cake where everything goes in the blender" (de Camargo *et al.*, 2024, p. 3632) and another categorically stresses the importance of time: "if something shows-up and I see that it takes a long time to prepare, it's too complicated, bye" (de Camargo *et al.*, 2024, p. 3633).

Conclusion

Gen Z's food styles are distinctively different from those of their parents and grandparents. It is a generation that values sustainability, global cuisines and reliable information; it is health conscious when it comes to food and expects quick and speedy meal options. The inclusiveness of Gen Z defies borders and political systems as an American Gen Z girl shares: "I think it's important to include everyone. I don't want to have a society that has only one ideal" (The Gen Z Reckoning, 2025, para. 35). Gen Z does however share the ideal about living in a more sustainable world in response to all challenges created by previous generations.

When we started researching on Gen Z, they were the young and upcoming generation. Now they have matured not only in terms of age, education, jobs and spending power. Soon they will also be shaping the political and business environments. They are imposing their expectations for a kinder world where the global human community can live happier, with less stress and in harmony with the natural world. Many challenges lay ahead of them for this to happen. With digital technology in their disposal, they need to overcome mental health conditions, respond to the transgressed planetary boundaries, deal with increasing societal and political disfunction, and find their way out of the polycrisis. They need to heal themselves and the world.

A male Gen Zer explains how his generation feels: "For young people, the world at the moment is a very scary place. We can't control any of that. But there are things we can do..." (Tapper, 2024, para. 21). Food is one of these areas where young people can have control and it is up to them to decide and act on what they consume.

References

Akbar, B. M. B., Prawesti, D. R. D., & Perbani, W. S. A. (2024). Big picture mental health of Generation Z in the world. *Jurnal Kesehatan Komunitas Indonesia*, 4(1), 1–20. https://doi.org/10.58545/jkki.v4i1.223

Albright, M. B. (2022). *Eat & Flourish: How Food Supports Emotional Well-Being.* Blackstone Publishing, Ashland, OR.

Ang, W. H. D., Shorey, S., Lopez, V., Chew, H. S. J., & Lau, Y. (2022). Generation Z undergraduate students' resilience during the COVID-19 pandemic: A qualitative study. *Current Psychology*, 41, 8132–8146. https://doi.org/10.1007/s12144-021-01830-4

Australian Bureau of Statistics (ABS). (2024a). Education and work in Australia. https://www.abs.gov.au/statistics/people/education/education-and-work-australia/latest-release#survey-material

Australian Bureau of Statistics (ABS). (2024b). How Australian generations spend their time on recreation and leisure. https://www.abs.gov.au/articles/how-australian-generations-spent-their-time-recreation-and-leisure

Bambridge-Sutton, A. (2024). Is Gen Z embracing the healthy eating trend? Food Navigator. https://www.foodnavigator.com/Article/2024/12/13/does-gen-z-embrace-healthy-food/

Bogueva, D. & Marinova, D. (2020). Cultured meat and Australia's Generation Z. *Frontiers in Nutrition: Nutrition and Food Science Technology*, 7, 148. https://doi.org/10.3389/fnut.2020.00148

Bogueva, D. & Marinova, D. (2022). Australian Generation Z and the nexus between climate change and alternative proteins. *Animals*, 12, 2512. https://doi.org/10.3390/ani12192512

Camphouse. (2024). Gen Z Trends 2025: What Marketers need to Know. https://camphouse.io/blog/gen-z-trends

Chan, C. K. Y. & Lee, K. K. W. (2023). The AI generation gap: Are Gen Z students more interested in adopting generative AI such as ChatGPT in teaching and learning than their Gen X and millennial generation teachers? *Smart Learning Environments*, 10, 60. https://doi.org/10.1186/s40561-023-00269-3

Content Science. (2025). Gen Z Content Consumption Fact Sheet. https://review.content-science.com/gen-z-content-consumption-fact-sheet

Cost, B. (2025). Gen Z is a failure at cooking — even whipping up a simple egg is 'too stressful,' they admit. News.com.au. https://nypost.com/2025/01/27/lifestyle/gen-z-is-a-failure-at-cooking-even-whipping-up-a-simple-egg-is-too-stressful-they-admit

Czernyszewicz, E. & Wiśniewska, M. Z. (2024). Credibility of food processing companies in terms of food safety: perspective of Generation Z consumers. *Central European Management Journal*, 32, 2, 320–335. https://doi.org/10.1108/CEMJ-10-2022-0175

DataReportal (2025). Digital 2025: Global overview report. Meltwater. https://datareportal.com/reports/digital-2025-global-overview-report

de Camargo, A. M., Botelho, A. M., Dean, M., & Fiates, G. M. R. (2024). Beyond catching a glimpse: Young adults' perceptions of social media cooking content. *British Food Journal*, 126(10), 2024, 3624–3643. https://doi.org/10.1108/BFJ-01-2024-0091

de Vargas, E. B., da Silva Dias, M., Schuch, I., Zanini Rosa, P. B., Marcos Fanton, M., & Canuto, R. (2024). The community food environment as an effect modifier of the relationship between racial discrimination and food insecurity among adults in Southern Brazil. *International Journal for Equity in Health*, 23, 224. https://doi.org/10.1186/s12939-024-02311-3

Dennington, A. (2021). Gen Ztressed: A new generation of college students. https://timelycare.com/blog/generation-z-college-students

Diekman, C., Ryan, C. D., & Oliver, T. L. (2023). Misinformation and disinformation in food science and nutrition: Impact on practice. *The Journal of Nutrition*, 153(1), 3–9. https://doi.org/10.1016/j.tjnut.2022.10.001

Ding, L., Jiang, C., & Qu, H. (2022). Generation Z domestic food tourists' experienced restaurant innovativeness toward destination cognitive food image and revisit intention. *International Journal of Contemporary Hospitality Management*, 34(11), 4157–4177. https://doi.org/10.1108/IJCHM-07-2021-0903

Djafarova, E. & Foots, S. (2022). Exploring ethical consumption of generation Z: Theory of planned behaviour. *Young Consumers*, 23(3), 413–431. https://doi.org/10.1108/YC-10-2021-1405

Duarte, F. (2025). Alarming Average Screen Time Statistics (2025). Exploding Topics. https://explodingtopics.com/blog/screen-time-stats#impact-of-screen-time

EIT Food. (2021). Our Food, Our Food's System. What Generation Z Wants from a Healthy Food System. https://www.eitfood.eu/media/news-pdf/Our_Food,_Our_Food_System_-_EIT_Food_report_.pdf

Electoral Integrity Assurance Taskforce (EIAT). (n.d.) Disinformation and Misinformation. https://www.aec.gov.au/About_AEC/files/eiat/eiat-disinformation-factsheet.pdfAustralianGovernment.

Feller, M. (2021). I'm a Black Dietitian — Here's What I Want You to Know About Food and Racism. Healthline. https://www.healthline.com/health/food-nutrition/black-dietician-food-and-racism

Fielding-Singh, P. (2021). How Racism Shapes Our Perception of Healthy Food. https://www.sciencefriday.com/articles/racism-healthy-food

Fisu, A. A., Syabri, I., & Andani, I. G. A. (2024). Urban dynamics and Gen-Z mobility: The influence of land use diversity and density on daily trip patterns in Indonesia. *Sustainable Futures*, 8, 100388. https://doi.org/10.1016/j.sftr.2024.100388

Forkes-Reece, C. (2024). Gen Z: A generation of health-conscious consumers. https://ahdb.org.uk/news/consumer-insight-gen-z-a-generation-of-health-conscious-consumers

Francis, T. & Hoefel, F. (2018). 'True Gen': Generation Z and its implications for companies. https://camphouse.io/blog/gen-z-trends

Fromm, J. (2023). The Gen Z Mindset Is Changing Coffee. Forbes. https://www.forbes.com/sites/jefffromm/2023/12/06/the-gen-z-mindset-is-changing-coffee/

FunFacts. (2023). Culinary Confessions: Cooking Habits of Gen Z & Millennials. https://www.homeruninnpizza.com/news-blog/fun-facts/cooking-habits-gen-z-millennials/

Ghouse, S. M., Shekhar, R., & Chaudhary, M. (2024). Sustainable choices of Gen Y and Gen Z: Exploring green horizons. *Management & Sustainability: An Arab Review*. https://doi.org/10.1108/MSAR-04-2024-0018

GilPress. (2024). Gen Z: Statistics, Data and Trends (2024). https://whatsthebigdata.com/gen-z-statis/

Grants Assist. (2025). Gen Z — What Social Issues Are Important to Gen Z? https://australiangrants.org/gen-z-what-social-issues-are-important-to-gen-z/

Hassoun, A., Beacock, I., Consolvo, S., Goldberg, B., Kelley, P. G., Russell, D. M. (2023). Practicing information sensibility: How Gen Z engages with online information. Proceedings of the 2023 CHI Conference on Human Factors in Computing Systems, Article No. 662, pp. 1–17. https://doi.org/10.1145/3544548.3581328

Haynes, M. (2024). Gen Z focuses on walkable neighbourhoods and public transportation. https://blog.kubapay.com/gen-z-focuses-on-walkable-neighbourhoods-and-public-transportation

Heo, M. L. & Jang, Y. M. (2023). Implications for health management through an analysis of changes in life patterns of Generation Z college students experiencing the COVID-19 pandemic: A qualitative study. *Journal of Multidisciplinary Healthcare*, 16, 1661–1670. https://doi.org/10.2147/JMDH.S415120

Hyer, H. (2024). What's cooking with Gen Z: 4 marketing tips from a fellow Gen Zer. https://www.supermarketperimeter.com/articles/10989-whats-cooking-with-gen-z-4-marketing-tips-from-a-fellow-gen-zer

Jennings, K.-A. (2023). A Quick Guide to Intuitive Eating. Healthline. https://www.healthline.com/nutrition/quick-guide-intuitive-eating

Kamenidou, I. C., Mamalis, S. A., Pavlidis, S., & Bara, E.-Z. G. (2019). Segmenting the Generation Z cohort university students based on sustainable food consumption behaviour: A preliminary study. *Sustainability*, 11(3), 837. https://doi.org/10.3390/su11030837

Katz, R., Ogilvie, S., Shaw, J. & Woodhead, L. (2021). *Gen Z, Explained: The Art of Living in a Digital Age.* The University of Chicago Press, Chicago, IL.

Kılıc, B., Bekar, A., & Yozukmaz, N. (2021). The new foodie generation: Gen Z. In Stylos, N., Rahimi, R., Okumus, B., & Williams, S. (eds) *Generation Z Marketing and Management in Tourism and Hospitality: The Future of the Industry*, pp. 223–248. Springer Nature, eBook. https://doi.org/10.1007/978-3-030-70695-1

Lepicier, R. (2023). Reaching Generation (Gen) Z Through Food and Flavor. https://www.linkedin.com/pulse/reaching-generation-gen-z-through-food-flavor-ryan-lepicier-mba/

Manley, A., Seock, Y.-K., & Shin, J. (2023). Exploring the perceptions and motivations of Gen Z and Millennials toward sustainable clothing. *Family & Consumer Sciences*, 51(4), 313–327. https://doi.org/10.1111/fcsr.12475

Marinova, D. & Bogueva, D. (2022). *Food in a Planetary Emergency.* Springer, Singapore.

Martinez, T. (2025). Tracking the Latest Gen Z Food Trends in 2025. https://pos.toasttab.com/blog/on-the-line/gen-z-food-trends-and-statistics

McCrindle. (2025). What comes after Generation Z? Introducing Generation Alpha. https://mccrindle.com.au/article/what-comes-after-generation-z-introducing-generation-alpha/

Moh, T. K., Hoang, T., Nguyen, V., Truong, T., & Nguyen, T. (2023). Cooking PaPa: An online application helps removing the barrier between modern generation and cooking. *Artificial Intelligence and Applications*, 1(3). https://doi.org/10.47852/bonviewAIA3202629

Nelson, A. (2024). Homemade made easy. https://digital.supermarketperimeter.com/sosland/sp/supermarket-perimeter-march-2024/index.php#/p/20

NielsenIQ. (2024). Spend Z: Gen Z changes everything. https://nielseniq.com/wp-content/uploads/sites/4/2024/12/NIQ_GenZ-Report-22FL_11.26.pdf

Noor, A. (2025). 20+ Gen Z Statistics for Employers. https://www.qureos.com/hiring-guide/gen-z-statistics

Nuttall, C. (2025). 12 characteristics of Gen Z in 2025. https://www.gwi.com/blog/generation-z-characteristics

Odoms-Young, A. M. & Marino, B. (2019). Examining the impact of structural racism on food insecurity: Implications for addressing racial/ethnic disparities. *Family and Community Health*, 41, S3–S6. https://doi.org/10.1097/FCH.0000000000000183

Our World in Data. (2023). Gender gap in primary, secondary and tertiary education, World. https://ourworldindata.org/grapher/gender-gap-education-levels

Park, E. & Widyanta, A. (2022). Food tourism experience and changing destination foodscape: An exploratory study of an emerging food destination. *Tourism Management Perspectives*, 42, 100964. https://doi.org/10.1016/j.tmp.2022.100964

Parker, K. (2025). Do Gen Z Even Care to Buy a Home Anymore? Is It All Just Real Estate Dystopia? https://kittyandmiles.com.au/do-gen-z-even-care-to-buy-a-home-anymore/

Parker, K. & Igielnik, R. (2020). On the Cusp of Adulthood and Facing an Uncertain Future: What We Know About Gen Z So Far. Pew Research Centre. https://www.pewresearch.org/social-trends/2020/05/14/on-the-cusp-of-adulthood-and-facing-an-uncertain-future-what-we-know-about-gen-z-so-far/

PDI Technologies. (2023). The Green Generation: Tracking Gen-Z Trends in Sustainable Purchasing Habits. https://pditechnologies.com/blog/tracking-gen-z-consumer-trends-sustainability/

Pickup, O. (2022). How Gen Z is driving the sustainable commuting movement. https://www.worklife.news/leadership/how-gen-z-is-driving-the-sustainable-commuting-movement/

Quad. (2025). Marketing to Gen Z: This generation's focus on sustainability trends with caveats. https://www.quad.com/insights/marketing-to-gen-z-this-generations-focus-on-sustainability-trends-with-caveats

Rand, S. (2025). Gen Z food trends: What today's young US consumers want. https://www.askattest.com/blog/research/gen-z-food-trends

Rismawati, R., Muhaimin, M. Y., Sanur, I. S., Ahmadin, A., Aba, A., & Khaer, F. (2025). Character education as an antidote to racism among Generation Z. *Jurnal Edukasi Terkini*, 2(1), 23–32. https://doi.org/10.70310/jet

Ritchie, H., Samborska, V., Ahuja, N., Ortiz-Ospina, E., & Roser, M. (2023). Global Education. Our World in Data. https://ourworldindata.org/global-education

Seemiller, C. & Grace, M. (2017). Generation Z: Educating and engaging the next generation of students. *About Campus*, 22(3), 21–26. https://doi.org/10.1002/abc.21293

Shaw, J. (2025). Why Gen Z Values Sustainability: Tips for Marketing to the Eco-Conscious Generation. Kadence International. https://kadence.com/why-gen-z-values-sustainability-tips-for-marketing-to-the-eco-conscious-generation

Shoup, M. E. (2022). From ube to gulab jamun... Gen Z and Millennials embrace 'diverse, unusual flavors' in desserts and confections. https://www.foodnavigator-usa.com/Article/2022/11/22/gen-z-and-millennials-embrace-diverse-unusual-flavors-in-desserts-and-confections

Statista. (2024). Share of Gen Z who avoid brands that are not anti-racism in the U.S. 2022. https://www.statista.com/statistics/1353165/us-avoid-anti-racist-brands-gen-z/

Su, C.-H., Tsai, C.-H., Chen, M.-H., & Lv, W. Q. (2019). U.S. sustainable food market Generation Z consumer segments. *Sustainability*, 11(13), 3607. https://doi.org/10.3390/su11133607

Tapper, J. (2024). What's up with Generation Z? The Guardian. https://www.theguardian.com/society/2024/mar/03/whats-up-with-generation-z

The Annie E. Casey Foundation. (2024). Social issues that matter to Generation Z. https://www.aecf.org/blog/generation-z-social-issues

The Gen Z Reckoning. (2025). A Generation at the Tipping Point. https://genzreckoning.com

West, H. (n.d.). Gen Z in Higher Education: How Their Values, Learning Styles, and Tech Use Are Changing Universities. https://www.goodcourse.co/post/teaching-gen-z-rethinking-education-for-a-new-generation

West, P. (2025). What Gen Z gets wrong about 'racist' Britain. The Spectator. https://www.spectator.co.uk/article/what-gen-z-gets-wrong-about-racist-britain/

Witherington, A. (2025). Gen Z: The Sober Curious Generation. https://www.mintel.com/insights/food-and-drink/gen-z-sober-curious-generation

World Health Organisation (WHO). (2025). Infodemic. https://www.who.int/health-topics/infodemic

World Population Review. (2025). Most Racially Diverse Countries 2025. https://worldpopulationreview.com/country-rankings/most-racially-diverse-countries#is-diversity-rising-or-falling-in-the-world

Yuan, Y., An, W., Li, J., Tang, H., & Lin, J. (2023). Cooking experience design for generation z in the digital age. In Marcus, A., Rosenzweig, E., & Soares, M. M. (eds) *Design, User Experience, and Usability*. HCII 2023. Lecture Notes in Computer Science, Vol. 14031. Springer, Cham. https://doi.org/10.1007/978-3-031-35696-4_26

Zuluaga, T. (2024). Gen Z And Millennial Restaurant Trends. Toast. https://pos.toasttab.com/blog/on-the-line/gen-z-food-trends

Chapter 4

The Plant-Based Revolution: Gen Z and the Rise of Green Eating

"If slaughterhouses had glass walls, everyone would be a vegetarian."

— Paul McCartney

This powerful statement by Paul McCartney, the legendary musician from The Beatles and outspoken advocate for plant-based living, underlines the ethical and environmental considerations that drive the shift toward plant-rich diets. His words highlight the disconnect between consumers and the realities of industrial animal agriculture, suggesting that greater transparency could inspire more compassionate and better dietary food choices. This message strongly resonates with Gen Z, a generation increasingly embracing plant-based eating for ethical, environmental and health reasons. Through years of conversations with Gen Z individuals from Australia and across the world, we continue to explore their evolving perspectives, shedding light on the factors driving their commitment to greener eating habits.

Before we delve into the conversations and Gen Z's perspectives, it is good to clarify the terms we use. The diet vocabulary is quite vast and constantly changing but here we focus only on some of the terms from a sustainability point of view without going into too much detail about the origins, popularity, prescribed quantities, pros and cons of the different dietary styles. First is healthy eating — a way of eating that is beneficial to human health. Green eating or a planetary diet is a way of eating that is beneficial for human health while not putting too much pressure on the planet's biophysical systems. Given all scientific evidence we have, making plant-based dietary choices has the co-benefits of being good for human health and for the planet; hence, it represents green eating. The broad categories of foods defined as plants are: fruits, vegetables, nuts, seeds, legumes, whole grains, roots and tubers. Another clarification that is worth noting is that there are two other groups of foods that are not covered in the plant category but can be part of green eating; they are fungi (e.g. mushrooms) and algae (e.g. seaweed). We dedicate a separate chapter (Chapter 7) discussing this, but for now, let's make the assumption (be it scientifically incorrect) that any plant-related diet also includes consumption of fungi and algae. Furthermore, some plants are consumed raw, others are processed and cooked to improve their value for the consumer. The types and levels of processing vastly vary as people have mastered food production throughout the history. Soybeans, for example, are legumes but they have traditionally been processed into tofu and more recently, into foodstuff that mimics or is considered analogous to the appearance, taste and nutritional value of animal-sourced foods. Other plants, such as nuts, peas and oats (as well as soybeans), are being processed into plant-based milks. The plant-based analogues and plant-based milks are also described as plant-based alternatives (to animal-sourced products), new protein alternatives or plant-based substitutes.

Going back to the diets, the terms plant-forward, plant-rich, plant-strong and plant-based choices have a high degree of similarity but there are important nuances from a green eating perspective. They are all an improvement from the standard western diet which is high in animal-derived foods. The term that is probably less commonly used, but is the most inclusive is plant-forward. It describes a dietary style that favours and encourages the consumption of plants without excluding other food groups. It can be described as "put more plants on your plate" (Have a plant, 2021, p. 6). It overlaps to a certain degree with a flexitarian diet but the latter emphasises reduction in meat consumption while plant-forward tackles the issue from the opposite perspective assuming that eating more plants would lead to less consumption of meat and other animal products.

A plant-rich (or plant-centric) diet is similar to a plant-forward diet but also accepts the consumption of plant-based alternatives. If people are reluctant to eat the new protein alternatives, their plant-rich diet essentially would be identical to a plant-forward diet.

The last term is plant-based diet. This is essentially a diet based entirely on the consumption of plants, including fungi and algae; it can also incorporate plant-based alternatives. Such a diet is quite restrictive of all foods of animal origin. In essence, a plant-based diet is identical to a vegan diet without the philosophical, ideological, emotional, political, animal welfare and other connotations that the latter has. The final comment is about a vegetarian diet which can also be associated with particular worldviews; it is similar to a plant-forward or plant-rich diet but excludes mainly meat, fish and seafood, and allows other food groups, such as dairy, eggs and honey. We can now put the focus on Gen Z and how they navigate the dietary landscape of green eating.

For many Gen Z individuals, adopting a plant-based diet started as a commitment to animal welfare but has since evolved into a broader responsibility to protect the planet. Research shows that Gen Z is at the forefront of embracing plant-forward lifestyles, such as vegetarian, vegan and flexitarian, driven by a deep-rooted commitment to sustainability and social responsibility (Chen, 2024; Zulkosky, 2023). This generation's dietary choices reflect a larger movement toward conscious consumption, where environmental impact and ethical considerations play a crucial role in shaping food preferences. Many Gen Z individuals recognise the environmental benefits of plant-based eating, such as reduced greenhouse gas emissions and lower resource consumption, aligning with their commitment to combating climate change (Marinova and Bogueva, 2022). As one of the Gen Z individuals explained, climate change has been a constant presence in their generation's upbringing, influencing their worldview and sense of urgency:

> *"Unlike our parents, we've grown up with climate change as a constant reality. It's been part of everything... everywhere... our conversations, education and news since we were born. Growing up with this constant reality has shaped how we think... our*

> *perspective in a way that makes any inaction feel impossible... You know... with our world... inaction just isn't an option. We're not only aware of this crisis... around us.... We're also surrounded by solutions and urgent calls to act, which has created a strong 'do something now' mindset in our generation. We don't have the luxury of ignoring the problem.... like our grandparents and even parents do. We've been taught by our parents that the future depends on us and the choices we make today, including with the food we consume."* (Female, born 2003)

In line with the urgent calls from the EAT-Lancet Commission (Eat Forum, n.d.) and Chatham House (Benton *et al.*, 2021) to adopt a plant-based diet to prevent environmental collapse, young people are embracing the message and adjusting their eating habits more rapidly than older generations. One Gen Z individual highlighted the growing trend of plant-based eating among young people while also acknowledging the lack of agreement between them and their parents or relatives regarding their food choices. This generational divide reflects differing perspectives on diet, sustainability and ethical morally just consumption, with Gen Z often prioritising ethical, environmental and health-conscious decisions that may not align with the views of older generations:

> *"It feels like the older generation just doesn't take our way of eating... my veganism seriously. They question it, doubt its benefits, and sometimes even see it as unnecessary and dangerous. I've been vegetarian for nine years since primary school and recently I've become fully vegan... and it's causing a lot of tension, especially with my parents and grandparents. They just don't get why I've made this choice,... they simply can't get it. I try not to mention food in our conversations as when we talk about food, this often turns into debates and even a fight. It's frustrating because,.... for me, it's about ethics, animal welfare and also to a great extent about sustainability and health... But to them, it's just a 'phase' of being young... or something really extreme. I wish there was more understanding instead of constant resistance."* (Female, born 2001)

The above comment reflects a personal struggle that aligns with the broader rise of green eating, highlighting how younger generations are embracing sustainable food choices despite resistance from older family members. A frustration with family scepticism underscores a common challenge faced by those adopting green eating habits — while they see it as a necessary shift for sustainability, health and animal welfare, others may perceive it as extreme or temporary. This generational divide highlights the cultural and social barriers that often accompany shifts toward more sustainable diets.

Notwithstanding this, young people are endorsing green eating and in this chapter we examine some of the trends and then move to analysing the barriers and challenges they have. Plant-based alternatives are replacements of livestock-based products introduced by the food industry. We comment on how Gen Z relates to them before concluding that these young people are prioritising health while looking for solutions for the future.

The Rise of Green Eating

Green eating is about making mindful food choices that support both personal health and the natural environment. It involves prioritising ethical production, whole, minimally processed foods, reducing reliance on resource-intensive products and actively working to cut down on food waste (Fong *et al.*, 2021). This approach also encourages choosing whenever possible locally sourced or organic ingredients, as these options often have a lower environmental footprint (Marinova and Bogueva, 2022). Beyond just what is on the plate, green eating reflects a broader commitment to sustainability, considering how food is produced, transported and consumed to minimise harm to the planet while promoting long-term food security.

By mapping the challenges and enablers of green-eating behaviours, our conversations with Gen Z provide valuable insights into the dynamics of their food choices. Understanding these factors, we believe, is essential for fostering a shift toward more sustainable eating habits and helping this generation navigate the complex relationship between health, ethics and sustainability. Many Gen Z individuals recognise that their generation is at the forefront of a shift toward healthier and more environmentally conscious eating habits.

One of the Gen Zers we spoke to reflected:

> *"I think there is a trend that has been changing recently among many young people I know. And many people are starting to eat healthier and healthier. And this starts with my generation (Gen Z), because these are the people who are much more sensitive to the topic of ecology, and also... a pro-environmental lifestyle. It's not happening with all people from my generation, I still shouldn't make generalisations about everyone, but I think the majority are".* (Female, born 2002)

This shift is not only about sustainability but also about personal things. Many Gen Zers see food choices as a tangible way to take control of their health and well-being. When we asked about healthy eating, and how they and their friends feel about the importance of mindful food choices, many shared that it is a priority in their daily lives:

> *"Certainly, quite a lot of my friends emphasise healthy eating... We are mindful of our food choices. This is in our interest to take the things that depend on us like.... things, such as what we eat, into our own hands."* (Female, born 2000)

Maintaining a healthy lifestyle requires a high degree of personal responsibility. Many Gen Zers feel that they have significant control over their eating habits, as this is one aspect of health that they can actively manage:

> *"You know, it's important to choose your food properly.... but also is important to accept the external factors that can influence my choices. I am vegetarian and I consume a predominantly plant-based diet, but often things such as access to affordable, cheap and healthy food are an issue.... Also, cultural influences, I mean now we are in Australia and eat all these beautiful foods but back in Germany we don't have such an abundant veggies' and fruits' variety you have here. Also, we must be mindful of time constraints, as sometimes even [if] you have the best intention to eat healthy, you may not have the time to cook or take the food from somewhere. All of these can impact the extent of our control with food. Despite all these challenges, I still believe that if I am taking ownership of my dietary decisions, [this] is a powerful step toward achieving my own healthier lifestyle."* (Female, born 2001)

Beyond simply choosing healthier foods, Gen Z also approaches eating as an exploratory and cultural experience. They are eager to experiment with different cuisines, valuing authenticity and traditional food cultures over ultra-processed alternatives:

> *"We love to explore food, to form our food preferences by trying. We like to discover other foods, other flavours. We would like to explore and try all the cuisines in the world... to dive into their traditional food culture, to eat real food, not to eat something that is artificially made out of not real ingredients you cook yourself, but something that is artificially produced... like the novel proteins... the plant-based meats."* (Male, born 1998)

However, this enthusiasm for green eating does not necessarily translate into unconditional support for food industry innovations. While Gen Z is curious about novel protein products, they often prefer to base their judgments on firsthand experience rather than industry claims. The plant-based revolution, driven in part by this generation, is reshaping not only how we think about food but also what kinds of food are acceptable. While they champion sustainability and conscious eating, many Gen Z consumers prefer whole, natural vegetables and fruits over highly processed plant-based meats. For them, eating green is not just about replacing animal products. It is about embracing simplicity, authenticity and minimally processed foods.

Gen Z's Commitment to Plant-based Diets: Motivations and Trends

In general, consumers' desire to transition toward a more plant-based diet and reduce their meat consumption is shaped by a complex network of psychological, cultural and environmental factors. One of the key barriers to this shift is attachment to consuming animal flesh because of habitual and identity-driven connections to meat and its taste (Graça *et al.*, 2015; Bogueva and Marinova, 2018; Bryant *et al.*, 2019). Many individuals perceive meat as an essential component of their diet, often associating it with tradition, pleasure and even social status, making it difficult to replace with plant-based options (Bogueva and Phau, 2016).

Food choice motivations play a crucial role in shaping dietary preferences as individuals prioritise different factors such as taste, convenience, cost, health benefits and ethical considerations when making food-related decisions (Graça *et al.*, 2019; Van Loo *et al.*, 2017). For some, the sensory appeal of meat remains a dominant factor, while others may be more influenced by health or ethical concerns (Bogueva *et al.*, 2021).

Perceptions about plant-based diets also significantly impact consumer adoption rates. While some individuals view plant-based eating as a healthier and more sustainable alternative, others perceive it as restrictive, nutritionally inadequate or lacking in variety (Fehér *et al.*, 2020; Adamczyk *et al.*, 2022). Such perceptions can either encourage or deter the transition to a plant-rich diet, depending on the level of awareness, exposure and access to plant-sourced food options.

Gen Z's approach to plant-based eating is characterised by a combination of commitment and flexibility. While a notable 72% of US Gen Z vegans intend to maintain their diet over the next five years, only 27% of the generation currently consumes plant-based foods most or all of the time (Zulkosky, 2023). Nonetheless, a considerable further 65% express a desire to adopt a more plant-forward diet, reflecting a growing inclination to reduce, rather than completely eliminate, meat consumption. This highlights a shift from strict veganism or vegetarianism to a more flexible, reducetarian or flexitarian approach.

Research suggests that Gen Z often opts for a vegan diet for health reasons, such as making them feel better, avoiding illnesses and improving digestion. This is the main motivator for 52% of the surveyed US Gen Z consumers (Zulkosky, 2023). Many (40%) also care about the environment, but only 17% say it is their top reason. Similarly, 17% are influenced by ethical concerns, including animal welfare, but these are less important than health. Social media, documentaries and influencers are helping raise awareness, which might make ethics a bigger factor in the future (Zulkosky, 2023). A Gen Zer explains:

> *"I first went vegetarian because I wanted to feel somehow healthier as most of the time I was lacking energy. I cut out all animal products and I think this really helped me with my digestion problems and energy levels. Eating a plant-based diet was just the right choice for me. You asked me about... whether I have chosen it because of considerations for my consumption's environmental impact and ethics reasons. These are also important to me, but honestly, my health was the biggest factor. I think a lot of people in my generation start this way, and then as we learn more from social media and documentaries, we care more about the bigger picture."*
> (Female, born 1999)

Another young adult from Gen Z explained that many of their friends, despite identifying as almost 99% vegetarian or vegan, occasionally make exception when specific dishes like fish or chicken are served. This behaviour could stem from cultural, social or personal factors, such as the significance of sharing meals during social gatherings, blending in with others, the belief that these exceptions do not strongly conflict with their ethical principles, or even simply a weakness for certain flavours or dishes they find hard to resist. This flexible approach may reflect the Gen Z's tendency to avoid rigid labels and embrace a more personal, pragmatic way of navigating their dietary choices. It also aligns with the flexitarian dietary options where the aim is to significantly reduce meat consumption without giving it up completely.

Barriers to Vegetarianism and Veganism among Gen Z: Challenges and Opportunities

There are numerous barriers that make adopting plant-based, vegetarian or vegan lifestyles challenging for many people. According to Perez-Cueto *et al.* (2022), who identified barriers to adopting plant-based foods across 10 European countries, challenges to such diets vary depending on individual eating habits with omnivores being more reluctant to make such a shift. Common obstacles include the belief that consuming animal-based meat is natural for humans and that we need animal products for adequate nutrition. Some concerns exist regarding the nutritional adequacy of plant-based foods, particularly doubts about satiety and their ability to provide sufficient energy and strength (Perez-Cueto *et al.*, 2022). There are also concerns about the taste of the new plant-based alternatives, in some cases combined with negative past experiences with these products. A study of 10 European countries revealed notable gender differences in the barriers to dietary change, with women scoring lower in all barriers to adopting plant-based foods (Perez-Cueto *et al.*, 2022). Factors such as portion sizes, food preferences and attitudes contributed to these differences, underscoring variations in health risks and susceptibility to disease (Perez-Cueto *et al.*, 2022).

Many studies have identified social norms, food neophobia, disgust (Onwezen *et al.*, 2021), preparation challenges and cooking skills, lack of information (Fehér *et al.*, 2020), enjoyment of the taste of meat and the belief that humans are meant to consume large amounts of meat, as significant obstacles (Rickerby and Green, 2024). Some barriers are based on misperceptions. For example, some consumers associate plant-rich eating with strict dietary rules or bland, unappealing food, which can deter them from giving it a try (Goldman *et al.*, 2025). Stigma and limited cooking skills further hinder the transition to plant-rich eating (Graça *et al.*, 2019). In fact, Rickerby and Green (2024) identified 40 barriers influencing the reluctance to adopt a plant-rich diet, including financial constraints, lack of knowledge, emotional attachments, distrust in the food system and suppliers, health concerns, convenience, social dynamics, enjoyment of meat, environmental considerations, accessibility, personal capability and media influence. Among the 40 identified barriers, concerns about nutritional intake and dietary requirements, grouped under the broader "health" theme, were perceived the biggest obstacle (Rickerby and Green, 2024) despite ample evidence to the contrary.

Specifically for Gen Z, notwithstanding the growing interest in plant-based diets, several barriers prevent widespread vegetarianism and veganism in the USA. Reluctance to adopt veganism stems from a mix of personal, nutritional and environmental concerns. Among non-vegans, 51% stated they would not give up meat (Zulkosky, 2023), showing how deeply ingrained western dietary habits are. Others — 21%, worry about potential nutritional deficiencies (Zulkosky, 2023). Opinions about veganism's environmental impact are divided with nearly half being sceptical of its benefits and ability to solve ecological challenges; only 40% believed that it helps protect the environment, and a smaller section — 12%, perceived mixed effects (Zulkosky, 2023).

These findings indicate the complexity of adopting veganism on a broader scale although transitioning to plant-rich diets may be easier. While individual dietary preferences and nutritional concerns are significant barriers, the scepticism about environmental benefits highlights the need for more robust education and transparent data about the impact and advantages of veganism. For many, the decision to adopt a plant-based lifestyle requires not only practical solutions, such as improved access to appealing alternatives, but also addressing misconceptions and providing clear evidence of its advantages for human health and the planet. Without tackling these concerns holistically, the shift toward plant-rich diets may struggle to gain widespread traction.

Although some think Gen Z leads the vegan trend, data shows that only 3% of them are strictly vegan, which is less than the 4% share of Millennials (Zulkosky, 2023). According to the same US survey, 79% of Gen Zers who are not vegan have no plans to adopt veganism in the near future (Zulkosky, 2023). Gen Z is interested in plant-based eating but prefers a flexible approach rather than being vegan. They focus on health, the environment and ethics, making small, lasting changes that could impact on the food industry.

While Gen Z's interest in plant-based eating signals a shift toward flexibility, it is important to note the diverse attitudes within this group. From our conversations with Gen Z, we noticed a clear divide, ranging from committed vegans to devoted carnivores. This observation aligns with findings from a study conducted across Greece, India and the UK, which also identified two distinct consumer groups: proponents and opponents of plant-based diets (Raptou *et al.*, 2024).

These groups showed statistically significant differences in their perceptions, ranging from the health benefits of plant-based diets to their attachment to animal-based proteins, dissatisfaction with plant-based alternatives' attributes, and concerns about adequate protein intake. The study further revealed a striking "homogeneity" in the factors influencing youth's intentions to adopt plant-based diets across the three countries (Raptou *et al.*, 2024). Key drivers included attitudes toward plant-based eating, meal preparation habits, perceived barriers to green eating and personal factors, such as knowledge of healthy eating and physical activity levels. These insights suggest that while Gen Z may appear divided, the factors shaping their choices — both obstacles and motivators, are often similar regardless of geography.

As one Gen Z individual explains, the trend is more about balance and flexibility than strict dietary choices:

> *"I'm all for eating more veggies and making better food choices, but that doesn't mean I have to give up meat completely. A lot of us in Gen Z are more about balance than extremes. We care about health, sustainability and the planet, but we also want flexibility in what we eat. You told me about the stats that most of us aren't planning to go fully vegan anytime soon, which is correct. But that doesn't mean we aren't making changes. Cutting back on meat and eating more plant-based foods is something I can get behind, but giving it up entirely? Not so much. It's about progress, not perfection, you know (Ha-ha-ha)"* (Female, born 2000)

While these challenges highlight why many Gen Z individuals hesitate to adopt veganism fully, they also present opportunities for change. By addressing cultural, nutritional and perception-based barriers through education, innovation and better communication, the shift towards plant-forward eating can become more accessible, sustainable and widely accepted.

> *"I can't eat differently. It's difficult to explain. Going plant-based isn't just about food for me. It's about kind of staying true to my way of being healthy, respecting sustainability, and my kindness to animals we have to kill to consume them. We need to consider that even small choices add up, and my choice to become vegan*

> *I believe helps create a better future for the planet and also the next generations coming after us."* (Female, born 2004)

This perspective reflects a deeply personal and value-driven approach to plant-based eating. For the individual, adopting a plant-based consumption style is not merely a dietary choice but an expression of health, environmental responsibility and compassion for animals. The belief in the cumulative impact of small, conscious decisions shows a commitment to creating a global sustainable future.

Plant-based Alternatives: Not a Major Part in Plant-based Eating

In the last few years, plant-based meat alternatives have tried to take the consumer market by storm. They were very positively received especially by the Millennials' generation who has long been viewed as catalysts in the adoption of plant-based meat alternatives, driven largely by their support for flexitarianism (Knaapila *et al.*, 2022), but were not so visibly liked by the representatives of Gen Z.

Despite the intention behind these alternatives being to represent a feasible solution to environmental challenges and global food insecurity, their broader acceptance remains uncertain due to concerns over processing, nutritional value and lack of consumer trust.

> *"I completely agree that we need to rethink the way we consume food in today's world. I truly hope that the dietary changes adopted by my generation can contribute to combating climate change. One effective approach to achieving this is through reducing meat consumption, which has the potential to significantly lower greenhouse gas emissions and promote sustainable food systems. However, I believe these changes should not come at the expense of compromising the quality and content of the food we eat. Personally, I don't see fake meat as the solution to this problem."* (Male, born 2000)

Mistrust toward plant-based meat alternatives seems common, with scepticism largely stemming from their heavily processed nature and ambiguous ingredient lists. These factors have raised doubts among consumers regarding the health benefits and authenticity of such products, further undermining their appeal (Fehér *et al.*, 2020). While reducing meat consumption aligns with the environmental goals of Gen Z, there is also a growing recognition that plant-based alternatives should uphold the integrity of natural, minimally processed foods. Some young consumers are wary of highly processed meat substitutes, questioning whether they truly align with the values of sustainability and health. Instead of embracing lab-engineered products, they advocate for a switch toward whole, natural foods which for them better reflect the essence of plant-based eating. As one respondent emphasised, plant-based diets should not just replace meat with ultra-processed alternatives but rather focus on unprocessed, nutrient-dense ingredients that align with both personal and environmental well-being:

> *"Plant-based doesn't mean eating alternatives, despite being labelled as plant-based. Many of these products are not really plant-based. They rely heavily on artificial ingredients and chemicals you can't pronounce rather than whole, natural foods. They are more chemical and artificial ingredients-based. These are not actual foods I would embrace. For me, true plant-based eating means embracing real, minimally processed foods, not heavily engineered substitutes."* (Female, born 2001)

The emphasis on authentic, minimally processed foods and the move toward reducing meat consumption share common ground in addressing sustainability and health concerns. Both perspectives highlight the desire to move beyond synthetic alternatives and embrace dietary solutions that feel real, natural and culturally resonant. Together, these viewpoints suggest Gen Z's generational demand for sustainable, minimally processed food solutions that align with environmental goals and the desire for genuine, healthful eating experiences. Such insights highlight a potential path forward fostering innovation in natural, plant-based foods that address these shared priorities.

While Gen Z admires technology, including food technology, and recognises its potential to address issues related to food scarcity or supply chain challenges, many also point out its limitations. Lack of naturalness, which this generation finds extremely important, is one such limitation. These young people are gravitating toward "back-to-basics nutrition", favouring minimally processed foods with clean labels and transparent ingredients. This also reflects a desire for traditional, nostalgic products that offer comfort and a sense of familiarity, as they allow to reconnect with simpler times (Moors, 2024). Hence, there is resentment towards plant-based meat alternatives.

Foods in the eyes of Gen Z are deemed "natural" when they lack human alteration, such as heavy processing, adding enhancements or removing undesirable elements (Rozin, 2005; André *et al.*, 2019). Examples of natural, non-meat protein sources for this generation often include lentils, pulses, soybeans as well as mushrooms, other fungi and algae in the form of sushi. In one of our conversations, a Gen Z individual shared:

> *"You know... a lot of my friends have gone vegan... They cut out animal products entirely... They do everything to distant themselves from animal-sourced food and products. They eat only veggies, beans, seaweed and fruits. They're wearing vegan shoes, drinking plant milks... oat or almond milk, and eat nuts and beans... But that doesn't mean they're okay with eating something that feels artificial or heavily processed, like some of the plant-based meats out there. We want food that not only feels real, but is real... not something that's been too processed or engineered in a lab... There's a big difference."* (Male, born 1998)

Even when plant-based meats happen to align with Gen Z's expectations, another significant challenge emerges — the high price of these alternatives, which can limit accessibility for many. Furthermore, for individuals who follow a vegan lifestyle for animal welfare reasons, there is an added reluctance. They prefer to avoid products designed to mimic the appearance and taste of meat, even when they are plant-based. A Gen Z's opinion explains:

> *"I find food technology incredibly fascinating, as it holds the potential to address major challenges like future food shortages. However, at this stage, these alternatives are often prohibitively expensive, making them inaccessible for the majority of people. Additionally, for many of us who embrace veganism for animal welfare reasons, the idea of eating something that mimics the look,*

> *taste or texture of meat feels incompatible with the values that led us to adopt this lifestyle".* (Female, born 2001)

This highlights a nuanced perspective within Gen Z, balancing an openness to innovation with a deep commitment to authenticity, affordability and ethical living. This opinion also explains a global market for plant-based meat alternatives which is not going as strong as originally predicted. In the USA, despite the meat alternatives being consumed now more than before the COVD-19 pandemic, they represent only 1% of the sales in the meat category and the strongest interest comes from Gen X and the Millennials (Circana, 2025). With a price gap of US$4.20 above total livestock meats, the new alternative proteins are significantly less popular among Gen Z. Following the initial surge in investment, challenges within the industry have led to the shutdown of certain companies and the discontinuation of specific product offerings.

The sector has faced difficulties in maintaining growth and with competition. Since 2020, the number of alternative meat brands in the USA has decreased by 28%, dropping from 116 to 83 (Mridul, 2024). This consolidation reflects a maturing industry where competition has intensified, and the space is no longer driven mainly by novelty. To thrive in this evolving market, brands must set themselves apart whether through ground-breaking innovation, compelling marketing strategies or strategic collaborations. The industry transformation features a shift from rapid expansion to a focus on long-term viability, where only the most adaptable and distinctive players will endure. However, consumer interest is a major factor in the maturing of the industry and there is a distinct reluctance of Gen Z to embrace these products.

One of the Gen Z individuals we had the opportunity to speak with shared their perspective, highlighting the challenges plant-based alternatives face in developing and maintaining their popularity. Initially, brands such as Beyond Meat and Impossible Foods captured attention due to their novelty, sparking widespread curiosity. Over time, however, enthusiasm waned as concerns around flavour, quality of ingredients and transparency began to surface. These factors led many to question the authenticity and value of such products, resulting in a decline in trust and lack of sustained interest (Bambridge-Sutton, 2024). A study in Austria found country of origin and price were the most important attributes for Gen Z consumers (Meixner *et al.*, 2024). Our Gen Zer explains:

> *"I remember when some of these brands first came out, everyone was so excited to try them. People were rushing to find places that sold Beyond or Impossible products, mostly out of curiosity. But among my friends and others I know, this trend quickly came and went. Honestly, it's not surprising. Some of these products just don't feel exciting — whether it's the taste, the ingredients or the overall quality. They lack authenticity and definitely don't feel worth the high price... Why should I stick to something that doesn't even feel like it's here to stay?! Many of the friends who used to buy these products stopped because they no longer felt good about eating them. For them, there was no compelling reason to continue, aside from their own health, but even that wasn't enough. What really put me off was the lack of transparency about the ingredients and the brands' missions... despite all the advertising. I'm not just looking for a meat replacement. I want a product that truly aligns with my values and makes me feel good about my choices. For me, plant-based meat products are not serving that purpose anymore. It seems like no one is buying them these days."* (Male, born 1997)

This insight captures the essence of the challenge and opportunity for the meat alternative market. It highlights a larger shift in consumer priorities. People are not just seeking substitutes, they are looking for food options that resonate with their personal values, promote well-being and justify their cost. The struggle of plant-based brands to meet these expectations underscores the need for the industry to evolve and focus on creating offerings that are not only sustainable but also meaningful and genuine to the people they aim to serve.

Innovation to Support Green Eating

Within the innovation literature, the emergence of the plant-based alternatives and the technologies associated with them have been described as a new technological innovation system in a sustainability transition towards more plant-sourced proteins (Tziva *et al.*, 2020). Marinova *et al.* (2017) describe these technologies as being a defining part of the latest and still emerging 7th Technological Wave when we are likely to see further clustering of research, development and application of new ways of producing food that are more sustainable and regenerative for the planet's health. What is particularly characteristic and a distinctive feature of these technological innovation systems is the role of consumers as a driving force in these processes (Tziva *et al.*, 2020). Gen Z consumers will play a deciding role for the future of plant-based alternatives.

To win over this observing and demanding generation, brands must move beyond merely imitating meat and focus on delivering authentic, meaningful products that resonate with Gen Z's values. In a competitive and maturing market, success will hinge on the ability to innovate with purpose and foster trust with consumers. Some companies are looking already into creating hybrid products or plant-based meats that include precision fermentation-made ingredients. Hybrid products combine animal meat with plant-based alternatives to achieve improved taste and better overall health and sustainability performance. Plant-based meats with precision fermentation are using microorganisms to produce key ingredients that replicate animal-based qualities. Companies are leveraging this technology to create dairy-like proteins and heme, enhancing the taste, texture and nutritional profile of plant-based alternatives (Boukid *et al.*, 2023). There are likely to be many more emerging innovative solutions in the alternative protein industry as economies around the world step firmly into the 7th Technological Wave.

Another aspect of the 7th Technological Wave is that human health shifts to the centre of innovative activities. While all previous waves were directed towards different ways of exploiting the natural resources, the 7th Technological Wave focusses on human health together with the well-being of the planet's biophysical environment. This shift is similarly driven by consumer demands and expectations. The perception of healthiness in plant-based alternatives is a crucial concern for Gen Z, particularly those who identify as health-conscious

Figure 4.1. Farmers' market in Sofia, Bulgaria.

eaters. Many in this generation prioritise whole, minimally processed foods that align with their active lifestyles and personal wellness goals (Marinova and Bogueva, 2022). Local farmers' markets (see Figure 4.1) are popular with Gen Z, particularly if they are conveniently located as this allows them to genuinely connect with food and aligns with a shared sustainability ethos (Aanyamittu, 2023).

One Gen Z individual shared the following perspective with us:

> *"I'm actually one of those people who eats pretty healthy. I enjoy cooking at home and choosing foods that I believe are best for my body. As someone who is quite athletic, I focus on nutrition that supports my lifestyle, and many of my friends feel the same way. We prioritise foods that promote the overall well-being and rarely indulge in fast food. When it comes to plant-based meats, I don't see them as a healthy choice — they often feel overly processed, and for me, they don't align with my idea of clean eating."* (Male, born 1997).

This sentiment reflects a broader scepticism among health-conscious Gen Z consumers, who may be drawn to plant-based options for sustainability reasons but remain cautious about ultra-processed meat alternatives. For them, healthy eating is not just about eliminating animal products but about making informed, balanced choices that truly benefit their bodies. This is part of Gen Z's focus on practical, day-to-day improvements in emotional and physical health (Moors, 2024).

Exploring Gen Z's attitudes toward plant-based alternatives reveals a complex interplay of curiosity and scepticism. While there is a desire for innovation, many plant-based meat alternatives fail to meet the health claims that align with Gen Z's preference for a healthier dietary lifestyle. Misconceptions and facts are often mixed casting shadow on both new and established products.

For example, ingredient composition plays a critical role in product acceptance. Soy, one of the most common and cost-effective alternative protein sources, faces notable resistance from this demographic, a finding echoed in research by Meixner *et al.* (2024). This aversion seems to stem from a mix of cultural perceptions and broader concerns about environmental and health impacts. While soy is valued for its nutritional profile, functional qualities and environmental performance, Gen Z often associates it with over-industrialisation, genetic modification and outdated practices. As one young person shared with us:

> *"I'll eat it if I have to, but I don't really trust this soy product. It feels not right. If I think soy, I think genetically modified. It's really outdated, and there are better options now."* (Male, born 1997).

Despite its numerous advantages, the negative perception about soy may undermine the appeal of such products among Gen Z consumers. Given this resistance, the industry needs to re-evaluate how it positions and communicates its products. Innovation can also help by moving to, for example, pea protein to replace soy which could be better received and gain a market traction with Gen Z. Innovation is needed not only at the product level, but also in marketing, packaging, distribution channels and points of contacts with this value-driven and health-conscious generation.

Driving Change

In conversations with Gen Z, an intriguing and recurring theme emerges — a keen awareness of and involvement in driving change, particularly in the food industry. Unlike previous generations, Gen Z does not just witness transformation, they actively shape it. This mirrors their broader ethos across various domains, from sustainability efforts spearheaded by figures like Greta Thunberg to the reimagining of consumer behaviour and industry practices. Gen Z consistently demonstrates a commitment to influencing change in everything they do.

According to a 2024 Deloitte survey, which collected insights from 23,000 respondents across 44 countries, 61% of Gen Z and 58% of Millennials believe they possess the power to drive change within their organisations (Deloitte, 2024). This sense of agency extends beyond workplaces and reflects broader expectations for the world at large (Deloitte, 2024). In a conversation, a Gen Z individual highlighted a striking dynamic — many people tend to underestimate the scale of change at its commencement and, more critically, fail to grasp the far-reaching impact it will eventually have. In the context of the food industry, this insight becomes especially relevant. Most food companies anticipate that Gen Z, now an influential generation, will enthusiastically embrace technological advancements in food, including alternative proteins and lab-grown innovations. However, it is important to note that Gen Z's relationship with food is shaped not just by novelty but by deeper values like sustainability, authenticity, transparency and ethical responsibility. As such, their support for these "fruits" of technology may depend on how well these innovations align with their principles. For instance, transparency, environmental impact and cultural relevance will likely influence the adoption of such products.

Millennials were among the first to embrace plant-based meat alternatives as innovations, drawn by their curiosity and values of sustainability and ethical consumption. Their enthusiasm helped integrate these products into mainstream grocery stores and restaurant menus, influencing even fast-food chains. For Gen Z, the adoption of plant-based alternatives was less of a surprise and more of an inevitability. While Millennials may have pioneered the initial acceptance of plant-based alternative products, Gen Z approaches them with a pragmatic lens, focusing on and questioning their authenticity, seeking transparency and alignment with their values (Marinova and Bogueva, 2022). One Gen Z representative reflected:

> *"The adoption of these products may seem like a trend to older generations, but for us, it's just part of a larger movement towards sustainable and conscious living. It's not about replacing meat with a fad... it's about reshaping how we think about food and its impact."* (Male, born 1996)

This perspective highlights the generational shift in how food technology is embraced — not merely as a novelty but as a step toward systemic change.

However, challenges persist. The barriers to adopting plant-rich diets, including cost, accessibility, taste preferences and cultural norms, remain significant. Scepticism about the heavily processed nature and lack of transparency of some plant-based alternatives further dampens enthusiasm, with many questioning their health benefits and authenticity. While Millennials helped bring these products into the spotlight, Gen Z pushes for improvements that better align with their priorities, such as sustainability, nutritional value, transparency and ethical integrity.

For food companies, this presents a critical challenge. They must navigate the expectations of a generation that not only recognises the need for change but actively demands it. The normalisation of plant-based foods demonstrates the power of generational influence in reshaping the food system, yet it also signals the need for industries to evolve continuously. Meeting the high expectations of Gen Z requires innovation that is not only sustainable but also meaningful, transparent and deeply aligned with their values. For Gen Z, the rapid normalisation of plant-based alternatives was simply a natural progression rather than an unexpected shift.

Rather than viewing the plant-based products as mere innovations or passing trends, Gen Z sees them as stepping stones in a broader journey toward aligning consumer habits with ethical and environmental values.

> *"You know, it's frustrating for my older Millennial sister how slow people are to take food trends seriously. My parents thought oat and soy milks were just a hipster thing — something for people chasing obscure food trends. But now they're the ones adding it to their coffee. I do too, sometimes, but when I do, I can't help thinking that as plant-based milks go mainstream, I'm the one who has become the hipster — the one embracing unconventional, alternative lifestyles that were once seen as counter to mainstream culture. When I think of soy milk, though, I also think about the deforestation in the Amazon rainforest. I think how human actions have cleared entire strips of magnificent, centuries-old trees to make way for soybeans. And not just any soybeans, but genetically modified ones, designed to withstand pests and other unforeseen circumstances. This same genetically modified soy that is free from any possible pests is being modified into something else like super-processed foods like plant-based meats. Maybe me and my friends, other Gen Zs like me, are starting to carry the stereotype of being overly concerned with appearing trendy or nonconformist in the eyes of our parents and older siblings, who now seem to belong to the plant-based meats and milks mainstream. Sometimes, I even blame my Gen X parents for not stopping this madness with food technology. Instead of enjoying simple, beautiful foods like vegetables, beans, legumes and fruits, we're being pushed toward something artificial. I'd rather consume nothing than eat this new food. It's complex, and it makes me really sad."* (Female, born 2003)

Furthermore, for these young consumers, origin and price emerge as the most significant factors influencing their purchasing decisions. This sentiment reflects a pragmatic mindset. As one Gen Z expressed:

> *"If it's locally made and affordable, I'm more likely to try it. But I'm not going to pay extra just because it's plant-based. It has to make sense for me and the planet."* (Male, born 2006)

While Gen Z values sustainability, they are not always willing to pay a premium for plant-based meat alternatives. Similar findings were reported in previous research (Meixner *et al.*, 2024). Interestingly, attributes such as fat content and organic production methods hold less influence for this generation (Meixner *et al.*, 2024). This may be due to the perception that processed foods, including many plant-based meat alternatives, are inherently less healthy than fresh, whole foods.

Another Gen Z questioned the whole process asking:

> *"What good can we expect from such a vicious cycle of food production disguised as ecological and sustainable food, something we all need in the future to meet the food or nutritional needs of a growing world population. Most of my friends are either vegan or vegetarian. Many of us are like that and we are like that not because we are concerned about the sustainable development of food, but simply about killing animals to satisfy our so great and especially in the western world limitless human gluttony for traditional meat."* (Male, born 2004)

These words encapsulate a broader generational sentiment. Gen Z is unwilling to accept products they perceive as overly-processed, nor are they willing to wait for the food industry to repeat the mistakes it made with climate change (Marinova and Bogueva, 2022), only to catch up later in an attempt to understand their needs. This generation recognises the urgency of implementing sustainable and ethical practices in their daily diets not as a distant ambition, but as a pressing necessity. For Gen Z, the shift extends far beyond the appeal of trendy products; it is about revolutionising the food system to place the planet, animal welfare and personal health at its core.

Their frustration often stems from watching older generations underestimate the magnitude of these changes in food safety, health, even traditions and cultures. One Gen Z added:

> *"It's like we see the wave coming. We just need everyone else to stop underestimating how big it is going to be — and start paddling with us into the same direction. We need our parents, the society to stop the slowly creeping food madness of all this unnatural food coming our way."* (Female, born 2002)

This insight aligns with trends we have observed over the last years. When changes in food systems take place, they first fulfill the excitement of their research and development (R&D) creators and venture-capitalist investors. They often arrive quietly, unnoticed by the consumers, who are following traditional meat-centred dietary practices, especially in Australia (Whitton *et al.*, 2021). However, once these changes gain momentum, their impact is expected to be profound and far-reaching.

Gen Z does not just adapt to these changes — they demand them. Their voices amplify the urgency, fuelled by the social media's ability to accelerate awareness and adoption. They are the first generation to grow up with an acute understanding of climate change and its implications for food security. At the same time, they are the first generation to be willing to stick to traditional and ethical food. This awareness drives their choices, from rejecting wasteful packaging in the same way as plant-based alternatives to championing locally sourced and sustainably produced foods.

What strikes us most is their optimism, even in the face of being surrounded by systemic inertia as they see their relationship with food. They believe in the power of collective action, of influencing change through their choices and voices. Their perspective reminds us all of the importance of not dismissing change when it seems distant or small; because when it comes, as they so clearly see it, its impact will be anything but insignificant. Gen Z is already paddling. The question is, are we ready to paddle alongside them? Are we all ready to be part of the green eating revolution?

References

Aanyamittu. (2023). Gen Z and the Farmers Behind Farmers' Markets. Medium. https://medium.com/@aanyamittu/gen-z-and-the-farmers-behind-farmers-markets-78d2d09f5c28#

Adamczyk, D., Jaworska, D., Affeltowicz, D., & Maison, D. (2022). Plant-based dairy alternatives: Consumers' perceptions, motivations, and barriers — Results from a qualitative study in Poland, Germany, and France. *Nutrients*, 14, 2171. https://doi.org/10.3390/nu14102171

André, Q., Chandon, P., & Haws, K. (2019). Healthy through presence or absence, nature or science? A framework for understanding front-of-package food claims. *Journal of Public Policy and Marketing*, 38, 172–191. https://journals.sagepub.com/doi/pdf/10.1177/0743915618824332?trk=public_post_comment-text

Bambridge-Sutton, A. (2024). Why do consumers trust — and distrust — meat alternatives? Food Navigator. https://www.foodnavigator.com/Article/2024/06/07/plant-based-meat-consumer-trust-and-distrust/

Benton, T., Bieg, C., Harwatt, H., Pudasaini, R., & Wellesley, L. (2021). Food system impacts on biodiversity loss. Three levers for food system transformation in support of nature. Chatham House. https://www.chathamhouse.org/2021/02/food-system-impacts-biodiversity-loss

Bogueva, D. & Marinova, D. (2018). What is more important perception of masculinity or personal health and the environment? In Bogueva, D., Marinova, D., Raphaely, T. (eds) *Handbook of Research on Social Marketing and its Influence on Animal Origin Food Product Consumption*, pp. 148–162. IGI Global, Hershey, PA.

Bogueva, D. & Phau, I. (2016). Meat myths and marketing. In Raphaely, T., Marinova, D. (eds) *Impact of Meat Consumption on Health and Environmental Sustainability*, pp. 264–276. IGI Global, Hershey, PA.

Bogueva, D., Marinova, D., & Raphaely, T. (2021). Influencing sustainable food-related behaviour changes: A case study in Sydney, Australia. In Bhattacharyya, J., Dash, M. K., Hewege, C., Balaji, M. S., & Weng, L. (eds), *Social and Sustainability Marketing: A Casebook for Reaching your Socially Responsible Consumers Through Marketing Science*, pp. 345–386. Routledge, New York, NY. https://doi.org/10.4324/9781003188186

Boukid, F., Hassoun, A., Zouari, A., Tülbek, M. Ç., Mefleh, M., Aït-Kaddour, A., & Castellari, M. (2023). Fermentation for designing innovative plant-based meat and dairy alternatives. *Foods*, 12(5), 1005. https://doi.org/10.3390/foods12051005

Bryant, C., Szejda, K., Parekh, N., Deshpande, V., & Tse, B. (2019). A survey of consumer perceptions of plant-based and clean meat in the USA, India, and China. *Frontiers of Sustainable Food Systems*, 3, 432863. https://doi.org/10.3389/fsufs.2019.00011

Chen, Z. (2024). Why Gen Z is embracing veganism: Health, environment, and ethics. Woke Waves. https://www.wokewaves.com/posts/gen-z-veganism-trends

Circana. (2025). The Current State of Meat Alternatives. https://www.circana.com/post/the-current-state-of-meat-alternatives

Deloitte. (2024). 2024 Gen Z and Millennial Survey: Living and working with purpose in a transforming world. https://www.deloitte.com/global/en/issues/work/content/genz-millennialsurvey.html

Eat Forum. (n.d.). The EAT-Lancet Commission on Food, Planet, Health. Can we feed a future population of 10 billion people a healthy diet within planetary boundaries? https://eatforum.org/eat-lancet-commission/

Fehér, A., Gazdecki, M., Véha, M., Szakály, M., & Szakály, Z. (2020). A comprehensive review of the benefits of and the barriers to the switch to a plant-based diet. *Sustainability*, 12, 4136. https://doi.org/10.3390/su12104136

Fong, B. Y. F., Chiu, W. K., Chan, W. F. M., & Lam, T. Y. (2021). A review study of a green diet and healthy ageing. *International Journal of Environmental Research and Public Health*, 18(15), 8024. https://doi.org/10.3390/ijerph18158024

Goldman, D. M., Nagra, M., Warbeck, C., Winings, E., & Landry, M. J. (2025). Plant-based diets: Nutritional considerations and public health implications. In Preedy, V. R., Patel, V. B. (eds) *Handbook of Public Health Nutrition*, pp. 1–27. Springer, Cham. https://doi.org/10.1007/978-3-031-32047-7_163-1

Graça, J., Calheiros, M. M., & Oliveira, A. (2015). Attached to meat? (Un)Willingness and intentions to adopt a more plant-based diet. *Appetite*, 95, 113–125. https://doi.org/10.1016/j.appet.2015.06.024

Graça, J., Godinho, C. A., & Truninger, M. (2019). Reducing meat consumption and following plant-based diets: Current evidence and future directions to inform integrated transitions. *Trends in Food Science and Technology*, 91, 380–390. https://doi.org/10.1016/j.tifs.2019.07.046

Have a Plant. (2021). The plant-forward eating guide: Insights, ideas & inspirations for plant-passionate professionals. https://fruitsandveggies.org/wp-content/uploads/2021/01/Have-A-Plant-Plant-Forward-Eating-Guide.pdf

Knaapila, A., Michel, F., Jouppila, K., Sontag-Strohm, T., & Piironen, V. (2022). Millennials' consumption of and attitudes toward meat and plant-based meat alternatives by consumer segment in Finland. *Foods*, 11(3), 456. https://doi.org/10.3390/foods11030456

Marinova, D. & Bogueva, D. (2022). *Food in a Planetary Emergency.* Springer, Singapore.

Marinova, D., Hong, J. Todorov, V., & Guo, X. (2017). Understanding innovation for sustainability. In Hartz-Karp, J. & Marinova, D. (eds) *Methods for Sustainability Research* (pp. 217–230), Edward Elgar, Cheltenham, UK.

Meixner, O., Malleier, M., & Haas, R. (2024). Towards sustainable eating habits of Generation Z: Perception of and willingness to pay for plant-based meat alternatives. *Sustainability*, 16(8), 3414. https://doi.org/10.3390/su16083414

Moors, R. (2024). How Gen Z and Gen Alpha are shaping the future of food. Global insights. https://insights.figlobal.com/foodtech/how-gen-z-and-gen-alpha-are-shaping-the-future-of-food

Mridul, A. (2024). Despite declines, plant-based meat is selling better than it did pre-pandemic. Green Queen. https://www.greenqueen.com.hk/plant-based-meat-alternatives-circana-report-sales-decline-retail/#:~:text=Since%202020%2C%20the%20number%20of,fresh%20fruit%20and%20salad%20vegetables

Onwezen, M. C., Bouwman, E. P., Reinders, M. J., & Dagevos, H. A. (2021). Systematic review on consumer acceptance of alternative proteins: Pulses, algae, insects, plant-based meat alternatives, and cultured meat. *Appetite*, 159, https://doi.org/10.1016/j.appet.2020.105058

Perez-Cueto, F. J. A., Rini, L., Faber, I., Rasmussen, M. A., Bechtold, K. B., Schouteten, J. J., & Steur, H. D. (2022). How barriers towards plant-based food consumption differ according to dietary lifestyle: Findings from a consumer survey in 10 EU countries. *International Journal of Gastronomy and Food Science*, 29, 100587. https://doi.org/10.1016/j.ijgfs.2022.100587

Raptou E., Tsiami A., Negro G., Ghuriani V., Baweja P., Smaoui S., & Varzakas T. (2024). Gen Z's willingness to adopt plant-based diets: Empirical evidence from Greece, India, and the UK. *Foods*, 13(13), 2076. https://doi.org/10.3390/foods13132076

Rickerby, A. & Green, R. (2024). Barriers to adopting a plant-based diet in high-income countries: A systematic review. *Nutrients*, 16(6), 823. https://doi.org/10.3390/nu16060823

Rozin, P. (2005). The meaning of food in our lives: A cross-cultural perspective on eating and well-being. *Journal of Nutrition Education and Behaviour,* 37, S107–S112. https://bpb-us-w2.wpmucdn.com/web.sas.upenn.edu/dist/7/206/files/2016/09/meaningfoodinlivesJNEB2005-26kuvut.pdf

Tziva M., Negro S. O., Kalfagianni A., & Hekkert M. P. (2020). Understanding the protein transition: The rise of plant-based meat substitutes. *Environmental Innovation and Societal Transitions*, 35, 217–231. https://doi.org/10.1016/j.eist.2019.09.004

Van Loo, E. J., Hoefkens, C., & Verbeke, W. (2017). Healthy, sustainable and plant-based eating: Perceived (mis)match and involvement-based consumer segments as targets for future policy. *Food Policy*, 69, 46–57. https://doi.org/10.1016/j.foodpol.2017.03.001

Whitton, C., Bogueva, D., Marinova, D., & Phillips, C. J. C. (2021). Are we approaching peak meat consumption? Analysis of meat consumption from 2000 to 2019 in 35 countries and its relationship to gross domestic product. *Animals*, 11(12), 3466. https://doi.org/10.3390/ani11123466

Zulkosky, C. (2023). Gen Z, veganism, and the future. The Food Institute. https://foodinstitute.com/focus/gen-z-vegans-and-the-future/

Chapter 5

Cultured Meat and Seafood: A Brave New World for Gen Z

"We shall escape the absurdity of growing a whole chicken in order to eat the breast or wing, by growing these parts separately under a suitable medium."

— Winston Churchill

Winston Churchill's remarkably forward-thinking vision for food technology reads today like a prophecy of cultured meat, also described as lab-grown, clean, cultivated, cellular, slaughter-free, synthetic, cell-based, cell-cultured or in vitro meat. In his 1931 essay *"Fifty Years Hence"*, published in *The Strand Magazine*, Churchill imagined a future where scientists could grow meat without raising entire animals, decades before biotechnology made such ideas possible (Churchill, 1931).

His timeline however was optimistic — he anticipated lab-grown meat would be commonplace within 50 years; the first cultured meat burger was only unveiled in 2013 (Merchant, 2013), and the technology remains in development more than a century later. Nevertheless, Churchill's foresight aligns strikingly with today's efforts to transform the food systems. Modern companies are pioneering cultured meat and seafood to address pressing issues such as animal welfare, greenhouse gas emissions and global food security.

Churchill's vision was not solely practical. He also projected how effortlessly these innovations would integrate into everyday life. As he boldly predicted: "The new foods will be practically indistinguishable from the natural product from the outset, and any changes will be so gradual as to escape observation." (Churchill, 1931)

His ideas broke away from the constraints of traditional farming and stand as a remarkable example of visionary thinking well ahead of its time. Yet, in the minds of Gen Z, these once sci-fi-like advancements are not always embraced with enthusiasm, reflecting a complex relationship between emerging technologies and consumer acceptance. The chapter discusses this in detail. In most places when we refer to cultured meat, we also include cultured seafood or cellular aquaculture as both are produced with the same core technology. In essence, cells are harvested from live commercially raised animals (cows, chickens, pigs or quails), fish or other seafood and are put under conditions where they can multiply (Mogi, 2023).

The Current Status of Cultured Meat Technology

Cultured meat is often described by researchers and industry experts as a revolution in progress, one with significant implications for the global value chain (Benussi and Samoggia, 2025; Treich, 2021). The industry stands at a pivotal juncture, balancing the groundbreaking potential it promises with the ongoing technological and regulatory challenges it faces. Since the Dutch scientist Professor Mark Post of Maastricht University unveiled the world's first lab-grown burger in 2013 (Good Food Institute, 2025), the field has made notable development. However, despite advancements, price, mass commercialisation and regulatory hurdles continue to challenge the path to mainstream adoption. At its core, cultured meat is produced by cultivating animal cells in a controlled environment, eliminating the need to raise and slaughter livestock. This process begins with the extraction of stem cells or other cells, such as skeletal, adipose-derived, from muscles and fibroblasts, which are then nourished in a nutrient-rich medium that allows them to multiply and form muscle tissue (Zandonadi *et al.*, 2025). Cultured seafood follows the same principles, using cells from fish, crab, shrimp or other seafood animals to create alternatives to conventionally caught or farmed seafood. These technologies aim to closely replicate the taste, texture and nutritional profile of traditional meat while offering a far more sustainable and ethical method of production (Goswami *et al.*, 2024).

The production process relies heavily on biotechnology and cellular agriculture, fields that harness biological systems to produce food. Scientists cultivate the extracted cells in bioreactors — specialised vessels that provide the ideal conditions for growth. The developing cells are fed a blend of amino acids, vitamins and growth factors to encourage their development into structured muscle tissue. Some companies are exploring the use of plant-based scaffolding to improve the texture and mouthfeel of cultured meat, thereby enhancing its consumer appeal (Sergelidis, 2019). Despite dramatic progress in these technologies, achieving scalability without sacrificing quality or affordability remains a formidable hurdle.

One young Gen Z consumer voiced a perspective to us that reflects broader generational ambivalence:

> *"I doubt that cultured meat can be developed to a point where it truly replicates the flavour of traditionally farmed meat. I'm not convinced it's worth paying a premium for something that still feels experimental, which I believe will be the case if it's released on the market. The science behind it still seems uncertain, and I don't see any clear personal benefits in choosing cultured meat over what I already eat, real meat."* (Male, born 2002)

This sentiment shows a central challenge how to convince consumers, especially younger ones, that cultured meat is not only safe and sustainable, but also desirable and comparable in quality to the meat they are familiar with.

To address such scepticism, many companies in the cultured meat industry have adopted strategic communication and product development approaches aimed at increasing acceptance, particularly among younger generations. Educational messaging that highlights environmental and animal welfare benefits is a common strategy. At the same time, some brands are employing "naturalness framing", presenting cultured meat as a continuation of traditional meat rather than a radical departure from it (Ruzgys and Pickering, 2020). Taste and texture are also at the forefront of research and development, as sensory appeal remains one of the most influential factors in food choice. Targeted marketing efforts are focused on Gen Z, a demographic often characterised by openness to innovation but cautious about what they perceive as artificial or experimental (Bogueva and Marinova, 2020).

Several countries are leading the charge with unique strategies. The California-based company Eat Just Inc., for example, received regulatory approval in Singapore for its cultured chicken for commercial sale and continues to promote its environmental and ethical credentials (Marsh, 2023). The Israeli Aleph Farms and Believer Meats (ex-Future Meat Technologies) are pushing boundaries in cost reduction and product realism, while the Netherlands-based Mosa Meat, the developer of the first cultured beef burger, continues to innovate in flavour and scalability. Furthermore, the Singaporean Shiok Meats is revolutionising seafood with lab-grown shrimp and crab, especially appealing to eco-conscious consumers in Asia.

Australia's Vow is charting a different course by developing new types of meat entirely, such as its Forged Parfait made from Japanese quail cells. Rather than replicating conventional meats, Vow's approach centres on crafting novel, gourmet-style cultured meats that provoke curiosity and signal culinary innovation. The company has even experimented with exotic options, such as kangaroo and crocodile, aiming to reframe cultured meat not as a compromise, but as an exciting evolution in food culture (Santoreneos, 2024; Vow, n.d.).

Several countries, including Singapore, USA, Israel, South Korea, Switzerland, Thailand, UK, New Zealand and Australia together with the European Union have started paving the way for cultured meat through regulatory approvals, although regulatory barriers, including around labelling, have not yet been finalised. The industry is keen to see this happening and is cooperating with regulators. For example, the Australian company Vow is already working with Food

Standards Australia New Zealand (FSANZ) to align its processes with safety and labelling guidelines (FSANZ, 2024).

Nevertheless, significant technological challenges remain. Processes, such as cell proliferation, differentiation, maturation and tissue assembly, require further optimisation, standardisation and consistency (Lee *et al.*, 2024). Key components, including scalable bioreactors and 3D scaffolds, are also still under active development. In addition, the energy demands of large-scale production could offset some of the claimed environmental benefits, particularly if renewable energy sources are not integrated (Zandonadi *et al.*, 2025).

In parallel with these technological hurdles, ensuring the safety and regulatory oversight of cultured meat production is equally critical. The successful development and commercialisation of these products depend on clear, up-to-date regulatory frameworks (Zandonadi *et al.*, 2025). Regulatory agencies around the world are actively exploring how best to govern this emerging industry. Cultured meat production presents unique food safety considerations, including potential genetic instability of cell lines, microbiological risks, chemical residues from the production process, concerns about allergenicity and nutritional uncertainties. Currently, there is no standardised testing protocol for assessing and evaluating these potential risks. As a result, robust food safety strategies, such as the implementation of the Hazard Analysis and Critical Control Points (HACCP) systems and adherence to best practices must be integrated throughout the production process to ensure consumer safety as well as build and earn public trust (Zandonadi *et al.*, 2025).

Further complicating matters, the industrialisation of cultivated meat remains in its infancy. While regulatory approvals are beginning to occur, key details about the technologies used and the nature of the final product remain unclear. Large-scale production is still in early stages, with significant gaps in standardisation, safety evaluation and transparency. Industrial scalability is hindered by proprietary technologies, incomplete infrastructure (e.g. bioreactors and 3D supports) and ongoing research needs (Lee *et al.*, 2024). Gaining consumer trust and preventing industry fraud will require more public information, regulatory oversight and independent scientific involvement.

Ultimately, the future of cultured meat hinges on more than regulatory approval or scientific breakthroughs. Its success will depend on winning the hearts, minds and taste buds of consumers, particularly Gen Z and the generations to come, namely Alpha and Beta. In this context, cultured meat is not merely a technological innovation but a cultural negotiation. Its success will be measured not only in environmental gains or production efficiency but in its ability to resonate with everyday eaters, to feel familiar, valuable and aligned with their beliefs and expectations. Whether the next generation embraces lab-grown meat as a new norm or rejects it as unnatural remains one of the defining questions for the future of food systems.

Cultured Meat: A Disruptive Force in Food Innovation

Cultured meat represents one of the most transformative alternatives to conventional animal farming. As Siegrist *et al.* (2018) explain, cultured meat is created through tissue-engineering techniques that allow animal cells to grow in a controlled setting, separate from the animal itself. This method eliminates the need to harm or kill animals. The final product closely resembles conventional meat in both texture and flavour (Siegrist *et al.*, 2018). Similarly, cultured fish and seafood follow the same principles, using fish cells to create seafood products without the need for fishing or aquaculture. Cultured seafood is created by cultivating muscle, fat and fibroblast cells from aquatic organisms, with or without genetic modification. Bioengineering replicates tissue structures found in traditional seafood, ensuring comparable nutrition (Chandimali *et al.*, 2024). Unlike traditional methods that depend on raising, slaughtering and killing animals, lab-grown meat, fish and seafood offer a more sustainable, ethical and cruelty-free way to produce protein. The aimed result is to create a product designed to mirror the taste, texture and nutritional value of conventional meat or seafood, without the environmental impact. Despite of these promises, one Gen Zer we spoke with, expressed scepticism about the relevance and appeal of lab-grown meat and seafood:

> *"Cultured meat feels like unnecessary technology to me. Cultured seafood is another nuisance. It feels like yet another challenge for my generation to confront. They are not something I want... neither are appealing to me... not even in a fantasy world...."* (Male, born 2005)

This kind of hesitation reflects one of many challenges faced by the industry. Another major hurdle is achieving industrialisation. Mass production remains in its early experimental stages, with no standardised methods in place (Lee *et al.*, 2024). A further complicating matter is the lack of methodology for optimisation of data sharing. Commercial competition has led many companies and research teams to guard their innovations, limiting knowledge-sharing and slowing collective progress. This lack of collaboration hinders efforts to develop efficient and scalable production systems.

In addition to technological challenges, cultured meat faces significant societal and cultural barriers (Siddiqui *et al.*, 2022). Consumer scepticism, high production costs and the need for long-term studies on health and sustainability continue to cause a delay in widespread acceptance (Lee *et al.*, 2024). While the industry promotes its potential to reduce environmental impact and improve food security, many consumers remain unconvinced. A particularly revealing viewpoint comes from a Gen Z consumer:

> *"I get that lab-grown meat is supposed to be better for the planet and more ethical, but honestly, I don't fully trust it. It feels like we're being sold this perfect solution, but the product just isn't there yet. If it's made in a lab and companies won't even share how they're making it, how are we supposed to know what's really in it? I'm not convinced it tastes or feels like real meat either. Everyone says it's a revolution, but to me, it still seems like an expensive experiment for people with moneys, investors..., not real food. I want food that's natural and transparent, for accountability..., not something that comes with more questions than answers."* (Female, born 2000)

This statement reveals a nuanced and critical Gen Z perspective on cultured meat. While acknowledging the ethical and environmental promises of lab-grown meat, the expressed view clearly shows distrust in the product's current form and development process and highlights a scepticism rooted in both, the lack of transparency and the perceived artificiality of the product. The mention of secrecy *"companies won't even share how they're making it"*, points to a broader discomfort with the commercial motivations behind food innovation, especially when such secrecy contradicts the values of openness and honesty that many Gen Z consumers demand.

Overall, Gen Z has a nuanced and critical perspective on cultured meat. While recognising its ethical and environmental potential, opinions like this reveal deep concerns about transparency, trust and the perceived artificiality of the product. The reference to secrecy highlights a broader unease with the commercial motivations driving food innovation, particularly when those motivations appear to conflict with Gen Z's values of openness and authenticity. Although Gen Z is often praised for its commitment to sustainability, this example shows that trust remains a crucial factor in food acceptance. Discomfort with the lab-based nature of cultured meat, coupled with the lack of transparency in its development, creates a credibility gap that the industry must address if it hopes to earn the support of younger generations.

The desire for food that is *"natural and transparent"* reflects a generational shift in priorities. It further suggests that trust in food systems, particularly new technologies, is heavily dependent on visible integrity and accessibility of information. For many young consumers, sustainability must be matched with integrity, control over personal health and sensory authenticity. The hesitation also touches on a sensory and emotional disconnect. Cultured meat, despite being engineered to replicate taste and texture, is not perceived as *"real food"*. It may fulfill ethical and environmental criteria, but unless it can also earn consumer trust and deliver a genuine eating experience, its potential remains limited.

As Gen Z continues to shape consumer trends through its demand for ethical, transparent and experiential food, their acceptance — or rejection — of cultured meat could play a decisive role in shaping future food systems (Marinova and Bogueva, 2022). Ethical intentions, while essential, may not be sufficient on their own. Success will depend equally on transparency and trustworthiness as well as a product that resonates on both sensory and emotional levels.

Trust, Transparency and Cautiousness

Beyond scientific advancements, consumer scepticism remains a crucial barrier to the mainstream adoption of cultured meat or aquatic products. Trust in both the products and their production methods and processes is still developing. To foster greater acceptance, companies must prioritise transparency, particularly in areas such as safety testing, nutritional profiling and food quality assurances. Although Gen Z consumers have demonstrated a strong interest in ethical and sustainable food alternatives, they also express caution regarding the taste, safety and cultural fit of lab-grown meat and seafood. As one Gen Zer explained:

> *"I think lab-grown meat is super interesting. You know, it feels like something out of a sci-fi movie... Probably it will be a super cool thing to make and I love the idea of a food system that's more ethical and sustainable. But at the same time, I can't help but feel uneasy... We are talking about it with you... like it's something like a potential food option, ... I mean not now but perhaps in the future... I still feel uneasy.... Hmmm... It's cool... and scary... Like, how do I really know it's safe? Or that it's not just another profit-driven product wrapped in green branding?... I want to believe in it, I really do, but there's this gap between what companies say and what I feel I can trust. Until there's more transparency, like real data, honest answers, maybe even a look inside how it's made... They can make a close-look movie about it... without this knowledge it's hard to fully embrace it. I'm curious, but I don't have these days to be anything else but cautious."* (Female, born 1999)

This opinion highlights the need for ongoing efforts to build consumer confidence through clear, evidence-based communication and inclusive dialogue with these young consumers.

Another Gen Zer was more positive about the prospects of cultured meat product consumption:

> *"I'm definitely open to trying lab-grown meat.... of course, if it were available at my local supermarket or on a menu, I'd be curious enough to give it a go. I am a vegan and I think the idea of reducing harm to animals and the environment with cell-culture technologies is really cool. For me, it's about that first experience... I want to know how it tastes, how it feels, and whether it actually lives up to the hype. Ha-ha-ha (laughing) I'm not saying I'll switch forever after one bite, but trying it at least once feels like the only real way to know if it's something I could get behind."* (Female, born 2001)

This view, although positive reflects a pragmatic curiosity common among Gen Z consumers, who are open to innovation and ethical progress, but grounded in personal experience as the key factor in shaping long-term acceptance. It is also reinforcing the idea that trust must be earned through clear communication and consistent evidence.

Crucially, further research into the scalability of production, long-term health effects and regulatory frameworks will shape the trajectory of this emerging sector. As scientific and technological breakthroughs continue, the success of cultured meat and seafood will depend not only on innovation but also on how effectively these developments are communicated and understood by the public. The difference between niche curiosity, which in most cases these novel foods are experiencing, especially from Gen Zs, and widespread acceptance may ultimately hinge on trust.

> *"I get why cultured meat is a big deal. It could potentially solve so many problems, if it is the right technology and it is accepted... mass accepted by others. But for me, food isn't just about science, it's about culture and comfort. I'm not against it, but I'd need to see how it fits into my everyday life, not just in theory but on my plate. If it looks, tastes and feels like real food, then maybe I'll come around. But I'm not rushing into it just because it's new."* (Male, born 2000)

The high cost and complexity of production may further entrench inequalities in food access, making these products available only to affluent consumers in the short term. Also, a lack of standardised regulatory frameworks across countries introduces uncertainty around quality control, labelling and consumer protection (Zandonadi *et al.*, 2025). Addressing these risks transparently is essential to ensure that lab-grown alternatives have some future and chances to evolve as truly sustainable and trusted options within the global food system.

Risks from Cultured Meat and Seafood

While lab-grown meat and seafood represent ground-breaking innovations with the potential to transform the food system, reduce land and water use and eliminate animals from the supply chain, these technologies are not without risks. Despite the optimism surrounding them, they pose biological, technological, environmental, economic and socio-cultural challenges that require rigorous scrutiny.

One of the most prominent concerns relates to long-term human health. While initial studies suggest that cultured meat can be safe and nutritionally similar to conventional meat, the evidence remains preliminary (Zandonadi *et al.*, 2025). Much of the current research comes from early-stage laboratory trials or simulations, and comprehensive longitudinal studies on human consumption are still lacking. As one Gen Z young person put it:

> *"I get that lab-grown meat is supposed to help the planet, and that's great, but I don't know exactly how this is going to happen. I still wonder what we don't know yet about cultured meat. What's actually in it? How will it affect my body after say... ten years of eating it? It feels like we're moving fast without having all the answers. Until there's more long-term research and clear labelling, I'm not sure if freely available I'm ready to make it even a small, occasional part of my regular diet."* (Male, born 1997)

This kind of fear highlights a crucial point: the lack of transparency and robust labelling. Consumers are increasingly aware of what goes into their food, and the production process of cultured meat, while technically fascinating, is still unclear to most people. Questions remain around the safety of the growth media, particularly if it contains animal-derived components such as foetal bovine serum

(FBS) or synthetic additives that may not yet be fully regulated or disclosed. Even newer, serum-free media formulations may contain recombinant proteins and bioengineered compounds, the long-term effects of which are not entirely understood (Zandonadi *et al.*, 2025).

In addition to potential biological risks, chemical and allergenic hazards must be considered. The exposure to novel ingredients, cross-contamination or bioaccumulation of trace substances during the culturing process could pose unforeseen health threats. Unlike traditional meat, where decades of epidemiological data exist, cultured meat lacks a historical consumption baseline, making risk prediction and communication more challenging for both scientists and regulators.

Another often overlooked dimension is energy use and environmental trade-offs. Cultured meat is widely praised for its potential to reduce greenhouse gas emissions and resource consumption. However, its actual environmental footprint depends heavily on the energy source powering its production (Marinova and Bogueva, 2022). Current models suggest that if renewable energy is not integrated into the biomanufacturing process, cultured meat could in some scenarios generate more carbon emissions than poultry or plant-based proteins. This concern becomes more pressing as production scales up from laboratory batches to industrial volumes (Zandonadi *et al.*, 2025).

From an economic and ethical standpoint, issues of access and inequality also appear significant. At present, the cost of producing cultured meat remains prohibitively high. Most commercial cultured meat prototypes are priced far above conventional meat products, which restricts their accessibility to affluent consumers or niche markets. This could reinforce existing inequalities in global food systems, where vulnerable populations may be left behind while wealthy markets enjoy the benefits of innovation.

Cultured seafood may pose unique ecological risks. As marine-based cell lines are developed, the potential for biohazards in aquatic ecosystems through unintended discharge, contamination or genetic drift raises questions about marine biodiversity and biosafety that have yet to be fully addressed (Ong *et al.*, 2021).

Compounding these concerns is the absence of unified, standardised regulatory frameworks across countries. Although some regions, such as Singapore and the United States have begun developing guidelines, most nations still lack specific policies governing the production, testing, labelling and marketing of cultured proteins. This regulatory vacuum not only poses risks to quality assurance and consumer protection, but also opens the door to inconsistent labelling standards, making it difficult for consumers to make informed choices (Zandonadi *et al.*, 2025).

Despite these valid concerns, many consumers are willing to embrace some level of risk in the name of progress. As one Gen Zer explained:

> *"I'm not naïve. Of course, there are risks, like with any new tech. But if we waited for everything to be 100% risk-free, we'd never make any progress. As long as it's regulated properly and the science behind it is solid, I'm willing to take a chance. Honestly, the risks of sticking with the current food system, climate change, factory farming, agriculture antibiotic use... seem way worse to me."* (Male, born 2000)

This sentiment reflects an emerging generational divide in risk perception. While older consumers may be more cautious about novel food technologies, younger cohorts, especially Gen Z tend to weigh risks relative to broader system-level harms, such as the ecological impact of industrial farming or antibiotic resistance. For some of them, lab-grown meat is not simply a scientific curiosity but a potential pathway to food justice, sustainability and innovation, if it is made right (Bogueva and Marinova, 2020). Addressing these risks requires transparency, inclusivity and proactive governance. Regulatory bodies must collaborate with scientific experts, civil society and industry leaders to create robust assessment protocols, ethical guidelines and comprehensive labelling laws. Public trust will depend not only on safety but also on openness about production methods, traceability and clear communication about both benefits and uncertainties.

Ultimately, cultured meat and seafood carry both promise and peril. To navigate this delicate balance, the industry must move forward with caution, humility and responsibility, ensuring that the technology evolves not just in the lab but in alignment with the values and needs of society, including Gen Z.

Consumer Acceptance of Cultured Meat

Despite its potential environmental, health, ethical and medical research benefits (Treich, 2021; Hague, 2025; Bryant, 2020), cultured meat faces several psychological, cultural and scientific barriers as well as technical challenges that could hinder widespread adoption. One of the primary challenges is consumer perception. Many Gen Z individuals struggle with the idea of eating meat grown in a lab rather than sourced from traditional farming. Cultural resistance further complicates adoption, as meat consumption is deeply embedded in culinary traditions and social practices worldwide. Some consumers view lab-grown meat as unnatural or artificial, making integration into mainstream diets difficult. Another significant hurdle is the lack of long-term studies on the health effects and environmental impact of cultured meat. While preliminary research suggests lab-grown meat could reduce greenhouse gas emissions and land use, more data is needed to confirm its long-term sustainability.

Gen Z is widely recognised for its strong stance on sustainability, ethical consumption and climate activism. Unlike previous generations, Gen Z consumers are more likely to question the origins of their food and seek alternatives that align with their values (Lera Blog, 2025). Studies indicate that Gen Z is more open to plant-based diets and alternative proteins, making them a crucial audience to win for cultured meat and seafood adoption (Good Food Institute, 2025). Their purchasing decisions are often influenced by concerns about animal welfare, environmental impact and food security, all of which are central to the cultured meat drive. However, despite their progressive attitudes, Gen Z also exhibits hesitancy toward lab-grown meat, citing concerns about taste, safety and cultural acceptance (Sergelidis, 2019).

Young people's opinions on cultured meat vary widely, reflecting a mix of curiosity, scepticism, cultural identity and ethical considerations. As one Gen Z individual explained:

> *"A lot of people I know... people from my friends' circle, still don't trust lab-grown meat. It feels strange or unnatural, and some even argue it's not 'real' meat at all. This makes many of us feel uncomfortable, frustrated even angry about it and question about why we need it in the first place... I saw a post recently about the Italian Prime Minister... the Lollobrigida's bill banning cultured meat, ... and honestly, I am proud of my government which is keeping the Italian traditions. The resistance to it here is strong. It's not just about how it's made... it's about how people feel. Food is deeply personal and tied to tradition. When something new comes along and disrupts food production... and changes how food is made, especially something grown in a lab, people worry. They find it risky for their health, for their kids' health, for our own health.... They question if it's safe, if it's necessary, or whether we're pushing technology too far. For many, it just doesn't feel right."* (Female, born 2002)

Similarly, another Gen Zer highlights the deep cultural meaning of meat and explains that meat carries strong symbolic value in many cultures, passed down from generation to generation. It represents something meaningful and deeply rooted, something that cannot easily be replaced by lab-grown or artificial alternatives.

> *"Cultured meat... seriously?... Why would I eat that when I have access to fresh meat here? For me eating meat isn't just about food. It's a core part of our Aussie culture, identity, and how we connect with each other. I was raised with meat... Meat is everywhere... in our traditions, family meals, national pride to put a sausage or a steak on a barbie (Australian barbecues), and even in ideas about gender and strength. For many people, meat means power, status or something special. Immigrants who move here embrace it too... Even though cultured meat is a smart, innovative idea, it doesn't align... it doesn't really fit into these traditions. It feels detached from the emotions and stories surrounding meat... It seems fake, not just in how it's made, but in what it means, what represents. There is no way for me to accept it. My family and friends would never accept it too."* (Male, born 2000)

This disconnect highlights a core challenge for food-tech innovators: while lab-grown meat replicates the product, it fails to capture the rituals and meanings associated with conventional meat. This young person identifies a profound gap between what cultured meat offers and what meat *means* in cultural terms. This points out a core challenge for food tech innovators: lab-grown meat may replicate the product, but not the rituals or meanings around it. Lab-grown meat lacks a cultural narrative. Gen Z, often characterised as being both progressive and reflective, may resist innovations that feel culturally hollow, even if they offer ethical benefits. Some worry that technology, such as lab-grown meat, could make food feel less genuine or natural, while others are excited by its potential and about the possibilities it offers for the future.

Another critic from Gen Z frames cultured meat as a form of techno-indulgence:

> *"You know, cultured meat is marketed as a solution to ethical and environmental problems, but at its core, it's still about satisfying a desire, wanting to eat meat without guilt. That doesn't resonate with me. I value food choices grounded in health, sustainability and necessity, not indulgence. Cultured meat doesn't feel essential to me... Honestly, it feels like a technological answer to a craving, not a genuine need. It's not about survival or wellness.... I think it's about preserving an experience that perhaps we need to evolve away from."* (Female, born in 2001)

This perspective reflects Gen Z's broader tendencies toward minimalism and mindfulness. Some individuals view cultured meat as reinforcing outdated desires rather than challenging them. Their scepticism aligns with a generational inclination to question consumerism and seek authenticity, where sustainability and health are seen not as marketing points, but ethical imperatives.

Ultimately, there is a clear tension between authenticity — staying true to traditional systems, and technological optimism — believing innovations can lead to better outcomes. While older generations may celebrate lab-grown meat as a "clean" solution, Gen Z voices often call for a deeper, value-based transformation in consumption. Instead of just substituting conventional meat with a lab-grown alternative, they ask: If we truly want to change food systems, shouldn't we also change ourselves?

A study by Ruzgys and Pickering (2020) examined the attitudes of college-aged Canadians toward cultured meat revealing a complex mix of scepticism and cautious optimism. While many youths perceive cultured meat as unnatural and unappetising, they also recognise its ethical and environmental advantages as an alternative to traditional meat. Notably, taste emerged as a key factor influencing acceptance with individuals exhibiting higher level of food disgust showing lower behavioural intent to consume cultured meat (Ruzgys and Pickering, 2020).

Similarly, previous research on Australia's Gen Z found that 72% of young adults in Sydney are not ready to accept cultured meat, despite acknowledging its potential benefits for sustainability and food insecurity (Bogueva and Marinova, 2020). Many young consumers feel that cultured meat needs to appeal to them both emotionally and intellectually before they consider eating it. Concerns about masculinity and Australia's reputation for high-quality animal meat also shape opinions, with some young people believing that switching to cultured meat would betray traditional Australian values. However, 28% of the respondents expressed willingness to try cultured meat, suggesting that environmental and health concerns could drive broader acceptance over time (Bogueva and Marinova, 2020).

Media Framing, Social Influence and the Cultural Stakes of Cultured Meat

The way Gen Z navigates food choices is unlike any generation before. Digitally native, ethically conscious, critical and purpose-driven consumers, Gen Z demands transparency and authenticity from brands and institutions. While sustainability and animal welfare are central concerns, these young consumers' attitude to lab-grown meat remains cautious, tempered by questions of safety, taste, cultural continuity and trust. Despite recognising cultured meat's ethical and environmental promises, many remain sceptical and hesitant to embrace it. This presents a significant challenge for food-tech innovators and policymakers — how can cultured meat be framed and communicated in a way that resonates with Gen Z's core values rather than clashing with them?

Social media platforms like TikTok and X (formerly Twitter), where Gen Z often encounters and debates ideas, including new food innovations, reflect a spectrum of opinions on cultured meat, both from curiosity to critique. How cultured meat is framed on these platforms significantly shapes consumer perceptions. While some posts highlight its benefits, others express scepticism, particularly regarding taste and safety (Kouarfaté and Durif, 2023). Posts emphasising ethical values (e.g., sustainability and animal welfare), intrinsic attributes (e.g., taste, flavour and nutrition) and giving clear scientific explanations (e.g., of how it's made, safety assurances and health implications) are far more persuasive than vague or fragmented messaging (Kouarfaté and Durif, 2023). Media framing plays a significant role in shaping attitudes, with ethical and informational messaging having the strongest influence on acceptance. Structured narratives, those that explain not only *what* cultured meat is but *why* it matters, have a greater impact. They resonate especially with Gen Z, who value clarity, ethics and authenticity (Kouarfaté and Durif, 2023). This aligns with earlier findings that effective communication with structured messaging must integrate moral, sensory and informational appeals (Goodwin and Shoulders, 2013; Dilworth and McGregor, 2015; Bryant and Dillard, 2019; Pilařová *et al.*, 2022). As one Gen Z noted:

> *"Okay, let's be real... one of the biggest questions about lab-grown meat is whether it can actually recreate the meats we know and bloody love,... or if we're just going to end up with a kind of mushy, meat-flavoured paste. I mean, can you grow a perfect A5 Wagyu beef in a lab? What about the crispy skin on chicken wings, the layered fat in pork belly, or the tenderness of slow-cooked lamb shanks? Or are we looking at a future where everything just turns into some version of a burger patty?"* (Male, born 1999)

This insight stresses that the challenge goes beyond nutrition or sustainability. For many, it is not just about whether lab-grown meat can match traditional meat in nutritional terms. It is about whether it can replicate the culinary experience that meat provides. The cuts,

Figure 5.1. Gen Z and cultured meat.

Source: https://chatgpt.com/c/684bc840-3e8c-8004-933b-407bb72c42fa.

textures, presentation and cultural significance of meat are crucial. Just burger patties are unlikely to be enough (see Figure 5.1). Gen Z wants a sensory and cultural match, one that replicates the ritual and richness of traditional meat:

> *"You know with food it's not just about taste or texture. It's about culture. Food is so much more than a needed source to develop a product. It's about how we connect with our roots, celebrate with family and pass traditions down through generations. Think of your grandmother's roast recipe, or that dish everyone expects at your holiday table. These aren't just meals, they're memories... If lab-grown meat can't replicate not just what we eat, but how we eat it... then a huge part of our culinary heritage could be lost. This is a good reason for non-acceptance."* (Male, born 1998)

These narratives highlight that meat holds cultural and emotional value. This concern points to something deeper — the cultural stakes of cultured meat. Meat is more than protein. It's a symbol woven into countless traditions across the globe. If lab-grown products cannot mimic the diversity and specificity of meat from A5 Wagyu to duck confit to Middle Eastern lamb roasts, they risk eroding the cultural richness attached to food. Lab-grown alternatives must speak not only to ethics and innovation but also to identity and tradition.

While technical achievements and safety assurances are vital, so too is storytelling. Gen Z wants to know: Who's making this? Why? How does it fit into their lifestyle and identity? Can it honour the past while moving us into a better future? As the food industry continues pushing the boundaries of innovation, success will depend on navigating not just scientific or regulatory hurdles, but also psychological, cultural and sensory barriers. Gen Z, despite its scepticism, has the potential to drive this transition, if the values they espouse are truly heard and reflected in the way cultured meat is developed and presented.

Gen Z's Meat Paradox: Affection Meets Disillusionment

Gen Z's relationship with meat is anything but uniform. It is a paradox marked by rising ethical awareness on one hand and experiencing cultural and emotional ties on the other. While a vocal minority embraces veganism and plant-based diets, the majority of Gen Z still consumes and often admires meat, seeing it not just as food but as a meaningful part of their identity and lifestyle. This internal divide plays a crucial role in shaping Gen Z's ambivalent stance toward food innovations, including cultured meat.

While some embrace plant-based or vegan lifestyles, the majority still consumes meat — often with a sense of pride and cultural attachment. This duality reflects a generation caught between ethical awakening and a meaningful connection to food traditions. Compared to previous generations, Gen Z is significantly more attuned to the ethical and environmental consequences of traditional animal agriculture and factory farming. A study of university students in USA, China, Indonesia and Thailand shows that 93% of Gen Z report high engagement with issues of animal and environmental protection, and 84% have made lifestyle changes to reflect these values (Sinclair *et al.*, 2025). Gen Z also believes that society is not doing enough to protect animals and the environment, with only 31%

and 34% of the respondents respectively feeling that current efforts are sufficient (Vegconomist, 2025). In response, Gen Z is driving change by supporting ethical brands, advocating for policy reforms and considering careers in sustainability (Sinclair *et al.*, 2025).

Specifically in China, 29% of educated Gen Z is willing to purchase cultivated meat products (Koosis, 2024). The study by Sinclair *et al.* (2025) also points out that Gen Z usually attribute the problems to individual actions rather than systemic factors, such as factory farming, with this trend being more explicit in USA and China. This indicates that there is less information available about industry practices in these countries. Factory farming is associated with cruelty, pollution, overuse of antibiotics and industrial overreach. Industrial-scale farming often confines animals in cramped conditions, leading to stress, disease and unnatural behaviours (Cruelty Farm, n.d.). Factory farming raises significant ethical and environmental concerns. It is likely that the more Gen Z becomes aware of the conditions under which animals are being raised for food consumption, the more likely it would be to see dietary changes (Schaller, 2022).

The ethical awakening of Gen Z does not necessarily translate into widespread rejection of meat consumption. In Australia, for example, where meat consumption is culturally ingrained, many Gen Z consumers express pride in the country's agricultural heritage and view locally sourced meat as part of a sustainable and respectful food tradition (Bogueva and Marinova, 2020). For them, eating meat is not a careless act, it is part of national identity, embedded in social rituals, quality, national identity and cultural pride (Marinova and Bogueva, 2022). Rather than turning away from meat completely, many Gen Z individuals seek more responsible consumption opting for local, organic or free-range alternatives.

While Gen Z is often portrayed as leading the charge toward plant-rich diets, numbers tell a more nuanced story (Tan, 2023). Despite widespread ethical concern, strict plant-based diets remain a minority trend. Gen Z are the majority meat-eaters, but not unquestioning consumers. Worldwide, only 2–5% of Gen Z identify as vegan (Millstein, 2024). In Australia, according to YouGov research

(Tan, 2023), only a small percentage of Gen Z consumers adhere strictly to plant-based diets, but many are open to flexitarian approaches. Ethical and environmental concerns drive these shifts, but the jump to full veganism or vegetarianism is often hindered by social, cultural and sensory barriers. Despite increasing discourse around sustainability, in Australia, 69% of Gen Z still identify as meat eaters, the highest percentage among all generations except Baby Boomers (75%) while among Millennials and Gen X these shares are lower at 58% and 59% respectively (Tan, 2023). This may initially seem at odds with Gen Z's reputation for progressive, eco-conscious values. In reality, many young Australians maintain strong cultural and emotional ties to meat. Even among those sympathetic to animal rights, meat is not only a dietary choice but continues to play a significant emotional and social role. It is a symbol of identity, celebration, family, comfort and social bonding, particularly in outdoor and family-centric settings common in Australian life. In this context, rejecting meat can feel like rejecting one's community or heritage. The dissonance between values and habits leads many to seek compromises rather than complete lifestyle changes. The rise in flexitarianism shows a willingness to adapt, but not a wholesale shift away from meat. This apparent contradiction is not a hypocrisy. It reflects a generation striving for balance. They want to eat ethically but also preserve culinary experiences that bring joy, connection and meaning. Whether this will translate in adopting cultured meat will depend on how successful the food products will be in offering the same experiences.

Gen Z's continued meat consumption does not imply indifference to animal welfare or climate change. Rather, it reflects a split mindset, a desire to be ethical and sustainable, coupled with a deep-seated appreciation for the role meat plays in their lives. This tension explains their hesitant or selective openness to alternatives like cultured meat, which often struggle to replicate not just the taste but the meaning of traditional meat. Cultured meat and seafood, despite their technological promise, often fail to address these emotional and cultural dimensions.

Emotional Attachment as a Barrier to Cultured Alternatives

The deep respect and appreciation for meat presents a major challenge for lab-grown meat and seafood alternatives. While Gen Z is curious and open to innovation, their love for *real* meat, particularly when sourced responsibly, makes them reluctant to replace it with something they perceive as synthetic or unnatural. Our research confirms that many Gen Z consumers do not see cultured meat as a one-to-one replacement (Bogueva and Marinova, 2020). Even when they are told that it eliminates animal suffering and drastically reduces environmental impact, they often express scepticism. Common concerns include taste, texture, health implications, the lack of "realness" and, most critically, a sense that lab-grown meat lacks the cultural and emotional significance of traditional meat.

In Australian settings, where meat is central to BBQs, national identity and outdoor gatherings, replacing traditional steak with a cell-cultured version can feel like tampering with something sacred. It will feel like a cultural loss. Concerns about texture, flavour, health impacts and authenticity dominate Gen Z's scepticism. This evidence strengthens the argument that cultured meat must do more than replace traditional meat nutritionally or ethically. Gen Z does not view cultured meat as a straightforward substitute. It must also speak to Gen Z's emotional and social experience of meat eating (Bogueva and Marinova, 2020). Despite being open to innovation, Gen Z's attachment to "real" meat remains a barrier, and their current dietary habits reflect a broader reluctance to abandon meat entirely, even when alternatives exist.

If cultured meat and seafood are framed purely as a scientific experiment or ethical advancement rather than a trusted food, this risks alienating those who associate meat with tradition and community. Framing, therefore, must evolve. The message should position cultured meat and seafood products not as synthetic or futuristic, but as a natural evolution of meat that respects heritage while aligning with modern values. Taste and authenticity are equally essential. To resonate with Gen Z, cultured meat and seafood must faithfully replicate the sensory experience of traditional meat and seafood and convincingly answer whether it is real enough to satisfy both the palate and tradition. Taste, texture, cultural function and emotional connection all matter. Merely marketing it as a cruelty-free or sustainable product is not sufficient. Importantly, transparent storytelling is the key. Producers must reveal *who* is behind the innovation, *why* it was created, and *how* it fits into consumers' values and aspirations. A value-driven aligning with Gen Z's strong preference for identity, ethics and purpose can influence their consumption choices. Marketing messaging that centres on climate action, animal welfare and food security, without moralising those who continue to enjoy conventional meat, has the best chance of winning Gen Z's trust.

The defining tension in Gen Z's approach to meat is that these young people are simultaneously ethical consumers and cultural eaters. On one hand, they are the ethically-driven consumers pushing for sustainable alternatives; on the other, they are those who view meat as part of their heritage, comfort and cultural expression. Most Gen Zs link both identities, wanting to do better but unsure that technology alone can offer what tradition provides. Rather than absolute rejection of cultured meat, their attitude is one of hesitant curiosity, coloured by a desire for transparency, authenticity and cultural alignment.

Implications for Cultured Meat Adoption

Understanding this split is essential for the future of the cultured meat and seafood sector. For it to gain traction among Gen Z, producers must go beyond the technological narrative of efficiency and ethics. They must also address questions of identity, tradition and emotional connection. Adoption will depend on more than nutritional equivalence or ethical merit. Gen Z wants food that aligns with its values *and* honours its lived experiences.

This generation is positioned to lead a food system transformation, but only if they feel culturally and emotionally connected to the alternatives (Bogueva and Marinova, 2020). Producers and policymakers must recognise that for Gen Z, meat is more than a commodity, it is symbolic and any alternative must be able to carry that symbolic weight. Cultured meat must not only be ethical and safe, but it must also be meaningful.

This does not mean that cultured meat is doomed to fail among Gen Z. On the contrary, this generation is uniquely positioned to lead a food transition, but only if that transition resonates with *both* their values and their lived experiences. The future of cultured meat hinges on addressing Gen Z's values holistically. Two-thirds of them already factor animal cruelty into their purchasing decisions (Youth Sense, 2022). Many experience "guilt" when eating meat, due to rising awareness of its environmental and ethical implications (Minett, 2024). However, the transition will not be seamless. It will require more than technical innovation — it will demand emotional resonance. Food must feel real, not just be real. Until cultured meat can deliver on identity, tradition and trust, its road to widespread acceptance among Gen Z will remain a complex one.

The tension between ethical and cultural values makes Gen Z a powerful but unpredictable demographic: they demand radical change, but on their terms. They want ethical clarity, environmental accountability and cultural sensitivity in every product they consume. Cultured meat, then, is not simply an alternative, it could be a test case for how food systems might evolve in response to generational demands.

This complex landscape offers both a warning and an opportunity. Traditional systems are losing legitimacy among Gen Z. However, for newer technologies to succeed, they must engage not just with logic, but with values, identity and the stories we tell about food.

Conclusion

The challenge for marketers, policymakers and innovators lies in recognising that Gen Z's scepticism toward cultured meat is not merely about unfamiliarity. It is deeply tied to cultural identity, emotional resonance and a broader generational shift toward purpose-driven choices. Traditional industry marketing strategies that emphasise ethical and environmental benefits may fall short unless they also create compelling narratives that align with Gen Z's values and lived experiences. Similarly, policy and advocacy efforts must acknowledge that resistance to lab-grown meat is not a rejection of science but a demand for meaningful change that feels authentic rather than imposed.

Gen Z's engagement with sustainability is often rooted in questioning consumer habits rather than simply replacing existing ones with technological alternatives. Therefore, successful innovation in cultured meat must go beyond reducing environmental footprints — it must also address cultural footprints by considering how food is integrated into traditions, personal identity and communal connections. Without this cultural bridge, even the most advanced food technologies may struggle to gain widespread acceptance.

References

Benussi, C. & Samoggia, A. (2025). A systematic literature review of cultured meat through the conceptual frameworks of the entrepreneurial ecosystem and global value chain. *Foods*, 14(5), 885. https://doi.org/10.3390/foods14050885

Bogueva, D. & Marinova, D. (2020). Cultured meat and Australia's Generation Z. *Frontiers in Nutrition*, 7, 148. https://doi.org/10.3389/fnut.2020.00148

Bogueva, D. & Marinova, D. (2022). Australian Generation Z and the nexus between climate change and alternative proteins. *Animals*, 12, 2512. https://doi.org/10.3390/ani12192512

Bryant, C. J. (2020). Culture, meat, and cultured meat. *Journal of Animal Science*, 98(8), skaa172. https://doi.org/10.1093/jas/skaa172.

Bryant, C. & Dillard C. (2019). The impact of framing on acceptance of cultured meat. *Frontiers in Nutrition*, 6, 103. https://doi.org/10.3389/fnut.2019.00103

Chandimali, N., Park, E. H., Bak, S. G., Won, Y-S., Lim, H. J., & Lee, S. J. (2024). Not seafood but seafood: A review on cell-based cultured seafood in lieu of conventional seafood. *Food Control*, 162, 110472. https://doi.org/10.1016/j.foodcont.2024.110472

Churchill, W. (1931). Fifty Years Hence, The Strand Magazine, https://teachingamericanhistory.org/document/fifty-years-hence/ (Also available at: Eschner, 2017, https://www.smithsonianmag.com/smart-news/winston-churchill-imagined-lab-grown-hamburger-180967349/)

Cruelty Farm. (n.d.). Factory Farming Exposed: Ethical Issues, Environmental Impact, and Health Risks Revealed. https://cruelty.farm/the-unseen-consequences-uncovering-the-ethical-dilemmas-of-factory-farming/

Dilworth, T., & McGregor, A. (2015). Moral steaks? Ethical discourses of in vitro meat in academia and Australia. *Journal of Agricultural and Environmental Ethics*, 28, 85–107. https://doi.org/10.1007/s10806-014-9522-y

Food Safety Australia and New Zealand (FSANZ). (2024). A1269 Cultured Quail as a Novel Food. https://consultations.foodstandards.gov.au/sas/a1269-cultured-quail-as-a-novel-food/

Good Food Institute. (2025). The science of cultivated meat. https://gfi.org/science/the-science-of-cultivated-meat/

Goodwin, J. N. & Shoulders, C. W. (2013). The future of meat: A qualitative analysis of cultured meat media coverage. *Meat Science*, 95, 445–450. https://doi.org/10.1016/j.meatsci.2013.05.027

Goswami, M., Ovissipour, R., Bomkamp, C., Nitin, N., Lakra, W., Post, M., & Kaplan, D. L. (2024). Cell-cultivated aquatic food products: Emerging production systems for seafood. *Journal of Biological Engineering*, 18, 43. https://doi.org/10.1186/s13036-024-00436-1

Hague, J. (2025). Lab-grown meat: You may find it icky, but it could drive forward medical research. The Conversation. https://theconversation.com/lab-grown-meat-you-may-find-it-icky-but-it-could-drive-forward-medical-research-253565

Kouarfaté, B. B. & Durif, F. (2023). Understanding consumer attitudes toward cultured meat: The role of online media framing. *Sustainability*. 15(24), 16879. https://doi.org/10.3390/su152416879

Koosis, A. (2024). Gen Z In China: Driving Sustainable Protein Adoption. Faunalytics. https://faunalytics.org/gen-z-in-china-driving-sustainable-protein-adoption/

Lee, S. Y., Lee, D.Y., Yun, S. H., Lee, J., Mariano, E. Jr., Park, J. . . . Hur, S. J. (2024). Current technology and industrialization status of cell-cultivated meat. *Journal of Animal Science and Technology*, 66(1), 1–30. https://doi.org/10.5187/jast.2023.e107

Lera Blog. (2025). The Future of Food: Lab-Grown Meat and Plant-Based Alternatives. https://lerablog.org/health/nutrition/the-future-of-food-lab-grown-meat-and-plant-based-alternatives/.

Marinova, D., & Bogueva, D. (2022). *Food in a Planetary Emergency*. Springer Nature, Singapore.

Marsh, N. (2023). Why Singapore is the only place in the world selling lab-grown meat. BBC. https://www.bbc.com/news/business-65784505

Merchant, B. (2013). Winston Churchill predicted synthetic lab-grown meat in 1931. Vice. https://www.vice.com/en/article/winston-churchill-predicted-synthetic-lab-grown-meat-in-1931/

Millstein, S. (2024). Is veganism really growing? Using data to track the trend. Sentient Media. https://sentientmedia.org/increase-in-veganism/

Minett, C. (2024). Gen Z's changing attitudes towards meat consumption highlight generational shift. https://www.express.co.uk/life-style/food/1983156/young-adults-concerned-about-meat

Mogi, T. (2023). Cellular aquaculture: High hopes for the seafood equivalent of cultured meat. https://healthist.net/en/food/1975/

Ong, K. J., Johnston, J., Datar, I., Sewalt, V., Holmes, D., & Shatkin, J. A. (2021). Food safety considerations and research priorities for the cultured meat and seafood industry. *Comprehensive Reviews in Food Science and Food Safety*, 20(6), 5421–5448. https://doi.org/10.1111/1541-4337.12853

Pilařová, L., Kvasničková Stanislavská, L., Pilař, L., Balcarová, T., & Pitrová, J. (2022). Cultured meat on the social network Twitter: Clean, future and sustainable meats. *Foods*, 11, 2695. https://www.mdpi.com/2304-8158/11/17/2695

Ruzgys, S. & Pickering, G. J. (2020). Perceptions of cultured meat among youth and messaging strategies. *Frontiers in Sustainable Food Systems*, 4, 122. https://doi.org/10.3389/fsufs.2020.00122,

Santoreneos, A. (2024). Inside Vow: The spaceship of the cultured meat industry. Forbes. https://www.forbes.com.au/covers/innovation/inside-vow-the-spaceship-of-the-cultured-meat-industry/

Sergelidis, D. (2019). Lab grown meat: The future sustainable alternative to meat or a novel functional food? *Biomedical Journal of Scientific & Technical Research*, 17(1), 12440–12444. https://doi.org/10.26717/BJSTR.2019.17.002930

Schaller, E. (2022). What Messaging Makes People Oppose Animal Farming? Faunalytics. https://faunalytics.org/what-messaging-makes-people-oppose-animal-farming/

Siddiqui, S. A., Alvi, T., Sameen, A., Khan, S., Blinov, A. V., Nagdalian, A. A.... Onwezen M. (2022). Consumer acceptance of alternative proteins: A systematic review of current alternative protein sources and interventions adapted to increase their acceptability. *Sustainability*, 14(22), 15370. https://doi.org/10.3390/su142215370

Siegrist, M., Sütterlin, B., & Hartmann, C. (2018). Perceived naturalness and evoked disgust influence acceptance of cultured meat. *Meat Science*, 139, 213–219. https://doi.org/10.1016/j.meatsci.2018.02.007

Sinclair, M., Stennett, J., & Chung, J. Y. (2025). Exploring Gen Z's attitudes towards animals and the environment. Faunalytics. https://faunalytics.org/exploring-gen-zs-attitudes-towards-animals-and-the-environment/

Tan, S. (2023). World Food Day: What are Australia's major dietary preferences — and most favourite cuisines? YouGov. https://au.yougov.com/consumer/articles/47583-world-food-day-australia-major-dietary-preferences-most-favourite-cuisines-2023

Treich, N. (2021). Cultured meat: Promises and challenges. *Environmental and Resource Economics*, 79, 33–61. https://doi.org/10.1007/s10640-021-00551-3

Vegconomist (2025). Study: Gen Z believes more action is needed to protect animals and the environment. https://vegconomist.com/studies-and-numbers/gen-z-more-action-needed-protect-animals-environment/

Youth Sense (2022). Gen Z's animal cruelty awakening. https://www.youthsense.com.au/youth-insights/gen-zs-animal-cruelty-awakening/

Vow. (n.d.). We create new meats that meat-eaters choose selfishly. https://www.eatvow.com

Zandonadi, R. P., Ramos, M. C., Elias, F. T. S., & Guimarães, N. S. (2025). Global insights into cultured meat: Uncovering production processes, potential hazards, regulatory frameworks, and key challenges — A scoping review. *Foods*, 14(1):129. https://doi.org/10.3390/foods14010129.

Chapter 6

Bug Bites: Gen Z and the Potential of Insects as Food

> *"To me, the attempts to normalise eating parasite-infested bugs proves that a predator class exists and they are deliberately mocking humanity. What is this insect eating telling you?"*
>
> — David Wolfe

This chapter explores the tensions, curiosity and resistance surrounding entomophagy, the practice of eating insects, through the lens of Generation Z. Raised during an era of climate urgency, digital saturation and sociocultural disruptions, Gen Z is both open to sustainability and sceptical of unfamiliar food technologies. Their responses to insect-based foods act as a litmus test for broader societal acceptance (Halicka *et al.*, 2025).

The chapter investigates key drivers and barriers to entomophagy among Gen Z consumers, including the influence of media, education and branding, the roles of sensory expectations and food neophobia, as well as the impact of trust, transparency and cultural narratives. It also discusses how artificial intelligence (AI) intersects with insect-based food systems, optimising production, enhancing nutritional profiling and tailoring marketing strategies. Ultimately, this chapter positions insects not only as a protein source but also as a symbol of food system transformation, driven in part by Gen Z's values and digital fluency and to a greater extent by consumption behaviours and how insects are envisioned as part of a sustainable food future. It attempts to answer the question posed by the American author in the field of nutrition David Wolfe — what is this insect eating telling us.

Curious, Cautious, Conscientious and Resistant Young People

Gen Z represents a paradox in food culture — hedonistic, values-driven, adventurous, but emotionally resistant (Duran and Bilim, 2025). This cohort of young people craves novelty, cultural diversity and technological innovation in their food choices and experiences, often guided by social media trends and online research. Their consumption is intentional but impulsive, and brand loyalty is notably low. Functionality, quality and uniqueness rank high on their list of priorities. Despite their general openness to new food experiences, edible insects remain firmly on the fringe of Gen Z's dietary landscape (Sogari *et al.*, 2019; Duran and Bilim, 2025).

Entomophagy offers benefits such as high-quality protein, a low environmental footprint and potential for scalable food security solutions. Yet psychological barriers persist. For Gen Z, resistance is not merely about taste but rooted in trust, identity and instinct:

> *"I get that insects are sustainable, but it still feels like a line I'm not ready to cross. It's not about being close-minded — it's more like, if I eat that, what does it say about who I am? There's a weird identity shift that I'm not sure I'm okay with."* (Female, born 1998)

This quote reflects the internal conflict Gen Z often experiences, balancing openness to sustainable innovation with a strong sense of personal and social identity and desire for authenticity and self-alignment. It adds depth by showing that their resistance is not simply a matter of aversion or ignorance, but rather, it is intricately bound to how food choices communicate values, social identity and cultural belonging.

Supporting this notion, a study among British Gen Z consumers aged 18–30 found that 93% of the participants perceived insect-based products as more appealing when the insects were not visible

(Willeke *et al.*, 2025). This suggests that concealment may mitigate the immediate "yuck" factor. However, even in these disguised forms, insect-based options continued to be rated lower than conventional products in terms of taste, sweetness and texture. This points to a deeper discomfort, one that cannot be overcome by aesthetics or sensory masking alone.

What emerges is a nuanced tension: while Gen Z is curious and environmentally aware, they also exhibit a protective stance over the symbolic meaning of food. For them, eating is not just about nutrition or sustainability. It is a personal statement. Insects though scientifically validated as nutritious and environmentally friendly, challenge established narratives of what is considered "real" food — especially within Western contexts where insect consumption is not normalised. Introducing insect-based foods into this ecosystem requires more than rebranding or reformulation. It demands a cultural reframing that aligns with Gen Z's desire for both ethical integrity and emotional resonance in their food choices.

In the remainder of the chapter, we first describe the benefits associated with entomophagy — from a sustainability and nutritional point of view, then outline the barriers — personal and industry-based, for making this a mainstream option for Gen Z. We finally discuss some of the ways to overcome these concerns and barriers before asking the question whether Gen Z will embrace eating insects.

Why Insects? The Sustainability Argument

Insects present a compelling case from an environmental perspective. They have been consumed for millennia in Asia, Africa and by the Australian Aborigines (see Figure 6.1 showing a traditional insect market in Cambodia). However, their importance as a food source is now changing. Futurists and sustainability advocates point to insects as a crucial solution to the ecological crisis associated with food. Their cultivation requires significantly less land, water and feed compared to traditional livestock. They emit fewer greenhouse gases and can convert organic waste into high-quality protein, supporting circular economy models and enhancing soil health (van Huis and

Figure 6.1. Insects market, near Phnom Penh, Cambodia (photo authors).

Oonincx, 2017; Balal, 2025). With the global population expected to reach 9 billion by 2050, insect-based proteins offer a scalable, efficient solution to meet rising demand for food. Insects represent a more sustainable alternative. Studies show that insect farming converts feed into protein efficiently, emits fewer greenhouse gases, consumes minimal resources, and requires less space and water compared to livestock, making it an ideal candidate for future food security (Cardona, 2024). Insects can be grown on organic waste streams which can contribute to nutrient recycling and waste minimisation (Lisboa *et al.*, 2024). They also have high fecundity with short reproduction cycles allowing them to reproduce in big numbers (Grimaldi, 2023).

Technologies such as Artificial Intelligence (AI), Internet of Things (IoT) and automation are accelerating the development of industrial-scale insect farming. Precision monitoring systems track environmental conditions, health metrics and yields, optimising production while minimising waste (Grimaldi, 2025). Emerging technologies such as machine learning-driven feed optimisation and robotic harvesting systems are enhancing efficiency, ensuring consistent quality and scalability. These advancements position insect farming as a viable solution for meeting global protein demands sustainably (Lange and Nakamura, 2023). AI-driven feed optimisation and robotic harvesting ensure consistency and efficiency, enabling insect farms to produce nutrient-rich, environmentally responsible protein at scale. Policy support and increasing consumer awareness could transform insects from niche novelties in the West into mainstream staples.

Nutritional and Health Benefits

Edible insects are not only sustainable, but they are also nutritionally dense. Packed with high-quality protein, fibre, healthy fats, vitamins, micronutrients and bioactive compounds, they represent a viable alternative to conventional meat (Bashir, 2024). They have potential health benefits, including improved gut health, immune support and cardiovascular wellness, and may even possess probiotic properties (Stull, 2021).

However, risks exist and the picture is not entirely rosy. Ruiz *et al.* (2025) caution about heavy metal accumulation, microbial contamination and allergic reactions, especially for those allergic to shellfish or crustaceans. While antioxidant and antimicrobial properties are promising, long-term health effects of entomophagy remain under-researched. Questions about metabolism, immune responses, allergenicity and microbial safety require further scientific investigation and regulatory oversight. As insects enter the human food chain in new forms, ensuring nutritional benefits without compromising safety will be critical (Cunha *et al.*, 2023).

Psychological and Social Barriers

Despite their impressive environmental and health credentials, edible insects face a deep-rooted psychological resistance. The instinctive "yuck" reaction remains a strong barrier to mainstream acceptance. Many still perceive insects as pests rather than potential food, making the concept of entomophagy difficult to swallow, literally and figuratively. An ethically oriented but emotionally resistant towards insects, Gen Zer shared:

> *"Look, I care about the planet. I'm vegan most days because I hate how animals are treated in factory farms. So yeah, the whole 'insects don't suffer like mammals' thing makes me pause. But emotionally? I just can't. Bugs gross me out on a deep level... It's like a hardwired reaction. I don't want to see them in my food or think about chewing one. I know it sounds irrational when you look at the data, but food is also emotional and cultural, right? I grew up thinking bugs are dirty or dangerous. It's gonna take more than facts to change how I feel."* (Female, born 2003)

Psychologists emphasise that food neophobia and disgust sensitivity are persistent deterrents. Even when insects are presented in an unrecognisable form in familiar formats, such as protein bars or baked goods, many consumers hesitate to consume them due to ingrained mental associations (Willeke *et al.*, 2025). Although people are generally acceptive for insects to be used as feed for livestock, they prefer this to be done for fish or poultry rather than beef and pork (La Barbera *et al.*, 2021).

Religious beliefs also play a role adding another layer of complexity. While some faiths permit the consumption of insects, others strictly forbid it, restrict it to specific species (e.g. locusts and crickets) or even give conditions when to be used as feed. For example, Muslims in general accept traces of insects in food but they should not be intentional or visible (Ramanujan, 2024). A study in Lebanon showed edible insects largely rejected based on perceptions of cultural impropriety and filth (Bou-Mitri *et al.*, 2025). Ethical debates persist within vegetarian and vegan communities highlighting concerns about consuming living creatures, further limiting acceptance. These challenges make it difficult for insects to break into everyday diets despite their potential as a sustainable protein source.

It may be possible to use education, transparency and innovation to shift consumer perception. Reframing insects as desirable, safe and aspirational foods could help overcome disgust and build trust. Could Gen Z, with its digital fluency and sustainability ethos, hold the potential to catalyse such a shift, if provided with the right narratives, platforms and experiences? The Gen Z youth we spoke with were not convinced and expressed their categorical stance against consuming insects:

> *"Nah, sorry, this whole bug-eating trend feels like a hard pass for me. I know people say it's the future of food, but honestly... It just feels forced. Like someone's trying to normalise it so we'll eventually accept it out of desperation. I'm not buying into the 'eat bugs' to save the planet's pitch when we haven't even fixed food waste or overconsumption. There are plenty of plant-based options that don't involve me eating beetles. I don't want to feel like I'm being nudged toward something just because it's trendy or efficient. If it ever comes to that, maybe the system is already broken."* (Male, born 2001)

The pushback from Gen Z members is because they see insect consumption as a top-down solution that lacks authenticity and ignores systemic food issues. The emotional resistance goes beyond the physical "ick" factor and reflects broader disillusionment with how food system transitions are framed.

Social marketing campaigns have been proposed as a strategic tool to shift consumer attitudes toward edible insects, addressing key psychological barriers such as unfamiliarity, preparation methods, visual appeal and access to reliable information (Bucea-Manea-Tonis *et al.*, 2023). The aim of campaigns is to leverage behavioural science principles to normalise entomophagy, using targeted messaging, educational initiatives and exposure strategies to enhance familiarity and reduce neophobia. One young person explains this generation's resistance to social marketing campaigns for edible insects:

> *"No amount of flashy marketing or educational campaigns will convince me that eating insects is normal. It's not about logic or nutrition. You know... it just feels unnatural. If sustainability means swapping out real food for bugs, then I'd rather find another way. There's something about it that crosses a line, and no amount of exposure is going to change that."* (Male, born 2000)

There are deep-seated emotional and psychological barriers that any marketing would need to overcome. While campaigns can aim to normalise entomophagy through familiarity in order to shift perceptions, many Gen Z individuals see insect consumption as fundamentally incompatible with their ingrained cultural perceptions about food, personal beliefs and tradition.

Disgust is a key deterrent to entomophagy, often intensified by cultural perceptions of insects as pests rather than food. Cultural narratives portray insects as dirty or disease-ridden, contributing to visceral aversions (Orkusz and Orkusz, 2024). A study by Bou-Mitri *et al.* (2025) among Lebanese consumers similarly found that the main deterrents to insect-based food adoption were feelings of disgust (60.7%), perceptions that insects are dirty (57.6%), weird (29.4%) or unsafe (28.4%). Urbanisation seems to have intensified the negative perceptions about insects, as people increasingly encounter them indoors rather than in natural environments and associate them with contamination and hygiene concerns (Science Japan, 2021). Furthermore, historical and religious influences in Western societies have contributed to the widespread disgust toward insects as food, framing them as undesirable or even harmful (Orkusz and Orkusz, 2024).

Unlike older generations who may view insect consumption as a novelty, Gen Z is generally sceptical and resistant to the idea. Social media have attempted to challenge Gen Z's resentment. Influencers, including chef Joseph Yoon (Brooklyn Bugs, n.d.; Foster, 2019), and platforms such as TikTok and Instagram have started normalising insects through aesthetically appealing meals, showcasing insect-infused gourmet dishes and viral food challenges (Ants Gourmet, 2024; Taste Atlas, 2025; Bugvivant, n.d.; Feed Spot, 2025). Despite this exposure, Gen Z's acceptance of insect foods remains modest.

Part of this reluctance may stem from cultural narratives portraying insects as not just unappetising, but also as symbols of manipulation or loss of agency. Insect consumption is perceived as a political and ideological agenda. Along this vein, David Wolfe, a controversial figure in the alternative health movement, challenges the dominant sustainability narrative surrounding insect-based dietary choices. He contends that the push for entomophagy, framed as a sustainability solution, is not merely about environmental benefits, but rather part of a broader societal manipulation. Wolfe suggests that the effort to normalise insect consumption signals a top-down manipulation by a so-called “predator class”, using entomophagy as a symbolic gesture of control over humanity. His argument taps into deeper cultural anxieties, questioning issues of autonomy, authenticity and hidden agendas behind corporate food policies. While his perspective may seem fringe to some, it resonates with younger consumers, particularly Gen Z, who have grown increasingly sceptical of institutional authority and the motivations behind corporate-driven food agendas (Bogueva and Marinova, 2025).

The reluctance toward insect-based food products reflects a broader distrust of institutional decision-making, highlighting the ongoing tension between public choice and engineered change. Similarly, conspiracy-driven narratives have gained traction in some circles. For example, some claim that the European Union is forcing its citizens to eat insects as part of a hidden political or corporate agenda. It is even suggested that insects would replace traditional foods or that they are being introduced to control populations (EDMO, 2023). Despite lacking evidence in support of such theories, the claims further amplify mistrust (EDMO, 2023). They feed into wider public uncertainty and complicate efforts to normalise insect-containing food products.

These narratives reveal that the future of food is not merely technical nor nutritional but deeply psychological and political. Advocates must bridge the gap between ecological urgency and public perception, promoting entomophagy as an informed choice rather than an imposed necessity. Edible insects should be reframed not as desperate substitutes but as desirable, safe and even aspirational foods. This shift will require a blend of transparency, education and product innovation, addressing psychological resistance while highlighting the practical benefits of entomophagy. Can Gen Z overcome the disgust factor and see insects as a legitimate component for a sustainable future, or will instinctual barriers and ideological concerns continue to slow adoption?

Promoting entomophagy cannot be purely nutritional or ecological. It must also be transparent, culturally sensitive and choice-based. Positioning insect consumption as a voluntary, informed and ethical

Figure 6.2. Fried tarantula at a market near Phnom Penh, Cambodia (photo authors).

act, rather than a forced transition, will be essential in reframing public opinion. Gen Z's acceptance of edible insects will depend on trust, transparency and messages that affirm autonomy, ethics and social identity.

Despite the fact that over 1500 species of insects are regularly sold and consumed by people (see Figure 6.2) from over 300 ethnic groups in 113 different countries (MacEvilly, 2000), the Gen Z consumer, particularly in the West, has a long way to go to accept them as food. The complexity of the task is even higher when aspects, such as threat to identity, social stigma and regulations, are added to the picture.

Threats to Identity and Masculinity

A particularly influential factor in shaping the attitudes towards edible insects is the role of identity, specifically gender norms surrounding food choices. An earlier study of Australian consumers (Sogari *et al.*, 2019) revealed that some men perceive eating insects as a challenge to traditional masculinity, reinforcing the notion that insect-containing foods are socially undesirable. Hence, expression of personal and social identity particularly through cultural narratives, are part of dietary choices.

By understanding the intersection of psychological barriers and gender norms, we can better evaluate the future of insects as food. As some Gen Zers advised, reframing insect-based foods as protein-rich and performance-enhancing might help overcome ingrained perceptions. Nevertheless, their popularity is likely to remain limited:

> *"I get why people are hesitant... it's weird at first. But if it helps the planet and provides good nutrition, why not? I think a lot of it is just a mental block. If we can change how we see food, maybe insects won't feel so out of place anymore. But honestly, they are not food for everyone to like and accept."* (Female, born 2002)

For many, red meat remains symbolically linked to strength and masculinity, making the idea of consuming insects seem incompatible with these established perceptions. This association raises important questions about how gender influences food acceptance and a Gen Zer is categorical:

> *"Eating bugs just doesn't fit with how I see food. I grew up associating protein with real food, like meat... Meat has always been a symbol of health, strength, something real and substantial. Switching to insects feels like giving up something bigger than just a meal, it's about tradition, masculinity, and what food is supposed to represent."* (Male, born 1999)

The masculinity-related barrier persists in contemporary attitudes toward edible insects and gender dynamics play a role in this. Analysing consumer responses can help determine whether generational shifts, increased exposure to sustainable food narratives and evolving masculinity ideals have altered perceptions. Understanding these dynamics is crucial for developing effective strategies to promote entomophagy as a viable and socially accepted food choice.

Lack of Familiarity and Social Stigma

Unlike plant-based products, which are now widely available, insect-containing foods remain a niche in Western markets. Most Gen Z consumers have never encountered edible insects in a familiar grocery shop or fast-food setting, making this food option feel foreign and unappealing. This unfamiliarity adds to Gen Z's scepticism (Kamenidou *et al.*, 2023). However, some Gen Zers are open-minded realising both, the practical advantages and psychological barriers:

> *"I see both sides, honestly. On one hand, insect protein makes a lot of sense. It's efficient, sustainable and already eaten in other cultures. On the other hand, I don't think it'll go mainstream anytime soon — not without serious rebranding. People aren't going to swap their chicken nuggets for mealworm nuggets overnight. But if the products are well-made and you don't even realise bugs are in them, like with cricket flour or whatever, I think younger people might be more open than we think. It just needs to feel... normal. Right now, it still feels like something you'd eat on a dare."* (Female, from 1999)

This reinforces the idea that food choices are not just based on taste or nutrition but are socially constructed, influenced by identity, cultural norms and generational attitudes. Although entomophagy is normal in many parts of the world, including Asia, Africa, Latin America and for Aboriginal people in Australia, where insects are celebrated as delicacies or essential protein sources (Mishyna *et al.*, 2020), Western Gen Z still largely stigmatises them as food. As Grimaldi (2025) argues, this disconnect reflects the broader challenge of integrating edible insects into food cultures that have long positioned them as contaminants rather than cuisine.

The stigma surrounding insect consumption remains one of the most entrenched psychological barriers to entomophagy among Gen Z consumers in the West. Despite being digital natives immersed in global discourses, Gen Z is not immune to the influence of social norms, particularly those inherited from older generations. Raised largely by Gen X parents, who themselves were not exposed to entomophagy in mainstream culture, many Gen Z individuals subconsciously absorb inherited aversions to insects as food.

Western narratives not only frame insects as dirty, but eating them is seen as demeaning or associated with desperation (Grimaldi, 2025). This framing is reinforced through media portrayals that link insect consumption to survivalist scenarios, poverty or sensationalist television — further distancing it from everyday food culture. These symbolic meanings make insects a harder sell than other alternative proteins. Unlike plant-based meat, which benefits from branding and alignment with tech-forward lifestyles, edible insects carry historical baggage that invokes feelings of embarrassment, deviance or even shame. As one Gen Z consumer shared:

> *"Okay, real talk. I used to think eating bugs was just some gimmick for reality TV or survival shows. But lately I've been seeing more articles and TikToks about how insects are packed with protein and way better for the environment than beef or chicken. Like, less water, less land, less emissions. That's cool, right? I'm not saying I'd snack on fried crickets every day, but if it's in a protein bar or like, ground up into a burger patty, I'd be open to trying it. As long as it doesn't have legs sticking out, I think I could deal. It's just... if we're serious about sustainability, maybe we need to get over the ick factor. Slowly, though... not all at once."* (Male, born in 2000)

Overall, Gen Z's enthusiasm for insects is low. Although Gen Z is generally open to novel dietary experiences, in this particular case, psychological aversion combines with food neophobia and reluctance to even try the new products. In a study in Greece, Kamenidou *et al.* (2023) found that 88.5% of Gen Z respondents were unwilling to replace meat with insect protein, and only 6.3% were willing to participate in sensory tests. Sogari *et al.* (2019) similarly found that in Australia psychological barriers stifled adoption. Even when Gen Z acknowledges the environmental and nutritional benefits of edible insects, these factors do not necessarily translate into actual consumption behaviour (Sogari *et al.*, 2019). This suggests that while Gen Z may be open to sustainable food alternatives, insect-based options still face significant hurdles in terms of cultural acceptance and consumer psychology. Gen Z might intellectually support sustainable diets but stop short of including bugs on their plates.

Hence, Gen Z experiences a growing cognitive dissonance: intellectually aligned with sustainability, yet emotionally anchored in longstanding cultural narratives. There may be a pragmatic openness, but only when consumption is visually and socially neutralised through processing, blending and aesthetic concealment. Such strategies help bypass the stigma without forcing a direct confrontation with deeply internalised taboos.

Social pressure, performative disgust and lack of relatable role models continue to shape reluctance among Gen Z. Destigmatisation must occur at both cultural and interpersonal levels. Sustainability messaging is slowly shifting the discourse, but the transition toward normalised entomophagy among Gen Z will likely depend on a blend of peer validation, emotional reframing and culinary innovation that reimagines insects as aspirational rather than abject.

Availability and Innovation

While cricket-based protein powders and mealworm snacks are on the market, broader adoption of foods with insect ingredients hinges on innovation. Could lab-grown insect proteins or genetically modified breeds create more palatable alternatives? Research in biotech and food engineering is only beginning to scratch the surface. Singapore is at the forefront of this shift, leading in regulatory approval for food innovation. After approving cultured meat and welcoming it as food to its consumers, the country has officially approved also 16 insect species for human consumption, including beetles, locusts, grasshoppers and mealworms, provided they are farmed in regulated facilities rather than harvested from the wild (Al Jazeera, 2024). This responds to concerns about the cleanliness of insects and aligns with a broader global trend of embracing entomophagy (Al Jazeera, 2024). By greenlighting both lab-grown meat and edible insect species, it sets a precedent for embracing unconventional proteins. This positions Singapore as a key enabler rather than a direct producer, signalling a future where alternative proteins may become mainstream in global markets.

In the rest of the world, even the most curious and sustainability-conscious Gen Z consumers face a practical barrier: insect-based foods are rarely accessible through conventional retail or hospitality channels. Particularly in most Western countries, these products remain confined to niche health food stores, specialty websites or experimental food festivals. This limited visibility reduces exposure and the opportunities for normalised, low-risk tasting experiences (Platta *et al.*, 2024). For a generation accustomed to convenience, food delivery apps and on-demand consumption, the absence of edible insects in mainstream grocery chains or restaurant menus acts as a major deterrent to trials and experiments. Without routine availability, insect-based foods remain a novelty rather than a legitimate dietary choice.

Affordability adds another layer of difficulty. Many edible insect products, especially those marketed in processed forms, such as protein bars or flours, are priced higher than their plant-based or conventional protein counterparts. For Gen Z, often juggling student loans or early-career salaries, price can be a deciding factor, even if the willingness to experiment exists as one Gen Zer noted:

> *"I am a bit nosy for new things... I wanted to try insect pasta I saw online, but it costs twice what I usually pay. If I'm taking a chance on something new, I'm not sure I want to also feel like I'm wasting money if I don't like it."* (Male, born 1999)

The economic reality reinforces existing consumption patterns and limits the normalisation of insect-based products in everyday eating. With increased demand, the prices are likely to fall down but for this to happen Gen Z needs to be convinced that at least their health is properly protected through policy, regulations and standards.

Policy, Regulatory and Standardisation Challenges

Globally, regulatory frameworks surrounding edible insects vary. Similar to Singapore, Europe has approved several species for consumption, but in many countries, insect-based products remain largely unregulated and policies are underdeveloped. Other major unknowns are emerging risks, such as hidden allergens or microbiological risks. They impact consumer well-being and raise safety questions. With limited data on long-term effects, policy discussions remain incomplete. This uncertainty poses a barrier not only for Gen Z consumers but also for entrepreneurs looking to innovate in the insect-based food industry. With no harmonised international standards, bringing insect-based products to market is a logistical and legal maze. Are there gaps in legislation hindering the ability to scale up insect products in the mainstream markets?

Transparency and trust are central to Gen Z's food choices. Beyond environmental concerns, they need to know what is in their food, where it comes from and how it is made. Insect-based foods currently suffer from a perceived lack of transparency. There are uncertainties around how insects are farmed, what they are fed, what their processing methods are, and what standards govern safety and hygiene as part of proper safety regulations. This uncertainty breeds hesitancy. Insect farming lacks the established oversight that plant-based or traditional animal protein sectors have developed over decades. Compared to traditional livestock or plant-based industries, edible insect farming lacks consistent regulatory oversight and labelling systems. Concerns about potential allergens, especially cross-reactivity with shellfish, and microbial contamination are frequently cited but remain under-researched (Kamenidou *et al.*, 2023).

Moreover, Gen Z often brings an ethical lens to food consumption. Questions around how insects are killed, whether they feel pain, under what conditions they are raised and the environmental integrity of farming practices remain largely unanswered in mainstream media or product information. For a generation concerned not only about health but also with the moral dimensions of food production, these gaps in transparency create friction and can erode trust. This generation is digitally literate and used to scanning QR codes, reading sourcing statements or investigating brands online. In the absence of robust, accessible information, scepticism arises, particularly among Gen Z, who are accustomed to investigating brands and products before purchase:

> *"I know it's supposed to be sustainable, but where's the info? With oat milk or tofu, I know the supply chain, the labels, even the carbon impact. With insects, it's all a bit murky, like, where are the receipts?"* (Female, born 2001)

Until foods containing insects meet the same transparency benchmarks expected of other sustainable products, many Gen Z consumers will hesitate to fully embrace them. The policy, regulatory and standardisation environment needs to win the trust of Gen Z, particularly as insects compete with other dietary options.

Preference for Familiar Food and Other Alternatives

Many Gen Z environmentally-oriented consumers prefer familiar alternatives and are already comfortable with plant-based products including tofu, tempeh, mycoproteins, legumes, real fruit and vegetables. These are widely available, versatile and supported by strong branding and environmental messaging aligned with broader health and ethical narratives that resonate well with Gen Z. Such alternatives also align with growing vegan and flexitarian identities popular among younger consumers. Unlike insects, these alternative products are already embedded in vegetarian and vegan food cultures, with which many Gen Z consumers engage, whether full-time or flexibly.

Insects, on the other hand, often feel foreign, not just in flavour and texture, but in cultural positioning. While some might be open to the idea of insects as future food, the sensorial unfamiliarity and absence of social validation (e.g. seeing friends or influencers eating them regularly) make adoption more difficult. As Platta *et al.* (2024) point out, the plant-based sector benefits from both structural support and consumer familiarity, creating a significant lead over entomophagy in the race for sustainable protein adoption.

In essence, while Gen Z values sustainability, their preferences are shaped by availability, price, perceived safety and social context. Without greater integration of insect-based foods into everyday food systems, including clearer safety standards and more accessible formats, their uptake will remain limited compared to the rapid normalisation of plant-based alternatives.

Overcoming Disgust

Research has extensively examined the "yuck" factor in insect consumption, yet strategies for permanently reframing perceptions remain underexplored. Could marketing, influencers immersive experiences, virtual reality (VR) or sensory training help shift consumer mindsets in a lasting way? While disgust serves as a significant psychological barrier, emerging techniques in behavioural psychology suggest that targeted exposure, education and sensory adaptation could help normalise insects as foods.

Social media are very popular with Gen Z and can be used to challenge existing perceptions about entomophagy. Influencers create viral content featuring insect-based dishes, from daring challenges to gourmet experiences using TikTok and YouTube platforms (e.g. Wright, n.d.). However, translating digital spectacle into lasting dietary habits remains uncertain. While influencers can generate curiosity and engagement around insect-based foods, the transition from online interest to real-world adoption is hindered by the same deep-seated psychological and cultural barriers.

Although influencers wield massive power over Gen Z's choices, authenticity is a key factor. Gen Z values transparency and genuine endorsements over clear advertising. If influencers merely promote insect-based foods as a trend without demonstrating personal engagement, scepticism arises. Successful campaigns instead should feature influencers who integrate insect-based products into their daily routines, emphasising taste, nutritional benefits and sustainability in a relatable way.

Another challenge is visual appeal. Visually appealing content can normalise insect-based foods, but only if it is backed by authenticity and quality. While some brands attempt to make insect-based foods more appealing, products that visibly resemble insects tend to be rejected. Gen Z consumers are more likely to accept insect protein when it is disguised in processed forms. As Willeke *et al.* (2025) highlight, processed formats like cricket flour are significantly more accepted than whole insects as they eliminate the visual and textural barriers that often trigger disgust responses. Subtle integration into existing food products, such as protein bars, pasta or baked goods, can enhance consumer acceptance.

Furthermore, social proof and peer influence play a vital role. Gen Z is more likely to try new foods if they see their peers engaging with them positively. Viral challenges, taste-test videos and interactive content can help normalise insect consumption, but only if they are framed in a way that feels organic rather than forced.

An approach, particularly relevant in Asian countries, involves gradual exposure and development of familiarity, where consumers are introduced to insect-containing foods in controlled environments, reducing initial aversion. Research indicates that repeated exposure to products in which insects have been used can diminish disgust responses over time, making individuals more open to trying them (Castro and Chambers, 2019). This principle aligns with findings in behavioural psychology, where sensory desensitisation has been shown to reduce negative emotional associations with unfamiliar foods (Tuorila and Hartmann, 2020).

To address these barriers, product form matters immensely. Consumers are far more likely to try insect-based products when integrated invisibly into familiar food formats, such as flours, protein bars or pasta, rather than containing visible insect pieces. This preference is widely observed in Western markets, contrasting with traditional insect-eating cultures like in Mexico or Thailand, where the visibility of whole insects does not present a psychological barrier (Castro and Chambers, 2019).

Promoting insect-containing products through popular, familiar formats may serve as a critical first step in overcoming disgust barriers and building product familiarity. Equally important is education, clarifying that not all insects are unhygienic and that farmed edible insects adhere to strict food safety standards. By dismantling inaccurate beliefs and reinforcing hygiene, nutritional and environmental benefits, education can neutralise disgust at its root (Castro and Chambers, 2019).

This aligns with behavioural psychology's broader toolkit: immersive learning, gradual exposure and reframing are proven methods in combating negative emotional associations, such as in the case of anxiety (Maltzeva, 2025). Creating interactive platforms or campaigns that allow consumers to see, experience and understand insect farming in a hygienic, scientific context could pave the way toward normalisation.

Education is a powerful tool for reshaping perceptions of edible insects, particularly in countries where entomophagy is not culturally embedded. Research in Congo suggests that individuals with higher levels of education are more likely to consume insects, as they possess a better understanding of their nutritional benefits (Mwisomi *et al.*, 2023). By incorporating engaging interactive experiences, consumers can begin to see insects as a normal and acceptable part of a meal, which may help reframe psychological associations that contribute to disgust (Mwisomi *et al.*, 2023).

Further research and product innovation aimed at embedding insect-based ingredients into popular food types may be the most effective strategy to break disgust barriers and build consumer familiarity. Also, educating consumers that not all insects are unhygienic and that edible insect farming adheres to safety regulations will be critical in eliminating erroneous concepts from the consumer mindset (Castro and Chambers, 2019).

Will Gen Z Stick with Insects?

When speaking with representatives from Gen Z, we found ourselves asking: is interest in entomophagy just a fad? Will they be open to such experiences as they transition into adulthood and parenthood, or could this simply be a passing trend fuelled by social media? Longitudinal data on generational food behaviour, especially regarding edible insects as alternative proteins, is scarce and most likely, non-existent. Understanding whether insect acceptance is viable over time will be critical for consumer acceptance. As Gen Z moves into new stages of life, including career-building, family life, parenthood and shifting socioeconomic contexts, their priorities and consumption habits are also likely to evolve.

One key factor in determining whether insect consumption will be part of Gen Z's diet is habit formation. If Gen Zers integrate insect-based foods into their diets early and perceive them as convenient, nutritious, normal and socially acceptable, adoption may continue and repeated use will be more likely. However, if insect consumption remains tied to novelty, dare, occasional social performance (e.g. TikTok challenges or curiosity-driven tastings) or is primarily driven by social media engagement, it risks fading as other food trends emerge. So far, this appears to be the case.

> *"I tried cricket chips at a sustainability fair, and yeah, it wasn't awful, but I don't see myself adding it to my weekly shopping list. It felt more like an experiment than something I'd actually crave to consume or cook with."* (Female, born 2000)

Such an opinion reinforces the fact that insect consumption currently leans more toward novelty than habit. While curiosity may drive initial engagement, maintained consumption requires familiarity, convenience and cultural integration. The quoted Gen Zer's experience suggests that insect-based foods are often framed as an experiment rather than a regular dietary option. This aligns with existing research indicating that novelty plays a significant role in Gen Z's willingness to try insect-based products, yet few transition from trial to habitual consumption (Kamenidou *et al.*, 2023).

Moreover, this reaction highlights sensory and psychological barriers associated with edible insects. While the product was not unpleasant, it lacked the craveability factor that motivates repeat purchases. This suggests that for entomophagy to become mainstream, brands must focus on product development, enhancing taste, texture and integration into familiar food formats. Additionally, social normalisation through peer influence, restaurant offerings and everyday accessibility may help shift perceptions from niche novelty to a legitimate food category.

Equally important are market accessibility and product innovation. Long-term uptake will depend on whether insect-containing ingredients can be integrated into familiar products, such as protein bars, pasta or snack foods, ideally in ways that de-emphasise the insect origin. A perfect example of this are the cochineal insects that are being used for red colour in a myriad of products and is even being marketed as a natural product (Harvard Museum of Science and Culture, 2025).

Mainstream grocery stores, quick-service restaurants and meal delivery platforms can help normalise insect consumption. Without this visibility, entomophagy risks remaining a fringe behaviour rather than becoming a staple. The role of accessibility and convenience is particularly relevant, as reflected in the following Gen Z's attitude:

> *"Honestly, if bug protein just came as part of a cool energy bar at my gym or was blended into a smoothie I already like, I'd probably be eating it without even thinking. But if I have to go to a weird website or a specialty store to find it, it's just not happening."* (Male, born 1999)

Cultural and generational dynamics play a formative role and shape long-term food preferences. Younger consumers are typically more open to experimentation, but broader social and familial norms play a role in solidifying food habits. Studies suggest that early exposure and positive experiences with insect-containing foods, such as at school or university meals, home cooking, or youth-focused campaigns, could be key to sustained acceptance.

> *"We grew up with plant-based everything... soy, almond milk, tofu, sushi. That stuff feels normal now. Bugs? Not so much. I'm not against the idea, but it doesn't feel like something that belongs on my plate... yet."* (Female, born 2000)

Still, the current landscape for entomophagy is marked by mixed signals. Despite some openness within the Gen Z cohort, deep-seated emotions like disgust, fear and unfamiliarity continue to dominate, particularly in Western cultures. Kamenidou *et al.* (2023) found that while Gen Z is not inherently food neophobic and 20.4% express curiosity, suggesting potential for conversion if the right messages, form and trust mechanisms are in place. Behavioural segmentation provides further nuance. The same study identifies distinct Gen Z segments including future potential consumers (29.1%), rejecters (26.7%), "disgusted, prefer to starve" consumers (22.2%) and inconsistent individuals (22.0%) (Kamenidou *et al.*, 2023). This segmentation underscores the psychological and cultural complexity of entomophagy acceptance and suggests that a "one-size-fits-all" strategy will be ineffective. Gender, cultural background, ethical orientation and food literacy levels all contribute to shaping these attitudes.

> *"I get that insects are sustainable, but let's be real, if I'm grossed out just looking at it, I'm not going to eat it, no matter how many influencers say it's cool."* (Male, born 2001)

These perceptions are not easily changed by information alone. They require experiential, social and emotional engagement, emphasising sensory adaptation and cultural reframing.

Conclusion

The situation in Western countries regarding entomophagy is very different from places where insects have been eaten by people for millennia. Gen Z represents both a challenge and an opportunity for the future of insects as food. These young people's sustainability-driven values, openness to innovation and digital literacy make them a natural audience for alternative proteins. Yet, their continued engagement with insect-containing foods hinges on whether the industry can address persistent sensory, psychological and accessibility barriers within a properly regulated policy and food safety environment. Unless insect-based foods are reimagined as desirable, accessible, integrated into familiar food formats, embedded in everyday food culture and daily routines, edible insects' future risks remaining a novelty rather than catalysing meaningful dietary change and a food system transformation.

To move beyond experimentation, stakeholders, including food producers, policymakers, educators and marketers, must invest in normalising entomophagy through targeted communication, transparency, robust safety standards and innovative product design. Bridging current research gaps, especially around consumer psychology, social acceptability and innovation, will be essential to generating and converting any curiosity into lasting behaviour.

If these conditions are met, Gen Z's appetite for sustainability could help transform insect consumption from a niche movement into a mainstream solution, at least in some of its forms (e.g. blended, bars, flours etc.), offering a scalable response to global food security and environmental challenges.

References

Al Jazeera (2024). Will eating insects save the planet? https://www.aljazeera.com/news/2024/8/4/will-eating-insects-save-the-planet

Ants Gourmet (2024). From Mild to Wild: 3 Insect Food Challenges for Every Appetite. https://antsgourmet.com/blogs/helpful-resources/from-mild-to-wild-3-insect-food-challenges-for-every-appetite

Balal, C. (2025). The Role of Insects in Sustainable Agriculture: A Complete Guide. Wikifarmer. https://wikifarmer.com/library/en/article/the-role-of-insects-in-sustainable-agriculture-a-complete-guide

Bashir, M. H. (2024). Innovative solutions for food security with edible insects. The Agricultural Economist. https://agrieconomist.com/innovative-solutions-for-food-security-with-edible-insects

Bogueva D. & Marinova D. (2025). Will Australians eat alternative proteins? *Foods*, 14(9), 1526. https://doi.org/10.3390/foods14091526

Bou-Mitri, C., Mashtoub, L. A., Issa, A., Doumit, J., & Karam, L. (2025). Consumers perception of insect-based foods in Lebanon: A novel sustainable food source. *Journal of Insects as Food and Feed.* https://doi.org/10.1163/23524588-bja10239

Brooklyn Bugs (n.d.). Education. https://www.instagram.com/brooklynbugs/?hl=en

Bucea-Manea-Tonis, R., Martins, O. M. D., Urdes, L., Coelho, A. S., & Simion, V.-E. (2023). Nudging consumer behaviour with social marketing in Portugal: Can perception have an influence over trying insect-based food? *Insects*, 14(6), 547. https://doi.org/10.3390/insects14060547

Bugvivant (n.d.). Awesome Edible Insect Recipes. https://bugvivant.com/edible-insect-recipes/

Cardona, M. (2024). Insects: The Food of the Future. Emerald Review. https://emeraldreview.com/2024/08/insects-the-food-of-the-future/

Castro, M. & Chambers, E. (2019). Consumer avoidance of insect containing foods: Primary emotions, perceptions and sensory characteristics driving consumers considerations. *Foods*, 8(8), 351. https://doi.org/10.3390/foods8080351

Cunha, N., Andrade, V., Ruivo, P., & Pinto, P. (2023). Effects of insect consumption on human health: A systematic review of human studies. *Nutrients*, 15(14), 3076. https://doi.org/10.3390/nu15143076

Duran, G. & Bilim, Y. (eds) (2025). *From Bugs to Food: Generation Z and the Perception of Edible Insects.* Özgür Publications. https://doi.org/10.58830/ozgur.pub715

European Digital Media Observatory (EDMO) (2023). EU with a side of insects: A delicious treat for disinformers. https://edmo.eu/publications/eu-with-a-side-of-insects-a-delicious-treat-for-disinformers/

Feed Spot (2025). 15 Best Edible Insects Blogs and Websites in 2025. https://bloggers.feedspot.com/edible_insects_blogs/

Foster, K. (2019). The Normalisation of Edible Insects: Interview with Joseph Yoon. Harvard Political Review. https://harvardpolitics.com/the-normalization-of-edible-insects-interview-with-joseph-yoon/

Grimaldi, D. (ed.). (2023). *The Complete Insect: Anatomy, Physiology, Evolution, and Ecology.* Princeton University Press, Princeton, NJ.

Grimaldi, D. (2025). Why Do Some People Hate Eating Insects? Entomology blog. https://blog.entomologist.net/why-do-some-people-hate-eating-insects.html

Halicka, E., Kaczorowska, J., Rejman, K., & Plichta, M. (2025). Investigating the consumer choices of Gen Z: A sustainable food system perspective. *Nutrients*, 17(3), 591. https://doi.org/10.3390/nu17030591

Harvard Museum of Science and Culture. (2025). Cochineal: A Product of Nature. https://hmsc.harvard.edu/online-exhibits/cochineal1/product-nature/

Kamenidou, I., Mamalis, S., Gkitsas, S., Mylona, I., & Stavrianea, A. (2023). Is Generation Z ready to engage in entomophagy? A segmentation analysis study. *Nutrients*, 15(3), 525. https://doi.org/10.3390/nu15030525

La Barbera, F., Amato, M., Fasanelli, R., & Verneau, F. (2021). Perceived risk of insect-based foods: An assessment of the entomophagy attitude questionnaire predictive validity. *Insects*, 12(5), 403. https://doi.org/10.3390/insects12050403

Lange, K.W. & Nakamura, Y. (2023). Potential contribution of edible insects to sustainable consumption and production. *Frontiers in Sustainability*, 4, 1112950. https://doi.org/10.3389/frsus.2023.1112950

Lisboa, H. M., Nascimento, A., Arruda, A., Sarinho, A., Lima, J., Batista, L.... Andrade, R. (2024). Unlocking the potential of insect-based proteins: Sustainable solutions for global food security and nutrition. *Foods*, 13(12), 1846. https://doi.org/10.3390/foods13121846

MacEvilly, C. (2000). Bugs in the system. *Nutrition Bulletin*, 25(4), 267–268. https://doi.org/10.1046/j.1467-3010.2000.00068.x

Maltzeva, I. (2025). Exploring VR therapy: A scoping review on virtual reality exposure for anxiety disorders. https://psytechvr.com/vr-therapy-for-anxiety-disorders

Mishyna, M., Chen, J., & Benjamin, O. (2020). Sensory attributes of edible insects and insect-based foods — future outlooks for enhancing consumer appeal. *Trends in Food Science and Technology*, 95, 141–148. https://doi.org/10.1016/j.tifs.2019.11.016

Mwisomi, E. M., Luminet, O., Chang, B., Kiumba, N. M., & Schmitz, M. (2023). Psychosocial determinants of intentions and behaviour towards edible insects in the South-Western part of the Democratic Republic of Congo. *Cogent Psychology*, 10(1), 2216053. https://doi.org/10.1080/23311908.2023.2216053

Orkusz, A. & Orkusz, M. (2024). Edible insects in Slavic culture: Between tradition and disgust. *Insects*, 15(5), 306. https://doi.org/10.3390/insects15050306

Platta, A., Mikulec, A., Radzymińska, M., Kowalski, S., & Skotnicka M. (2024). Willingness to consume and purchase food with edible insects among Generation Z in Poland. *Foods*, 13(14), 2202. https://doi.org/10.3390/foods13142202

Ramanujan, K. (2024). Insect-based food: sustainable, nutritious — but not religious. Cornell Chronicle. https://news.cornell.edu/stories/2024/03/insect-based-food-sustainable-nutritious-not-religious

Ruiz, F. E. A., Ortega Jácome, J. F., Tejera, E., & Alvares- Suavz, J. M. (2025). Edible insects as functional foods: Bioactive compounds, health benefits, safety concerns, allergenicity, and regulatory considerations. *Frontiers in Nutrition*, 12, 1571084. https://doi.org/10.3389/fnut.2025.1571084

Science Japan (2021). Modern people's hatred of bugs is "caused by urbanisation," the University of Tokyo reveals. Science Technology Innovation Japan, Japan Science of Technology Agency. https://sj.jst.go.jp/stories/2021/s0726-01p.html

Sogari, G., Bogueva, D., & Marinova D. (2019). Australian consumers' response to insects as food. *Agriculture*, 9(5), 108. https://doi.org/10.3390/agriculture9050108

Stull, V. J. (2021). Impacts of insect consumption on human health. *Journal of Insects as Food and Feed*, 7(5), 695–714. https://doi.org/10.3920/JIFF2020.0115

Taste Atlas (2025). Top 23 Insect Dishes in the World. https://www.tasteatlas.com/best-rated-insect-dishes-in-the-world

Tuorila, H. & Hartmann, C. (2020). Consumer responses to novel and unfamiliar foods. *Current Opinion in Food Science*, 33, 1–8. https://doi.org/10.1016/j.cofs.2019.09.004

van Huis, A. & Oonincx, D. G. A. B. (2017). The environmental sustainability of insects as food and feed. A review. *Agronomy for Sustainable Development*, 37, 43. https://doi.org/10.1007/s13593-017-0452-8

Willeke, M., Tsiami, A., & Lara, S. W. (2025). Tasting the future: Sensory evaluation and perception of insect-based products among Gen Z and Millennials. *Gastronomy*, 3(1), 2. https://doi.org/10.3390/gastronomy3010002

Wright, Z. (n.d.). Zackkwright. TikTok. https://www.tiktok.com/@zackkwright/video/7507309745791880478

Chapter 7

Fungi, Algae and the Future of Food

In 2023, Dr Michael Greger, the founder of the influential Nutrition Facts webpage (nutritionfacts.org) wrote the book *How Not to Age* (Greger, 2023), which became instantly a New York Times best seller. The book uses more than one thousand literature sources to synthesise the available scientific evidence about slowing down and preventing diseases associated with ageing through dietary and lifestyle choices. Dr Greger shared in an interview that when he was researching the materials for the book, he was astonished about the health power of mushrooms and that they should be part of every person's regular diet. Gen Z are probably still too young to think about how not to age, but as everybody else they need to know about the benefits of mushrooms and other fungi as well as the potential seaweed and algae in general have as a food for today and the future.

Before we delve into the place of fungi and algae in Gen Z's dietary preferences, it is worth explaining what these two sources of food are. Despite these organisms belonging to two different taxonomy kingdoms (Encyclopedia Britannica, 2025), they have some similarities which make people as well as science associate them. They both are not plants (or animals), but are assumed to be part of a plant-based diet or dietary options because they are positioned lower than animals in the tree of life (LibreTexts, n.d.) and are at the base of the food chain. A common feature is the lack of vascular tissues that transport water, sugars, nutrients, hormones and other signalling molecules in plants. As in plants and animals, fungi and algae have eukaryotic cells with membrane-bound nucleus, organelles and mitochondria. These cells metabolise nutrients, provide their own energy and replicate themselves to produce succeeding generations (Britannica, 2025b). The reproduction of both algae and fungi as organisms occurs asexually by fragmentation (although some fungi may propagate sexually). Most importantly, a variety of fungi and algae can be sources of human food.

Algae generally belong to the kingdom Protista, although their taxonomy is evolving as more information about them is discovered. An important characteristic of algae is that they are not parasitic, their cells contain chlorophyl and they can synthesise their own food through photosynthesis. Hence, they are autotropic and do not need to consume other organisms for their survival. Algae live predominantly in aquatic environments and range in size between microscopic dimensions (many consist of one cell only) to dozens of metres in length (and millions of cells). Seaweed is macroalgae while phytoplankton is considered microalgae. Different algae are not closely related evolutionarily and some appear to be closer to fungi than to other algae but they all have the ability to photosynthesise (Britannica, 2025a). Fungi are different and taxonomically they form their own kingdom. Their mode of nutrition is heterotrophic which means they cannot make their own food like algae and plants, and obtain nutrients by absorbing organic compounds from the environment. They can exist feeding on other organisms — as parasites, on dead and decaying matter — as saprophytes, or in a symbiotic relationship with other organisms — as mycorrhizae (Hausman, 2025). The fungi kingdom is vastly underexplored with only 10–20% of the species described (Whitmee *et al.*, 2015). In addition to mushrooms, included in the large fungi kingdom are yeasts, moulds as well as spores. They grow in diverse habitats, including in extreme environmental conditions, such as high temperatures, dry places, deep sea habitats, under UV and cosmic radiation. Algae and fungi can also co-exist symbiotically forming lichens that can grow on almost any surface and are essential for nutrient cycling (Sisson and Vigus, 2023).

The importance of fungi and algae is multi-dimensional as they support the entire plant and animal world, including humans. Often people do not pay attention to these forms of life dismissing them as inferior or irrelevant. In fact, they are the foundation of a healthy planet and can also be used for direct human consumption. Below we explore in further detail the role of fungi and algae for planetary health and specifically for human health within the understanding that the latter is embedded in the former. The concept of planetary health is defined as "the achievement of the highest attainable standard of health, wellbeing and equity worldwide through judicious attention to the human systems — political, economic, and social — that shape the future of humanity and the Earth's natural systems that define the safe environmental limits within which humanity can flourish. Put simply, planetary health is the health of human civilization and the state of the natural systems on which it depends" (Whitmee *et al.*, 2015, p. 1978). Algae and fungi are important actors first for planetary health, and then some can be used to feed human population with the aim not only to improve public health, but also to reduce the environmental footprint of food and further contribute to better well-being of all species on this planet. The use of algae and fungi is discussed here only as human food despite opportunities for them to be converted or added into feed and other economic applications, such as for oil and energy.

Fungi and Health

Fungi have long been used as a source of food with traditional knowledge helping people to distinguish between edible and poisonous species. While some may still be looking in the forests for truffles — the fungus tubers that grow under the soil in a symbiotic relationship with the roots of some trees and are very difficult to cultivate on a farm, nowadays it is rare for people to forage for edible wild mushrooms. There are thousands of mushroom varieties in the vast fungi kingdom and it is estimated that only 5 to 10% of them have been scientifically described and named (Mushroom, n.d.). A study of 2,786 mushroom species across 99 countries identified 2,189 to be edible, of which 2,006 can be consumed safely and a further 183 require some form of pre-treatment for safe consumption or were

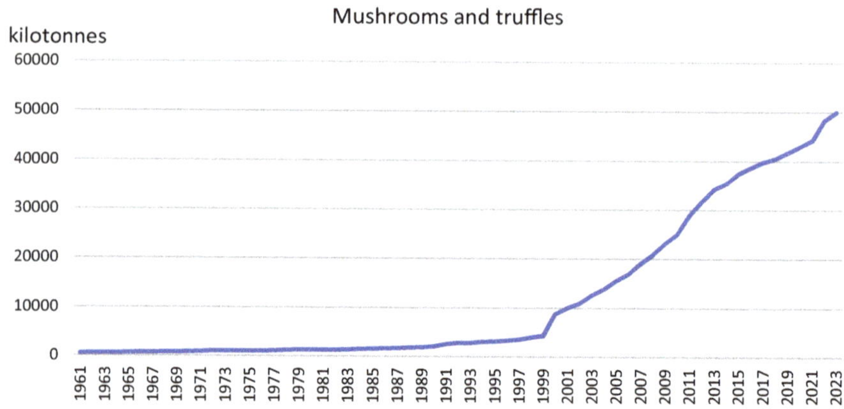

Figure 7.1. Global production of mushrooms and truffles (in kilotonnes).
Source: FAOSTAT.

associated with allergic reactions (Li *et al.*, 2021). However, only 30 mushroom species are grown commercially for food (Li *et al.*, 2021) despite global mushroom consumption rising 38-fold over the last 50 years (see Figure 7.1). The most common varieties are button, chanterelle, enoki, beech, black trumpet, hedgehog, oyster, king oyster, lion's mane, maitake, porcini, shiitake and field (or meadow) mushroom (Denenberg and the Editors of Epicurious, 2023). In Australia, the crop list of edible fungi includes 27 entries, including truffles (Australian Government, 2025).

Some mushrooms are grown commercially for their medicinal properties but overall, there is huge potential to expand into other edible species as a source for human food given the many benefits fungi have. Owing to the presence of the aforementioned properties and rich phytocomposition, mushrooms are being used in the production of nutraceuticals and pharmaceuticals.

Planetary Health

Ecologically, fungi are the principal decomposer of organic matter on the planet and contribute to nutrient cycling. This is a very important role they play as part of the natural environment. It is less known that in addition to being decomposers, fungi can serve as bioindicators for the health of the ecological systems "due to their ubiquitous distribution, diverse ecological roles, remarkable biological diversity, and high sensitivity to environmental changes" (Warnasuriya *et al.*, 2023, p. 1). Well before threats to human health emerge, fungi can be used to predict environmental changes and levels of contamination resulting from industrial processes, agricultural practices, waste disposal and natural disasters. Based on their prevalence, diversity, distribution and response to stress, fungi (and lichens) act as bioindicators for changes in the ecology that traditional environmental indicators may not be able to measure (Warnasuriya *et al.*, 2023). They can be indicators for soil, water and air quality as well as for the health of entire aquatic, marine and terrestrial environments. Similarly, they can act as bioindicators for areas surrounding industrial operations, such as ethanol-rich locations. Some fungi are pathogenic and represent a threat to the health of plants, animals and people and are bioindicators for disease-causing environments (Warnasuriya *et al.*, 2023). Fungi also absorb CO_2 contributing to fighting climate change, particularly when they grow in soils and forests (Averill *et al.*, 2018). It is estimated that fungi and their necromass — dead organic material stored in soil and often inside aggregates of soil particles, absorb one-third of the annual global emissions from burning fossil fuels, equivalent to 13.12 billion tonnes of CO_2 per year (Frew *et al.*, 2023). Any land clearing, including for growing food, affects life not only on the surface but also disrupts the fungal networks underground.

In addition to the ecological role of fungi, what is particularly important from a food perspective is how their environmental footprint compares to that of other dietary options. It is well-established that meat, and particularly beef, has a very high environmental footprint because of cattle being ruminant animals and their digestive system emitting high quantities of methane as well as the land and water requirements associated with providing feed. Edible mushrooms, which are macrofungi of the phylum *Basidiomycota* with large multicellular structures (fruiting bodies), are the most commonly consumed fungi (Wei *et al.*, 2022). They are grown commercially all year around, often in indoor environments controlled for temperature, humidity, light and air flow on special compost (substrate). Mushrooms require significantly less land, water and energy to grow compared to meat and the associated greenhouse gas emissions are much lower. The carbon footprint of commercially grown fresh mushrooms is 0.17 kg CO_2e per kg compared to 1.74 kg CO_2e per kg of farmed seaweed and a whooping 31.1 kg CO_2e per kg for beef (CarbonCloud, 2025). Replacing any meat with mushrooms is beneficial from a planetary health perspective and this is why fungi are also a desired option when looking for new alternative proteins.

The most popular meat replacement option based on fungi is Quorn developed in the 1960s based on the naturally occurring *Fusarium venenatum* which converts carbohydrate into protein-rich mycoprotein through a fermentation process (Finnigan *et al.*, 2019). With Quorn products being sold in supermarkets across 16 countries, they are a widely available option for consumers across the globe, including in Australia, UK and USA. These products compare favourably from an environmental point of view not only with beef, pork and chicken, but also with plant-sources options based on soybeans and tofu as well as with other alternative proteins, such as soy protein isolate (SPI) and textured vegetable protein (TVP). Table 7.1 provides some comparisons based on the ratio (as a percentage) between the footprint of a Quorn product and other dietary options. The figures for Quorn mince — the product that consumers can use directly, are very similar to Quorn mycoprotein. In all cases Quorn has only a fraction of the environmental footprint of the other food options, as low as 4–8% in the case of beef.

In addition to the favourable nutritional profile on their own, fungi are also used to improve the quality of meat products because of their antimicrobial and antioxidant properties (Pérez-Montes *et al.*, 2021). This is part of the trend of replacing some of the content of traditional meat products with different plants, cereals, pulses, seeds, algae and fungi with the benefit of improved taste, as well as inclusion of dietary fibre, missing in animal-based products (Das *et al.*, 2020). The ecological benefits from such a replacement can be significant as consumers are not challenged to embrace completely new products, but can continue to enjoy familiar, and even improved,

Table 7.1. Ratio between the average footprint of UK Quorn and other dietary options.

	Carbon equivalent (%)	Land use (%)	Total water use (%)
Ratio of Quorn mycoprotein vs soybean	68	12	31
Ratio of Quorn mycoprotein vs tofu	94	92	29
Ratio of Quorn mycoprotein vs soy protein isolate (SPI)	3	4	2
Ratio of Quorn mycoprotein vs textured vegetable protein (TVP)	3	4	2
Ratio of Quorn mince vs average beef mince	4	6	8
Ratio of Quorn pieces vs average chicken breasts	23	24	37
Ratio of Quorn sausage vs average pork primal cuts	11	21	39
Ratio of Quorn vegan ham vs pork ham	63	26	24

Source: Carbon Trust (2023).

taste of foodstuff whose environmental footprint is much lower. The example with Quorn (see Table 7.1) shows the significant difference between fungi and animal-based products but also soybean (and tofu) which is the best environmentally performing plant.

Human Health

Mushrooms have appealing taste, flavour and texture which combine with nutritional and medicinal value. They contain various essential amino acids that represent building blocks of vital proteins for humans, and are a rich source of phytonutrients, such as polysaccharides, dietary fibres and other micronutrients, including copper, iodine, iron, potassium, selenium, zinc and vitamins B, C and D while being low on sodium (Anusiya *et al.*, 2021). In addition to nutritional contribution, the therapeutic values of mushrooms are significant as they "offer a wide range of health benefits with a large spectrum of pharmacological properties, including antidiabetic, antioxidative, antiviral, antibacterial, osteoprotective, nephroprotective, hepatoprotective etc." (Anusiya *et al.*, 2021,

p. 11239). They also have cardiovascular, antitumour, anticancer, immunomodulatory, neuroprotective, anti-inflammatory and anti-ageing functions (Anusiya *et al.*, 2021; Kumar *et al.*, 2021). The polysaccharides in mushrooms act as prebiotics on the gut microbiota in the digestive systems.

The health and nutritional value of mushrooms is beyond doubt. Their umami flavour due to the high content of glutamic acid, sends to the brain the same savory taste and signals of satisfaction as when humans consume meat (Diepeveen *et al.*, 2022), without its negative consequences. Adding mushroom powder, for example, increases consumer acceptance of food products by improving mouth feel and juiciness (Hamza *et al.*, 2023; Kour *et al.*, 2022). Mushrooms and other edible fungi are important low-calorie, low-fat protein-rich dietary sources. They are considered superfoods and their global consumption is expected to rise at a compound annual growth rate (CAGR) of 9.7% until 2030 due to more people looking for better dietary options away from diets rich on animal-sourced foods (Grand View Research, 2021). Many of the new alternative proteins incorporate mushrooms as a main ingredient contributing to more sustainable diets with reduced environmental impacts and improved health benefits.

Despite the numerous health and planetary benefits of consuming mushrooms and recent increase in commercial production, they remain an underutilised dietary source with a wide nutritional and bioactive potential (Kumar *et al.*, 2021). The interest and attitude of Gen Z towards mushrooms and truffles as food items on their own or ingredients to other foodstuff are shaping the future of this superfood. This emerging appreciation is echoed in the way Gen Z talks about food and health with a distinctive mix of curiosity, functionality and flavour consciousness:

> *"Mushrooms are just the best, hey. They're tasty, good for your gut, packed with protein, and they don't mess up the planet while they grow. Low-key, they're like nature's shortcut to doing things right, heaps sustainable, heaps versatile. I chuck 'em in stir-fries, wraps, pasta, you name it. Plus, they're way better for the environment than half the stuff in the fridge. Honestly, they just make sense.*

Tasty, healthy, climate-friendly, what's not to love?" (Female, born 1998)

This candid perspective captures how Gen Z connects multiple values such as taste, health and sustainability into their everyday food choices. Their enthusiasm is not just about chasing food trends, they are genuinely excited by ingredients that are versatile, packed with nutrients and fit their lifestyle values. Mushrooms, with their low environmental impact and gut-friendly benefits, are fast becoming a favourite. Given the rising demand for mushrooms and creative use across Gen Z kitchens, they might just be the unofficial symbol of a generation choosing to eat well and live with a purpose. There is also the opportunity to venture into further parts of the fungi kingdom, including the thousands of other edible species that are currently underexplored.

Algae and Health

Despite algae, especially seaweed being traditionally consumed for millennia, particularly in East Asia, it is not included in the data and food balanced sheets of the Food and Agriculture Organisation of the United Nations. This is indicative of how underutilised these species are as food sources as well as of the opportunities that exist. Seaweed in particular grows in saline waters and does not compete for freshwater with other agricultural crops (FAO, 2024). It does not require fertilisers or pesticides to grow either and reduces the pressure on land. The numerous species of seaweed can be categorised as (LeRoy, 2022): green or chlorophytes (e.g. sea lettuce, umibudo and spirulina) estimated at 7,349 species in the Algaebase of the International Union for Conservation of Nature (IUCN), brown or phaeophytes (e.g. kelp, kombu, wakame, arame and hijiki) assessed at 2,065 species, and red or rhodophytes (e.g. nori, dulse and Irish moss) assessed at 6,734 species (Brakel *et al.*, 2021). They all have umami flavour and are gradually penetrating western menus. This subtle but persistent rise in popularity has not gone unnoticed by Gen Z, who are especially attuned to food choices that intersect with sustainability and personal wellness:

> *"Mate, seaweed's got that sci-fi vibe, but it's as ancient as the ocean itself. It's kind of wild when you think about it... Can you imagine that this humble sea veggie has fed coastal cultures for centuries?... Cool... and now it's rocking the spotlight in modern diets. It's packed with iron, iodine, vitamins and that umami hit we all chase, plus it grows fast without chewing up land or fresh water. Super cool!.... What I love most is how it ticks all the boxes I have. It's good for the planet,... good for your body,... and just bloody cool to cook with. My girlfriend chops it into stir-fries, blends it into smoothies, even sprinkle it on avo [avocado] toast. Eating seaweed makes me feel like I'm doing right by my health and the environment, without going full wellness influencer. It's ancient wisdom with a modern twist and it's just getting started."* (Male, born 1999)

This statement reflects how Gen Z often blends nostalgia and futurism in their food narratives, seeking options that honour tradition while aligning with environmental urgency and health priorities. The rising global seaweed industry intersects neatly with these values, offering low-impact, nutrient-dense foods without the baggage of industrial agriculture. As seaweed shifts from novelty to mainstream, Gen Z's openness and advocacy could be the cultural catalyst needed to integrate algae more deeply particularly into Western diets.

Many varieties of seaweed are edible, but the popular species consumed across the world are: nori — used in sushi, wakame — used in miso soup and seaweed salad, konbu (or kombu) — used to make favourable broths and tea, dulse — used in salads and sandwiches or as a snack, arame — for stir-fried, noodles and salads, and umibudo (or seaweed grapes) — on their own or as a garnish to other dishes (LeRoy, 2022).

Although coastal communities around the world continue to collect seaweeds from oceanic and marine environments, the vast majority of the supply, namely 97%, now comes from aquaculture — controlled cultivation in aquafarms (FAO, 2024). Seaweed cultivation is practised in 54 countries around the world with the industry

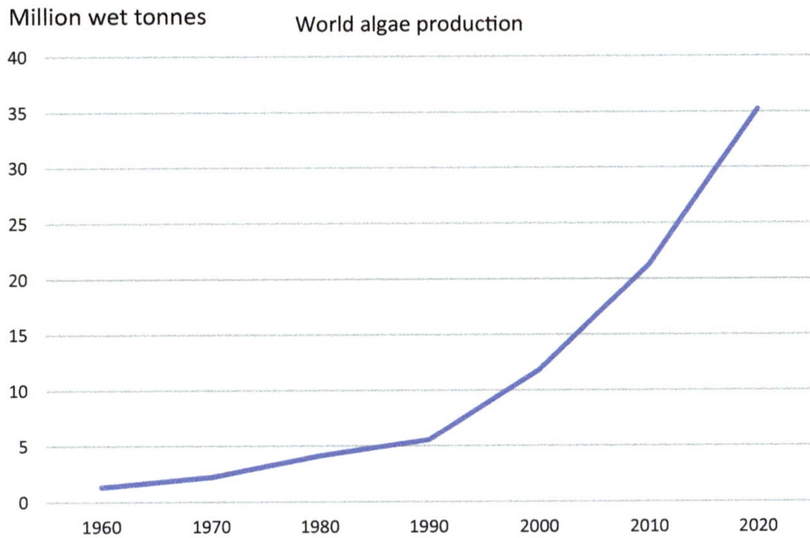

Figure 7.2. Global production of cultivated and wild algae (in million wet tonnes).

Source: FAO (2021).

experiencing exponential growth in the last 50 years (Cottier-Cook *et al.*, 2022), reaching 33 million tonnes wet weight in 2022 (Seaweed Insights, 2022), a value of US$ 17.14 billion in 2023 and projected to grow at a CAGR of 8.20% until 2032 (Fortune Business Insights, 2025). While the wild collection of seaweed remains unchanged since the 1960s at around 1.1 million tonnes per year, cultivation has drastically increased (FAO, 2021). The fast growth of algae production (see Figure 7.2) is driven by a range of factors with the food industry being one of them together with demand for animal feed, pharmaceuticals, fertilisers, fibre, biofuels, biostimulants for the soil and bioplastics. Traditionally practised in South-East Asia in countries, such as the Philippines, China, Indonesia, South Korea, Japan and Malaysia, seaweed cultivation is becoming a fast-growing aquaculture sector across the globe, including in USA,

Norway, Tanzania and Chile. Out of the thousands of species of seaweed, five varieties account for 95% of global cultivation: undaria — a genus of kelp or wakame, pyropia or nori, gracilaria or Irish moss, eucheumatoids farmed for carrageenan used in the food industry for gelling, thickening and stabilising mainly of dairy and meat products, and saccharina or kombu (Seaweed Insights, 2022).

Planetary Health

As a by-product of photosynthesis, algae produce oxygen and are of great ecological importance for the planet. It is estimated that 30 to 50% of the net global oxygen available to humans and animals come from algae (Britannica, 2025a). Within oceanic and sea environments, algae are essential for the existence of marine life; they are the equivalent to plants for land animals. Harvesting of wild seaweeds, and kelp in particular, needs to be carried out in a sustainable balanced way that allows them to regrow without affecting the ecology for the many species who depend on the sea forests for food, shelter and as breeding grounds, such as fish (e.g. herring), sea otters, octopuses, crabs, sea urchins, sunflower stars and other invertebrate (WWF, 2023; Bricker, 2024). Seaweed farming is becoming a widespread way to reduce the pressure on wild habitats, particularly as it is easy to do and species like kelp grow very fast (harvesting is possible within six weeks). Larger areas of seaweed also contribute to sequestering carbon dioxide and producing oxygen, making this an overall planet-positive food source with promising prospects for the future. By absorbing nitrogen and phosphorous, seaweed is also helping combat oxygen-deprived dead zones in marine environments and negate the effects of coastal eutrophication triggered by excess runoffs of fertilisers (Office of Communications, 2025). The carbon footprint of farmed seaweed is one of the lowest among dietary options and its eutrophication impact is negative, in other words, seaweed helps alleviate eutrophication (see Table 7.2). Depending on the methods of cultivation, seaweed is farmed predominantly in marine environments (mariculture) and if practiced on land, it uses degraded areas, including desert (Zhang *et al.*, 2022).

Table 7.2. Environmental indicators for farmed seaweed and other food products.

	Greenhouse gas emissions (in kilograms per kilogram of food product)	Eutrophying emissions (in grams of phosphate equivalents — PO_4 eq per kilogram of food product)	Land use (in square metres per 1000 kilocalories)
Apples	0.4 kg	1.45 g	1.3 m^2
Nuts	0.4 kg	19.1 g	2.1 m^2
Potatoes	0.5 kg	3.5 g	1.2 m^2
Bananas	0.9 kg	3.3 g	3.2 m^2
Seaweed farmed	1.1 kg	−11.0 g	∼0 m^2
Bivalves farmed	1.4 kg	−32.4 g	∼0 m^2
Wheat and rye	1.6 kg	7.2 g	1.4 m^2
Maize	1.7 kg	4.0 g	0.6 m^2
Tomatoes	2.1 kg	7.5 g	4.2 m^2
Tofu	3.2 kg	6.2 g	1.3 m^2
Silver bighead carp farmed	3.9 kg	9.3 g	4.7 m^2
Rice	4.4 kg	35.1 g	0.8 m^2
Eggs	4.7 kg	21.8 g	4.4 m^2
Salmon farmed	5.1 kg	33.5 g	4.7 m^2
Chicken	8.3 kg	48.7 g	6.6 m^2
Pork	12.3 kg	76.4 g	7.3 m^2
Cheese	23.9 kg	98.4 g	22.7 m^2
Beef (dairy herd)	33.3 kg	301.4 g	15.8 m^2
Lamb and mutton	39.7 kg	97.1 g	116.7 m^2
Beef (herd)	99.5 kg	365.3 g	119.5 m^2

Note: Eutrophying emissions calculated with Redfield's 16:1 (nitrogen to phosphorous) conversion ratio from https://ourworldindata.org/grapher/nitrogen-emissions-seafood and https://ourworldindata.org/grapher/phosphorous-emissions-seafood).
Source: https://ourworldindata.org/environmental-impacts-of-food; https://ourworldindata.org/grapher/eutrophying-emissions-per-kg-poore; https://ourworldindata.org/land-use-diets.

Another very sustainable food option is farmed aquatic molluscs (see Table 7.2) — edible bivalves or clams (mussels, oysters and scallops) as they sequester carbon, nitrogen and phosphorous in their shells and bodies.

However, the rapid expansion and globalisation of demand for seaweeds combined with the effects of climate change on temperature increase, rising seawaters and ocean acidification, are having negative impacts on seaweeds which are typically grown in shallow coastal waters. This has led to increased outbreaks of diseases, such as ice-ice disease and epiphytic filamentous algae, and pest infections, including viruses, bacteria, protists and eukaryotic endophytic algae (Cottier-Cook *et al.*, 2022). Compared to land-based crops, the genetic resources, breeding concepts and resilience of seaweeds against such stressors are underexplored (Brakel *et al.*, 2021). Calls are also being made to provide international mechanisms to safeguard aquatic environments and aquaculture from invasive species, pathogens, pests, diseases and wider environmental hazards, including the need for monitoring and to develop the ability for rapid detection and diagnosis in order to protect this valuable food source (Cottier-Cook *et al.*, 2022). Understanding the complexity of the various seaweeds, their lifecycles and the role they play within the marine ecology can help maintain diversity and avoid overexploitation for farming purposes (FAO, 2024). The importance of traditional knowledge in navigating the challenges of crop selection and breeding should be acknowledged when searching for sustainable solutions (Dumilag *et al.*, 2022).

Cultivating seaweed does not require the use of any fertilisers or pesticides and creates new habitats for marine wildlife that support the biodiversity of the coastal ecosystems (Vernik, 2024a). Seaweed acts as a water filter absorbing nitrogen and phosphorus from excess nutrient runoffs as well as serves as a buffer zone against storms protecting from beach erosion.

Seaweed is described as holding numerous superpowers (Vernik, 2024b). This multidimensional value of seaweed resonates strongly with Gen Z, particularly for those who often seek food options that are regenerative, not just sustainable:

> *"Eating seaweed feels like doing something good for the planet and my body at the same time. It's low-impact, full of nutrients and just makes sense. If it's healing the Earth while nourishing me — why wouldn't I be all over that?"* (Female, born 2000)

This sentiment captures Gen Z's ecological ethos, a drive to consume with a purpose. For many in this generation, seaweed offers more than a novel flavour or nutritional boost. It represents a tangible action towards reversing ecological damage. The appeal lies in its ecosystem services, its low-input cultivation and its cultural renewal ability. It has the potential to be a healthy and tasty nutritious food choice and at the same time contribute towards restoring aquatic environments, reducing carbon dioxide in the atmosphere, adding to oxygen generation and protecting coastal communities. Coastal populations, including deprived communities and particularly women, can benefit from seaweed farming economically and socially adding another sustainability dimension of these activities.

Human Health

Seaweed is a nutritious, low-fat, high-fibre, low-calorie, rich of micronutrients healthy dietary option that is increasingly becoming popular in western diets in the form of sushi, biscuits and as additive in another foodstuff. Seaweed is packed with minerals (calcium, iron, iodine, magnesium, phosphorus, potassium, zinc, copper, manganese, selenium and fluoride), amino acids, vitamins (A, B_1, B_2, B_9, B_{12}, C, D, E and K), dietary fibres, antioxidants and polyphenols beneficial for humans as well as other animals (Lomartire *et al.*, 2021). Nutritionists are arguing that seaweed may support thyroid, heart and gut health as well as immune function and balance blood sugar levels protecting against Diabetes Type 2 (Davis, 2024; Williams, 2024).

The benefits of seaweed have led to an abundance of new products based on whole seaweeds, extracts and supplements. Seaweeds have found their way in many nutraceutical, biomedical and pharmaceutical applications. In agriculture, seaweeds are used to improve soil fertility and as organic fertilisers; as animal feed, they are used for livestock such as cattle, lambs and pigs as well as fish (Lomartire *et al.*, 2021).

However, some seaweeds may have high levels of iodine and potassium, and hijiki is known to accumulate potentially harmful inorganic arsenic (LeRoy, 2022). The presence of heavy metals can also be problematic and the labelling of food products should provide needed information for the consumers (Cherry *et al.*, 2019). Seasonal variations can also impact the quality of seaweed (Cotas *et al.*, 2021) as well as the geographical location where it is being farmed. Overall, seaweed is grown in environments which are less controlled and less regulated. Harmonising food safety standards across the globe and more research are needed to prevent some negative effects of unchecked seaweed consumption, particularly when humans eat animals that have already been fed seaweeds.

Despite recent interest in algae and its versatile applications, including as a source of food, some health concerns may overshadow the numerous environmental and nutritional benefits seaweeds have. Their superpower will remain underutilised if Gen Z are not prepared to make seaweeds a mainstream component in their diets. Currently Asia accounts for 97% of total (cultivated and wild) production (FAO, 2021) with seaweeds continuing to hold a lot of potential for the other parts of the world and its young people.

Gen Z and Fungi

The bedroom of a friend's Gen Z daughter has numerous pictures and figurines of mushrooms. She explained to us that they are just very appealing and as it turns out, mushrooms fit within the aesthetics of most Gen Zers (see Figure 7.3). They "can be whimsical or weird, cute or scary, benign or deadly" and their duplicity has been part of human fascination for centuries (Worrel, 2025, para. 6). Gen Z are decorating their first homes with images of mushrooms (Demopoulos, 2023) which appears an innocent trend until microdosing on hallucinogens, such as magic mushrooms (Glover, 2024), starts to raise concerns about youth's recreational habits. The jury is still out as to whether hallucinogens are having a positive effect on Gen Z, who is also described as "generation stay at home" (Hinsliff, 2024) or "the anxious generation" (Haidt, 2024), but these young people started to use them during the COVID-19 pandemic (Daly, 2021) with no signs of slowing down (Glover, 2024). As one of them explains: "I think Gen Z has a very different understanding of not only mental

Figure 7.3. Room decoration made by an Australian Gen Z girl.

health, which has been tragically stigmatised for so long, but also of self-awareness, introspection, emotional intelligence, and spiritual health . . . I am however concerned that many young people have an overly casual attitude about these powerful tools. As psychedelics keep going mainstream, I hope to see education leveraged more than fear as a harm reduction tactic" (Holyanova, 2023, para. 7).

Notwithstanding what the long-term acceptance and impact of magic mushrooms are going to be, edible fungi are a completely different agenda, and we want to understand how these young people relate to them. Edible mushrooms, in contrast to psychedelics, are emerging as a symbol of curiosity, health consciousness and environmental engagement for Gen Z:

> *"Mushrooms? Absolute legends. They're protein-packed, planet-friendly and just a bit quirky — like nature's overachievers. They pop up in everything from burgers to brekkie [breakfast] wraps, and somehow nail flavour without messing with the Earth. I love*

> *that they grow fast, don't need heaps of resources and still manage to feel kind of mystical. They're not trying to be trendy, they're just naturally cool. Honestly, they're the kind of weird that makes sense."* (Male, born 2004)

For this generation, mushrooms represent more than just a meat alternative, they embody a playful and serious interest in sustainable living and mindful nutrition. Studies are yet to be conducted about young people's consumption of fungi and all the evidence we have is based on shared personal opinions. Nevertheless, such insights are quite important as they describe Gen Z's general attitudes to this superfood. Below are some examples of young people's fascination with mushrooms not only as a food but also as a window to a different world of opportunities.

Gen Z entrepreneurs in USA are leading a youth movement by using mushrooms and other fungi for improved health, education and environmentalism (Johnson, 2024). This movement is as much about identity and empowerment as it is about fungi:

> *"Fungi taught me to slow down, to observe. There's something revolutionary about growing your own food and medicine from something people used to ignore. I received a mushroom growing kit from my grandparents and I was fascinated by what number of mushrooms I can grow. They are like a miracle food."* (Female, born 2001)

The humble mushroom, once relegated to niche cuisines or forest floors, is now a medium through which Gen Z expresses autonomy, resilience and ecological awareness. Being digital natives, they are using Google Scholar, youtube and other social media to teach themselves how to cultivate mushrooms and establish communities of followers, organising mycology clubs and creating a mycological movement. One of these young people explains: "... this was the one thing I could dive into. No mentors, no guides, just me learning and growing. And the more I learned, the more I realised: mushrooms are like nature's superpowers. Each one has something unique to offer, from boosting the immune system to improving brain function. It's more than just growing — it's about connecting with nature in a whole new way" (mycosymbiote, 2025).

Foraging for wild mushrooms is an activity that has started to attract young people who are rediscovering contacts with nature as an adventure added by technology. Young Chinese Gen Z see it as an immersive tourist experience organised through social media and guided by experience professionals that allows them to find rare edible species (Li and Fan, 2024). One of them explains that despite dense foliage and difficult paths: "When I picked one exotic mushroom after another, I couldn't stop. I became more enthusiastic and increasingly fond of it" (Li and Fan, 2024, para. 7). Swedish youth prefer to use smartphone applications to identify wild mushrooms rather than rely on advice by more experienced people: "I love to try out new types of fungi. We make cuts in them and identify them through books and smartphone applications. I never ask people on Facebook for advice" (Svanberg and Lindh, 2019, p. 17). This resurgent interest in wild fungi can have a very positive influence on increasing the consumption of mushrooms, including cultivated species.

The more young people learn about mushrooms, the more likely they will be to endorse the environmental and health benefits as a constant component of daily menus. Mushrooms are like a thread that ties people through time and cultures "a bridge from our ancestors to us and to our descendants into the future" (Stamets, 2020, p. 12).

Gen Z and Algae

A 2022 study of Australian Gen Z shows that two-thirds of these young people are not aware of the link between climate change and people's food choices (Bogueva and Marinova, 2022). It is not surprising that they also vastly reject the new alternative proteins, such as cultured meat as well as those based on algae, insects and plant-sourced. Young Australians however are increasingly attracted to seaweed, particularly when used in products such as snacks, crackers and sushi (Young *et al.*, 2022; UniSC News, 2023). They are aware of the nutrient and health benefits of seaweed, particularly its fibre content as many Australians fail to meet the recommended dietary intake from fruit and vegetables. Gen Z's interest in seaweed is shaped by its nutritional profile and the way it fits with their fast-paced, snack-oriented lifestyle.

> *"I started eating seaweed chips because they're crunchy and tasty — but then I found out they're good for your gut and packed with fibre. Win-win."* (Female, born 2002)

This convergence of convenience, health and sustainability makes algae-based snacks especially appealing to young consumers seeking quick yet meaningful food choices. Despite Australia cultivating brown seaweed — 1,923 tonnes in 2019 (FAO, 2021), most of the seaweeds consumed are imported (Deepika *et al.*, 2022). Consequently, the price of foods with algae can be prohibitive for young people who also prefer the packaging used to be more sustainable (Young *et al.*, 2022).

The Australian Government is actively encouraging domestic production, including new product development, with ambitious plans to establish a 1.5-billion-dollar industry by 2040 (Maksan *et al.*, 2025). Alongside environmental performance and economic opportunities, many health benefits, including Australian brown seaweed being a source of essential dietary minerals and fibre, contribute towards these commitments (Skrzypczyk *et al.*, 2024). It is also working towards establishing food safety programs and management systems specifically for seaweed for human consumption to better regulate this booming industry (Winkel, 2023).

Although not aimed at young people specifically (their share was around 20% of the participants), a comparison of seaweed consumption between Australia, UK and Croatia distinguishes between food enthusiasts keen to try new products and food conservatists who are reluctant to do so (Maksan *et al.*, 2025). The share of Australian food enthusiasts was much higher at 60% compared to 41% in Croatia and 22% in UK. Respectively, the share of food conservatists was the highest in UK — 78%, compared to 59% in Croatia and 40% in Australia. Food conservatists were also older and more inclined to eat red meat, pork or chicken rather than fish or seafood. Also, Australia had the highest share of people who have consumed seaweed. Factors, such as food innovativeness, food neophobia, food involvement, food responsibility, health consciousness and symbolic value, influenced people's attitudes towards seaweed consumption. These findings show that age, culture and exposure to alternative new products influence the acceptance of edible algae, and this is where young people have an advantage. Indeed, many Gen Zers view seaweed not only as a novelty, but as a symbol of the kind of food system they want to see flourish:

> *"Seaweed just makes sense, hey. It grows quick, doesn't hog land or water, and even helps clean up the ocean — pretty amazing when you think about it. Compared to something like beef, it's way easier on the planet. It's stacked with nutrients, that salty goodness, and you can chuck it in anything. But I'm not saying it's the magic fix for everything. Like any trend, it's gotta be grown and sourced responsibly, no overharvesting or dodgy farming. It's one of those foods that could help us do better, if we treat it with care. More of that, yeah, but with a bit of common sense too."* (Male, born 1999)

This type of grounded enthusiasm, tempered with caution, illustrates how Gen Z navigates new food trends, not with blind acceptance, but with a mindset that blends curiosity, conscience and a call for accountability.

Consumer Acceptance of Fungi and Algae

Despite their numerous health benefits, both fungi and algae are rarely mentioned in dietary guidelines around the world with the exception of Canada, Brazil and the EAT-Lancet Report (Willett *et al.*, 2019). They are not part of the current Australian Dietary Guidelines and assumed to fall within the category of vegetables and legumes/beans. Increasingly calls are being made for this to change with fungi being recognised as a special category during the regular updates of the national dietary advice (e.g. Beckett *et al.*, 2021; Australian Mushroom Growers' Association, 2023). For example, edible mushrooms are seen as a sustainable food-based source suitable for tackling vitamin D deficiency and they should be recognised and promoted in dietary guidelines (Starck *et al.*, 2024). Algal protein meets or exceeds FAO's protein quality requirements for human nutrition and resembles sources, such as soybeans and eggs (Wu *et al.*, 2023), but seaweeds are similarly not included in dietary guidelines. Dietary guidelines are used by health and medical professionals and if they do not emphasise the benefits of fungi and algae, it is not surprising then that consumers are not including regularly these options as whole foods in their daily diets.

It appears that consumers are less cognisant about the environmental credentials of fungi and algae but are more aware of the health benefits and are keen to explore them as nutritional supplements and functional foods. This is driving the demand for mushrooms (Ess Team, 2025) as well as cultivated seaweed (Khan *et al.*, 2024). The versatility of fungi and algae and their potential to replace animal-based foods and food ingredients offer a new area of research and development in addition to direct consumption. From an environmental point of view, any reduction in raising animals for food is good as it will reduce the pressure on the land and decrease pollution. Eating fungi and algae instead is a very positive step and young people should be at the forefront of such changes. In addition to many valuable micronutrients and vitamins, proteins derived from algae and fungi are of similar high-quality as animal sources but more sustainable. They are metabolically potent and can contribute towards reducing oxidative stress, fight obesity and diabetes, prevent cancer, cardiovascular disease (especially hypertension) and neurodegenerative diseases, while having negligible adverse effects (Yimam *et al.*, 2024).

Are consumers then likely to accept and switch to these foods? Climate change is a top environmental concern for university Gen Z in Australia (Bardales Salguero *et al.*, 2024) and around the world (Bogueva *et al.*, 2024). The environmental profile of foods based on fungi and algae is much better, particularly compared to animal-based options. Food however is a very complex issue with a suite of sensory, health, psychological, marketing and environmental factors influencing individual choices.

Although we are yet to fully understand which are the main triggers influencing young people's choices in the case of fungi and algae, the following investigation sheds some light. A study in

Italy tested consumer preferences for truffles using the theory of planned behaviour that states that human behaviour is determined by individual intentions which are in turn influenced by attitude, subjective norms and perceived behaviour control (Laganà *et al.*, 2024). The researchers added consumers' food-specific curiosity to the above three constructs. Only 20% of the survey sample were Gen Z participants but they contributed to the overall understanding of the factors that influence consumer intention to use truffles. Personal experiences, organoleptic characteristics, such as aroma and taste, health and food safety, trust in the product provider, traceability information, the aspect of green products and, more generally, the protection of the environment and forests came as influencing factors in consumer interest and acceptance of truffles.

Although truffles are a very specific niche product, the findings that consumers are interested in green and sustainable options with smaller environmental impact in order to protect the natural environment are quite significant. So are also the expectation for the food products to be healthy, safe with reliable information about their nutritional values and provenance. These are all important requirements to satisfy for fungi and algae to have a future as a mainstream and regular dietary option preferred by young people.

Conclusion

Edible fungi and edible algae are special categories of organisms on this planet that are outside the plant–animal classifications. They offer immense opportunities but remain underexplored and underused dietary treasure trove. Whether Gen Z will be willing to consume them directly as whole foods or have them as ingredients replacing animal-sourced components, it is important that consumers acknowledge the numerous co-benefits for human health and planetary well-being of these superfoods holding superpowers to change the current dietary patterns. Supported by dietary and environmental guidance, increased consumption would also offer more sustainable food production opportunities which do not rely on the exploitation of a limited small number of species. Demand

and commercial cultivation of fungi and algae are on the rise driven by increasing understanding of the numerous benefits they offer. Armed with digital technologies, Gen Z is embracing these opportunities igniting a food movement for a more gentle and sustainable world.

References

Anusiya, G., Gowthama Prabu, U., Yamini, N. V., Sivarajasekar, N., Rambabu, K., Bharath, G., & Banat, F. (2021). A review of the therapeutic and biological effects of edible and wild mushrooms. *Bioengineered*, 12(2), 11239–11268. https://doi.org10.1080/21655979.2021.2001183

Australian Government. (2025). Edible fungi. Australian Pesticides and Veterinary Medicine Authority. https://www.apvma.gov.au/crop-groups/edible-fungi

Australian Mushroom Growers' Association. (2023). Scientific basis for a mushroom food group in the Australian Dietary Guidelines (MU22001). https://www.horticulture.com.au/growers/help-your-business-grow/research-reports-publications-fact-sheets-and-more/mu22001/

Averill, C., Dietze, M. C., & Bhatnagar, J. M. (2018). Continental-scale nitrogen pollution is shifting forest mycorrhizal associations and soil carbon stocks. *Global Change Biology*, 10, 4544–4553. https://doi.org/10.1111/gcb.14368

Bardales Salguero, R., Bogueva, D., & Marinova, D. (2024). Australia's university Generation Z and its concerns about climate change. *Sustainable Earth Reviews*, 7, 8, https://doi.org/10.1186/s42055-024-00075-w

Beckett, E., Cardwell, G., Fayet-Moore, F., Fuller, J., Kalotas, A., & Tapsell, L. (2021). Fungi Foods: Where do they fit in a plant-based diet? https://australianmushroomgrowers.com.au/wp-content/uploads/2021/10/NRAUS-AUSTRALIAN-MUSHROOMS_2021_Expert_Roundtable_Whitepaper-1.pdf

Bogueva, D. & Marinova, D. (2022). Australian Generation Z and the nexus between climate change and alternative proteins. *Animals*, 12, 2512. https://doi.org/10.3390/ani12192512

Bogueva, D., Marinova, D., Wäechter, N., & Tekiner, I. H. (2024). Societal concerns. In Seemiller, C. & Grace, M. (eds) *Gen Z around the World: Understanding the Global Cohort Culture of Generation Z*, pp. 119–127. Emerald Publishing, Leeds, UK.

Brakel, J., Sibonga, R. C., Dumilag, R. V., Montalescot, V., Campbell, I., Cottier-Cook, E. J., Ward, G., . . . Gachon, C. M. M. (2021). Exploring, harnessing and conserving marine genetic resources towards a sustainable seaweed aquaculture. *Plants People Planet*, 3, 337–349. https://doi.org/10.1002/ppp3.10190

Bricker, K. (2024). Six Ocean Animals That Rely on Kelp for Survival. Ocean Wise. https://ocean.org/blog/six-ocean-animals-that-rely-on-kelp-for-survival/

Britannica. (2025a). Algae. https://www.britannica.com/science/algae

Britannica. (2025b). Eukaryote. https://www.britannica.com/science/eukaryote

CarbonCloud (2025). Explore emission factors for 57688 food products. https://apps.carboncloud.com/climatehub/

Carbon Trust (2023). Quorn Footprint Comparison Report. https://www.quorn.co.uk/assets/files/content/Carbon-Trust-Comparison-Report-2023.pdf

Cherry, P., O'Hara, C., Magee, P. J., McSorley, E. M., & Allsopp, P. J. (2019). Risks and benefits of consuming edible seaweeds. *Nutrition Reviews*, 77(5), 307–329. https://doi.org/10.1093/nutrit/nuy066

Cotas, J., Pacheco, D., Araujo, G. S., Valado, A., Critchley, A. T., & Pereira, L. (2021). On the health benefits vs. risks of seaweeds and their constituents: The curious case of the polymer paradigm. *Marine Drugs*, 19(3), 164. https://doi.org/10.3390/md19030164

Cottier-Cook, E. J., Cabarubias, J. P., Brakel, J., Brodie, J., Buschmann, A. H., Campbell, I., Critchley, A. T., . . . Bondad-Reantaso, M. G. (2022). A new Progressive Management Pathway for improving seaweed biosecurity. *Nature Communications*, 13, 7401. https://doi.org/10.1038/s41467-022-34783-8

Daly, M. (2021). Psychedelic Use on the Rise As Gen Z Seeks To 'Enhance Connectedness'. https://www.vice.com/en/article/psychedelic-use-on-the-rise-as-gen-z-tries-to-enhance-connectedness/

Das, A. K., Nanda, P. K., Madane, P., Biswas, S., Das, A., Zhang, W., & Lorenzo, J. M. (2020). A comprehensive review on antioxidant dietary fibre enriched meat-based functional foods. *Trends in Food Science & Technology*, 99, 323–336. https://doi.org/10.1016/j.tifs.2020.03.010

Davis, S. (2024). Health Benefits of Seaweed. https://www.webmd.com/diet/health-benefits-seaweed

Deepika, C., Wolf, J., Moheimani, N., Hankamer, B., von Herzen, B., & Rao, A.R. (2022). Utilisation of seaweeds in the australian market — commercialisation strategies: Current trends and future prospects. In Ranga Rao, A., Ravishankar, G.A. (eds) *Sustainable Global Resources of Seaweeds*, Volume 1. Springer, Cham. https://doi.org/10.1007/978-3-030-91955-9_15

Demopoulos, A. (2023). Squiggly mirrors and mushrooms: how are gen Z decorating their first homes? The Guardian. https://www.theguardian.com/lifeandstyle/2023/nov/07/gen-z-interior-tiktok-trends-squiggle-mirror-pastels

Denenberg, Z. & The Editors of Epicurious. (2023). 15 Types of Mushrooms and How to Cook with Them. https://www.epicurious.com/ingredients/types-of-mushrooms

Diepeveen, J., Moerdijk-Poortvliet, T. C. W., & van der Leij, F. R. (2022). Molecular insights into human taste perception and umami tastants: A review. *Journal of Food Science*, 87(4), 1449–1465. https://doi.org/10.1111/1750-3841.16101

Dumilag, R. V., Crisostomo, B. A., Aguinaldo, Z. Z. A., Hinaloc, L. A. R., Liao, L. M., Roa-Quiaoit, H. A., . . . Roleda, M. Y. (2022). The diversity of eucheumatoid seaweed cultivars in the Philippines. *Reviews in Fisheries Science & Aquaculture*, 31(1), 47–65. https://doi.org/10.1080/23308249.2022.2060038

Encyclopedia Britannica. (2006). Taxonomy: The five-kingdom system of classification. https://kids.britannica.com/students/assembly/view/195075

Ess Team. (2025). Mushroom Consumption Trends by Region: What's Driving Demand? https://essfeed.com/mushroom-consumption-trends-by-region-whats-driving-demand/

FAOSTAT. (2025). Crops and livestock products. Food and Agriculture Organisation of the United Nations. https://www.fao.org/faostat/en/#data/QCL

Finnigan, T., Wall, B. T., Wilde, P. J., Stephens, F. B., Taylor, S. L., & Freedman, M. R. (2019). Mycoprotein: The future of nutritious non-meat protein, a symposium review. *Current Developments in Nutrition*, 3(6), nzz021. https://doi.org/10.1093/cdn/nzz021

Food and Agriculture Organisation of the United Nations (FAO). (2021). Global seaweeds and microalgae production, 1950–2019. https://openknowledge.fao.org/server/api/core/bitstreams/97409d09-2f8e-4712-b11e-60105d89959b/content

Food and Agriculture Organisation of the United Nations (FAO). (2024). The State of World Fisheries and Aquaculture 2024: Blue Transformation in Action. Rome. https://doi.org/10.4060/cd0683en

Fortune Business Insights. (2025). Commercial Seaweed Market Size, Share & Industry Analysis, By Type (Red Seaweed, Brown Seaweed, and

Green Seaweed), Form (Flakes, Powder, and Liquid), End-uses (Food & Beverages, Agricultural Fertilizers, Animal Feed Additives, Pharmaceuticals, and Cosmetics & Personal Care), and Regional Forecast, 2024–2032. https://www.fortunebusinessinsights.com/industry-reports/commercial-seaweed-market-100077 https://www.fortunebusinessinsights.com/industry-reports/commercial-seaweed-market-100077

Frew, A., Aguilar-Trigueros, C. A., Powell, J., & Weinberger, N. (2023). Hidden carbon: Fungi and their 'necromass' absorb one-third of the carbon emitted by burning fossil fuels every year. The Conversation. https://theconversation.com/hidden-carbon-fungi-and-their-necromass-absorb-one-third-of-the-carbon-emitted-by-burning-fossil-fuels-every-year-206674

Glover, E. (2024). Why are so many young people taking hallucinogens? https://www.dazeddigital.com/life-culture/article/62279/1/why-are-so-many-young-people-taking-hallucinogens

Grand View Research. (2021). Mushroom Market Size, Share & Trends Analysis Report By Product (Button, Shiitake, Oyster), By Form, By Distribution Channel, By Application (Food, Pharmaceuticals, Cosmetics), By Region, And Segment Forecasts, 2022 — 2030. https://www.grandviewresearch.com/industry-analysis/mushroom-market

Greger, M. (2023). *How Not to Age: The Scientific Approach to Getting Healthier as You Get Older*. Flatiron Books, New York, NY.

Haidt, J. (2024). *The Anxious Generation: How the Great Rewiring of Childhood is Causing an Epidemic of Mental Illness.* Alen Lane, London.

Hamza, A., Ghanekar, S., & Kumar, D. S. (2023). Current trends in health-promoting potential and biomaterial applications of edible mushrooms for human wellness. *Food Bioscience*, 51, 102290. https://doi.org/10.1016/j.fbio.2022.102290

Hausman, T. (2025). Mode of Nutrition of Fungi: Overview & Examples. https://study.com/academy/lesson/fungi-mode-of-nutrition-overview-examples-mushrooms-food.html

Hinsliff, G. (2024). Meet generation stay-at-home: 'You don't need to pay to go clubbing: you can sit at home and watch it on your phone'. The Guardian. https://www.theguardian.com/lifeandstyle/2024/feb/10/meet-generation-stay-at-home-dont-pay-to-go-clubbing-watch-it-at-home-on-your-phone

Holyanova, M. (2023). Revolutionizing the Conversation: The Impact of Millennials and Gen Z-ers on Psychedelic Media and Culture. https://maps.org/news/bulletin/the-impact-of-millennials-and-genz-on-psychedelics/

Johnson, R. (2024). The youth movement leading the mushroom renaissance: Five self-educated eco-entrepreneurs under 30 are changing the fungi game. https://www.rollingstone.com/culture-council/articles/movement-leading-mushroom-renaissance-educated-entrepreneurs-changing-fungi-1235139097

Khan, N., Sudhakar, K., & Mamat, R. (2024). Macroalgae farming for sustainable future: Navigating opportunities and driving innovation. *Heliyon*, 10(7), e28208. https://doi.org/10.1016/j.heliyon.2024.e28208

Kour, H., Kour, D., Kour, S., Singh, S., Hashmi, S. A. J., Yadav, A. N., Kumar, K., Sharma, Y. P., & Ahluwalia, A. S. (2022). Bioactive compounds from mushrooms: Emerging bioresources of food and nutraceuticals. *Food Bioscience*, 50(Part B), 102124. https://doi.org/10.1016/j.fbio.2022.102124

Kumar, K., Mehra, R., Guiné, R. P. F., Lima, M. J., Kumar, N., Kaushik, R., Ahmed, N., Yadav, A. N., & Kumar, H. (2021). Edible mushrooms: A comprehensive review on bioactive compounds with health benefits and processing aspects. *Foods*, 10(12), 2996. https://doi.org/10.3390/foods10122996

Laganà, V. R., Lombardi, F., Di Gregorio, D., & Nicolosi, A. (2024). Consumption preferences for truffles and truffle-based products: An application of the PLS-SEM model. *Sustainability*, 16(12), 5002. https://doi.org/10.3390/su16125002

LeRoy, M. (2022). A guide to edible seaweed. https://www.webstaurantstore.com/blog/3940/types-of-edible-seaweed.html

Li, Y. & Fan, Y. (2024). Growing Wild: Why Young China Can't Get Enough of Mushroom Hunts. https://www.sixthtone.com/news/1015861

Li, H., Tian, Y., Menolli Jr., N., Ye, L., Karunarathna, S. C., Perez-Moreno, J., Rahman, M. M., ... Mortimer, P. E. (2021). Reviewing the world's edible mushroom species: A new evidence-based classification system. *Comprehensive Reviews in Food Science and Food Safety*, 20(2), 1982–2014. https://doi.org/10.1111/1541-4337.12708

LibreTexts. (n.d.). Kingdom Classification According to Whittaker. https://bio.libretexts.org/Courses/Prince_Georges_Community_College/PGCC_Microbiology/04%3A_Microscopy_Staining_and_Classification/4.03%3A_Classification_and_Identification/4.3.01%3A_Kingdom_Classification_According_to_Whittaker

Lomartire, S., Marques, J. C., & Gonçalves, A. M. M. (2021). An overview to the health benefits of seaweeds consumption. *Marine Drugs*, 19(6), 341. https://doi.org/10.3390/md19060341

Maksan, M. T., Matulić, D., Mesić, Ž., & Memery, J. (2025). Segmenting and profiling seaweed consumers: A cross-cultural comparison of Australia, the United Kingdom and Croatia. *Food Quality and Preference*, 122, 105305. https://doi.org/10.1016/j.foodqual.2024.105305

Mushroom. (n.d.). How many mushrooms are there? Mushroom: The Journal of Wild Mushrooming. https://www.mushroomthejournal.com/how-many-mushrooms/

mycosymbiote. (2025). TikTok videos. https://www.tiktok.com/@mycosymbiote/video/7421593000049921323?lang=en

Office of Communications. (2025). Seaweed Aquaculture. National Oceanic and Atmospheric Administration (NOAA) Fisheries. https://www.fisheries.noaa.gov/national/aquaculture/seaweed-aquaculture

Pérez-Montes, A., Rangel-Vargas, E., Lorenzo, J. M., Leticia Romero, L., & Santos, E. M. (2021). Edible mushrooms as a novel trend in the development of healthier meat products. *Current Opinion in Food Science*, 37, 118–124. https://doi.org/10.1016/j.cofs.2020.10.004

Seaweed Insights. (2022). Global Production Overview. https://seaweedinsights.com/global-production/

Sisson, L. & Vigus, P. (2023). *Fungi of Aotearoa: A Curious Forager's Field Guide.* Penguin Books, Auckland, New Zealand.

Skrzypczyk, V. M., Callahan, D. L., Francis, D. S., & Bellgrove, A. (2024). Australian brown seaweeds as a source of essential dietary minerals. *Journal of Applied Phycology*, 36, 797–809. https://doi.org/10.1007/s10811-023-03074-0

Stamets, P. (ed.) (2020). *Fantastic Fungi: How Mushrooms Can Heal, Shift Consciousness, and Save the Planet.* Simon and Schuster, New York, NY.

Starck, C., Cassettari, T., Wright, J., Petocz, P., Beckett, E., & Fayet-Moore, F. (2024). Mushrooms: A food-based solution to vitamin D deficiency to include in dietary guidelines. *Frontiers in Nutrition*, 11. https://doi.org/10.3389/fnut.2024.1384273

Svanberg, I. & Lindh, H. (2019). Mushroom hunting and consumption in twenty-first century post-industrial Sweden. *Journal of Ethnobiology and Ethnomedicine*, 15, 42. https://doi.org/10.1186/s13002-019-0318-z

UniSC News. (2023). Seaweed consumption among young Australians. University of the Sunshine Coast. https://www.usc.edu.au/about/unisc-news/news-archive/2023/january/seaweed-consumption-among-young-australians

Vernik, D. (2024a). Is seaweed nutritious? And 5 other facts about seaweed. WWF. https://www.worldwildlife.org/stories/is-seaweed-nutritious-and-5-other-facts-about-seaweed

Vernik, D. (2024b). Why seaweed is a jack-of-all-trades in the fight against the climate crisis. WWF. https://www.worldwildlife.org/stories/why-seaweed-is-a-jack-of-all-trades-in-the-fight-against-the-climate-crisis

Warnasuriya, S. D., Udayanga, D., Manamgoda, D. S., & Biles, C. (2023). Fungi as environmental bioindicators. *Science of The Total Environment*, 892, 164583. https://doi.org/10.1016/j.scitotenv.2023.164583

Wei, Y., Li, Liu, Y., Xiang, S., Zhang, H., Yi, L., Shang, Y., & Xu, W. (2022). Identification techniques and detection methods of edible fungi species. *Food Chemistry*, 374, 131803. https://doi.org/10.1016/j.foodchem.2021.131803

Whitmee, S., Haines, A., Beyrer, C., Boltz, F., Capon, A. G., de Souza Dias, B. F., Ezeh, A., . . . Yach, D. (2015). Safeguarding human health in the Anthropocene epoch: Report of The Rockefeller Foundation-Lancet Commission on planetary health. *Lancet*, 386, 1973–2028. https://doi.org/10.1016/S0140-6736(15)60901-1

Willett, W., Rockström, J., Loken, B., Springmann, M., Lang, T., Vermeulen, S., Garnett, T., . . . Murray, C. J. L. (2019). Food in the Anthropocene: The EAT–Lancet Commission on healthy diets from sustainable food systems. *The Lancet*, 393(10170), 447–492. https://doi.org/10.1016/ S0140-6736(18)31788-4

Williams, J. (2024). The health benefits of seaweed. GoodFood. https://www.bbcgoodfood.com/health/nutrition/health-benefits-seaweed

Winkel, C. (2023). Development of a seaweed food safety program. AgriFutures Australia. https://agrifutures.com.au/product/development-of-a-seaweed-food-safety-program/

Worldwide Fund for Nature (WWF). (2023). From Sea to Table: Wild Harvested Kelp Versus Farmed Kelp. https://www.worldwildlife.org/blogs/sustainability-works/posts/from-sea-to-table-wild-harvested-kelp-versus-farmed-kelp

Worrel, A. (2025). The Mushroom Home Decor Craze is Delightfully Terrifying. https://amityworrel.com/2024/05/23/the-mushroom-home-decor-trend-is-delightful-and-terrifying-why-we-love-the-mushroom-trend

Wu, J. Y., Tso, R., Teo, H. S., & Haldar, S. (2023). The utility of algae as sources of high value nutritional ingredients, particularly for alternative/complementary proteins to improve human health. *Frontiers in Nutrition*, 10, 1277343. https://doi.org/10.3389/fnut.2023.1277343

Yimam, M. A., Andreini, M., Carnevale, S., & Muscaritoli, M. (2024). The role of algae, fungi, and insect-derived proteins and bioactive peptides in preventive and clinical nutrition. *Frontiers in Nutrition*, 11. https://doi.org/10.3389/fnut.2024.1461621

Young, M., Paul, N., Birch, D., & Swanepoel, L. (2022). Factors influencing the consumption of seaweed amongst young adults. *Foods*, 11(19), 3052. https://doi.org/10.3390/foods11193052

Zhang, L., Liao, W., Huang, Y., Zhang, L., Liao, W., & Huang, Y. (2022). Global seaweed farming and processing in the past 20 years. *Food Production, Processing and Nutrition*, 4, 23. https://doi.org/10.1186/s43014-022-00103-2

Chapter 8

Beyond Dairy: Gen Z and Alternative Milks

A survey of Australian Generation Z young people we conducted in 2024 showed that 98% of them have tried plant-based milk — 173 out of 177 respondents (Bogueva and Marinova, 2025). By comparison, just above half (namely 55%) of the Australian consumers have done so. A more important difference about young people is also manifested in their intention to continue to consume plant-based milks on a regular basis or occasionally (74% compared to 43% for the Australian population). The exact numbers are: 44% of Gen Z are planning to use plant-based milks regularly (compared to 23% for Australians generally) and a further 30% for Gen Z (compared to 20% for Australians generally) are keen to do so occasionally (Bogueva and Marinova, 2025).

Gen Z are curious and trend-setting with their food choices and the acceptance of alternative milks is widespread. This is particularly interesting for a country like Australia which prides itself with clean and quality dairy products and where a litre of cow's milk in the supermarket is cheaper than a litre of water. Plant-based milks, on the other hand, are generally more expensive — up to 3 times more expensive (Dairy Australia, 2022). Australian Gen Z, however, is not an exception. Evidence from the USA also shows that Gen Z consumers are willing to pay higher prices for products they perceive as more sustainable (Ramsing *et al.*, 2023). Their willingness to pay more is often driven by a combination of health, ethics and climate concerns, and some even see it as a way to align daily habits with personal values. As one young Gen Z representative from Australia explained:

> *"I grew up with dairy like most Aussies, but these days I'm all about oat or almond — just feels better, you know? Yeah, they're a bit pricey, but I'd rather fork out more for something that's lighter on the planet and doesn't mess with my gut. Plus, half my mates are off dairy anyway, so it's kind of the new normal."* (Female, born 2005)

Such sentiments are echoed across coffee shops, university campuses and social media, where dairy-free alternatives are no longer a niche preference but a cultural marker of Gen Z identity. The availability of soy, almond, oat, rice, corn, quinoa, pea, lupin, hemp, cashew, hazelnut, coconut, pistachio and macadamia milks, to name some among others, is making such choices easier. Almost any coffee shop in the West offers plant-based milk choices for the trendy coffee, tea and chai drinkers (see Figure 8.1). What is also lurking in the laboratories around the world is synthetic milk and other lab-grown dairy products.

Despite the evidence that Gen Z is pressing ahead replacing livestock products with plant-based alternatives, there are many aspects in this transition that require unpacking. Is this just a trend? Does taste matter? What are the nutritional and health benefits? Do plant-based dairy alternatives have lighter environmental footprints? Is it more ethical to consume plant-based alternatives to cow's milk? With all additives and processing involved, are plant-based alternatives ultra-processed foods? Does it matter how the alternative milks are packaged and labelled? What is the role of social media in popularising or condemning these food choices? Are young people's preferences influenced by gender? Gen Z does not have the answers to these questions; in fact, there are still many unknowns as the technology develops and new options reach commercialisation. Amid this uncertainty, young people are relying on personal experience, peer influence and digital platforms to make sense of their evolving food values. As one Gen Zer explained:

> *"It's not just about swapping milk for the planet... there's a whole mix of reasons. For me, it's about gut health, sure, but also the way*

Figure 8.1. Plant-based milk options in a coffee shop in Fremantle, Western Australia.

> *animals are treated and what's actually in the stuff we're drinking. You see something on TikTok one day, then your mates are all trying oat milk the next. We're figuring it out as we go, yeah?"* (Male, born 2000)

This ongoing experimentation with food identity is especially visible in Western cultures where dairy has long been central to people's diet. Also, some of these questions are more relevant for the western consumer. People in Asian countries, such as China, Japan, Malaysia and Singapore, have been widely consuming soy milk for hundreds of years with the oldest evidence dating back to 25–220 AD (Chen, 2023). A main reason for using soy beans grinded in water to produce a nutritious beverage, is the lactose intolerance prevalent in people from East Asian decent (MedlinePlus, 2023). Two thirds of the Australian Asian population nowadays have problems digesting lactose (Dairy Australia, 2024) and dietitians estimate that one in four (namely 27%) of Australian households have members who are lactose intolerant (Dynan, 2020). The situation is similar in other countries with migrant populations from East Asia. For many young Asian-Australians, plant-based milks, especially soy are not just alternatives, they're cultural staples. As one Gen Zer pointed out:

> *"Soy milk's always been around at home — it's not some new health kick for us. My parents drink it with breakfast, and me, a second-generation Chinese-Australian, I grew up thinking it was just normal. I reckon now it's kind of funny seeing everyone treat it like a trend, when for us it's just part of who we are."* (Female, born 2002)

This dual familiarity with both traditional and emerging food systems puts Asian-Australian Gen Zers in a unique position to influence and bridge cultural food narratives. Is this what is facilitating the wider acceptance of plant-based milks? Let's try to examine some of these issues in more detail and through the eyes of Gen Z.

Taste of Non-Dairy Alternatives

As with any food items, taste experience is a leading consideration for young consumers. Older generations may prioritise health concerns above taste and although wellness is a strong motivator for Gen Z, taste and sensory experience are essential for the uptake of any new foods. All other factors, such as psychological, environmental, marketing or animal well-being, may be a priority for some sections of Gen Z but not across this entire cohort. In fact, people continue to consume their favourite foods even when they are aware of negative consequences. For example, an Australian study of Sydney residents showed that a large majority of people aware of the negative consequences of meat consumption for planetary health, continue to consume it with high frequencies (Marinova and Bogueva, 2019) driven by its sensory qualities. At the other end of the spectrum, a study of vegan Gen Z in US shows that for more than half of them, the dominating reason for these dietary preferences is the health benefits from a plant-based diet (Vegconomist, 2023). The biggest obstacle for non-vegans in that study is meat cravings (Vegconomist, 2023) showing how important taste sensation is.

Gen Z's milk preferences are shaped by personal experiences, dietary needs and cultural rituals. The diversity of plant-based milks poses two taste-related trials to consumers. One is related to similarity of experience compared to cow's milk and the other is the richness of flavours and textures available. It is impossible to say which plant-based milk tastes the best as it is a matter of personal preferences. For many Gen Zers, it is a process of discovery and adaptation. As one of them shared:

> *"Not all plant milks hit the same, hey. Some are way too watery or have this weird aftertaste, but then you find one, like oat or macadamia, for me that actually tastes good in coffee or with brekkie [breakfast]. It's all trial and error, but once you land on your fave [favourite], there's no going back."* (Male, born 2000)

A study by Arla Foods found that nearly half of Gen Z in the UK felt ashamed to order dairy in public, and 29% admitted to only ordering dairy alternatives at cafés while still consuming dairy at home (Lawrence, 2022). This suggests that social identity and peer influence play a role in shaping milk choices, especially in public settings when ordering chai or coffee. Many young consumers are experimenting with different types before settling on their favourites and personal experiences often shape these choices.

Soy milk is perceived to taste the closest to cow's milk, mainly because of its high protein content, but other plant-based milks also have specific aromas, pleasant tastes, colours, smoothness and thickness that appeal to consumers (Lutze, 2024). One Gen Zer expressed how taste preferences and cultural habits shape choices:

> *"I'm lactose-intolerant, so I never really had cow's milk anyway. Almond and coconut were my go-tos, but oat or soy milks are my new fave [favourites]... they're smooth and don't mess with my chai. I don't think it's just a health thing for us anymore; taste is everything."* (Female, born 2005)

This view highlights the intersection of dietary needs, cultural drink rituals, such as chai, and sensory satisfaction in shaping Gen Z's food decisions. For many in this cohort, especially women and culturally diverse groups, taste and emotional connection are just as vital as health benefits. As a result, brands are increasingly tailoring products to appeal to specific flavour traditions and preferences (Insights Symrise, 2024).

Another popular use of plant-based milks is in coffee shops which Gen Z like and frequently enjoy but strongly oppose the surcharge they have to pay for non-dairy alternatives (Halabi *et al.*, 2024). As one Gen Zer put it:

> *"It's wild that I have to pay extra just because I want oat milk in my flat white. It's 2025... we shouldn't be punished for making better choices."* (Male, born 1999)

This sentiment is echoed widely among Gen Z, who feel that choosing sustainable or health-conscious options should not come with financial penalties. At the same time, individual preferences remain diverse. Proving individual preferences, a study of Gen Z university students across Greece, India and UK found that 33% were of the opinion that plant-based products, which include milk, have an inferior taste while 40% said that some of them tasted good (Raptou *et al.*, 2024). Generation Z continues to be the major driver of plant-based options, including veganism and flexitarianism, with the majority of those who have made the switch wanting to educate other people about the benefits of such food choices (Zulkosky, 2023).

Aftertaste is another major consideration for all consumers of alternative milks and dairy products. It can play a decisive role in people's preferences. We are yet to see investigations about what aftertaste is preferred by Gen Z and how different their preferences are from the mainstream consumer who is looking for a sensory experience that mimics livestock-based products (Amyoony *et al.*, 2023).

Health and Nutritional Qualities of Non-Dairy Alternatives

For those who are lactose-intolerant, that is 76% of the world population (Zechmann-Khreis, 2024), the first health benefit from plant-based milks is that they do not contain lactose. Many Gen Zers cite digestive issues, lactose intolerance and gut health as reasons for avoiding dairy (Yehl, 2024). This is more than just a passing preference, it is rooted in lived experiences:

> *"I used to get stomach cramps after cereal as a kid, and no one knew why. Mum swapped me to rice milk and it was a game changer, no more pain. Now I'm all about rice or oat milk, they feel lighter but still give me that full feeling."* (Female, born 2001)

This story points out how health awareness, parental influence and personal well-being shape Gen Z's long-term dietary habits. It also reflects a broader reality, where non-dairy alternatives are more than trends, they are practical solutions for everyday life.

Many people are concerned about allergies associated with soy, nuts or other plants used to produce plant-based milks. This is true but many substances can cause allergic reactions, including the proteins casein and whey, contained in cow's milk, which are some of the common allergens (Zuberbier, 2016). Soy allergy, for example, is relatively uncommon compared with milk allergy (Allergy and Anaphylaxis Australia, 2025). Being transparent about the ingredients on the part of manufacturers and reading the labels on the part of consumers are important to avoid any unpleasant allergic reactions. This sense of vigilance is something Gen Z takes seriously, especially when juggling dietary needs and social food trends. As one Gen Zer shared:

> *"I've got a mate who's allergic to nuts, and I'm gluten-sensitive, so we're always triple-checking the fine print on cartons. It's annoying, but you sort of get used to it... brands that are clear about what's in their stuff earn our trust."* (Female, born 2004)

This focus on transparency and ingredient clarity reinforces why Gen Z increasingly values ethical and health-conscious food choices. Beyond allergens, the discussion also extends to the nutritional profile of plant-based non-dairy alternatives.

Related to human health are the nutritional qualities of plant-based non-dairy alternatives. Although generally plant-based milks do not equal cow's milk in terms of nutrient content, they are commonly fortified to match the latter's nutritional profile (Ramsing *et al.*, 2023). Differences in the content of protein, zinc and potassium remain depending on the base plant ingredient and specific product (Ramsing *et al.*, 2023).

Although there may be some differences and concerns about the micronutrients contained in plant-based milks, the reality is that these products are generally perceived as healthier than their dairy counterparts because of the lower content of saturated fat and cholesterol (Paul *et al.*, 2020), but also because they are not exposed to contaminants associated with the production of cow's milk, such as hormones and antibiotics (Welsh *et al.*, 2019). An analysis of 148 plant-based milks across Australia, USA and Western Europe showed that they do not contain high levels of sodium, saturated fat or calories and that most were fortified with calcium to levels making them equal to or greater than dairy milk (Craig and Fresán, 2021).

Table 8.1 shows a comparison between the nutritional contents of cow's milk and fortified soy, rice, almond, oat and pea milks based on Australian data. Food essentially is a source of energy and cow's milk supplies the highest number of kJ per unit of liquid. By comparison, soy has 96% of the energy supply while almond has only 23% (see Table 8.1). A higher energy intake is good to maintain a healthy weight but when there is widespread overweight and obesity, plant-based milks can be better options. Better digestive health because of higher fibre content is also a preference consideration when it comes to plant-based milks (Sethi *et al.*, 2016).

Gen Z consumers are embracing the benefits of plant-based milks considering them healthier and more sustainable but they also do not want to compromise on texture and taste as well as nutrition.

Table 8.1. Nutritional content of cow's and different fortified plant-based milks per 250 ml.

	Cow's milk	Soy milk	Rice milk	Almond milk	Oat milk	Pea milk
Calcium [% of RDI]	34%	37%	23%	22%	36%	38%
Phosphorus [% of RDI]	23%	18%	5%	4%	22%	25%
Riboflavin [% of RDI]	28%	4%	0%	3%	0%	29%
Vitamin A [% of RDI]	15%	2%	0%	0%	0%	18%
Zinc [% of RDI]	7%	3%	2%	2%	2%	N/a
Magnesium [% of RDI]	8%	17%	8%	6%	6%	N/a
Energy [kJ]	698	620	670	160	583	345
Total fat [g]	8.8	7.0	2.5	3.8	2.0	5.0
Protein [g]	8.5	9.5	1.0	1.2	3.5	8.3
Total sugar [g]	13.8	6.8	9.8	0	8.0	0.3
Added sugar [g]	0	4.5	0	0	0	0
Free sugar [g]	0	4.5	0	0	0	0

Note: RDI — recommended dietary intake; N/a — not applicable.
Source: Dairy Australia (2022).

Production methods for plant-based milks are constantly technologically improving with the aim to increase the bioavailability of microelements, such as calcium and vitamins, as well as preserve their valuable components, including phenolic compounds, antioxidants, unsaturated fatty acids, phytosterols and isoflavones (Aydar *et al.*, 2020). Consumer perception also plays a big role with those who perceive plant-based milks as a natural product more willing to consume them than those who see them as highly processed and artificial, less likely to do so (Martínez-Padilla *et al.*, 2023). Clear marketing strategies and good product package labelling can help Gen Z navigate among the multitude of offered products.

Environmental Benefits

While there may be some uncertainty around the nutritional qualities of plant-based milks, their environmental benefits are well-established and beyond doubt. There are however differences in the environmental footprints of the most commonly used plant-based milks, namely almond, oat, rice, soy and oat (see Figure 8.2),

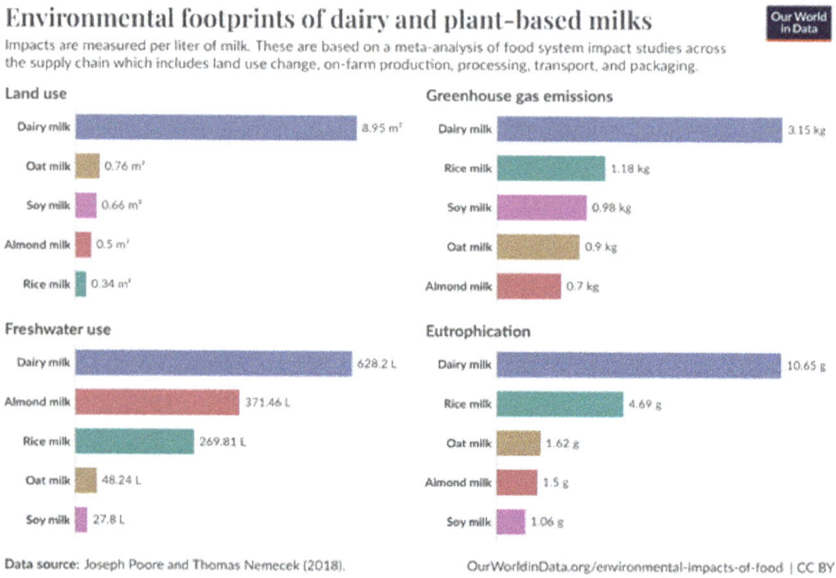

Figure 8.2. Environmental footprints of milks.

with soy and oat milks being relatively good performers in all aspects. Each plant-based milk has its own environmental and nutritional credentials as well as its story of production. For example, macadamia milk is of particular interest for a country like Australia as these nuts come from ancient native species commercialised in the 1960s (Australian Macadamias, n.d.). While the macadamia trees store carbon and help binding soil together avoiding erosion and degradation, their origins are from tropical areas and hence require a lot of rain to grow (ZeroSmart, 2020). Macadamia milk is nutritious, creamy and considered hypoallergenic for those allergic to nuts or soy, but it is also more expensive. Young and other people on a tight budget need to factor in the price of the alternative milks, particularly as currently there are no government subsidies for them while dairy production is heavily subsidised (Métraux, 2025).

Notwithstanding the differences, the comparison with dairy milk shows how environmentally superior plant-based alternatives are and consequently, how poorly dairy milk performs in terms of land use, freshwater consumption, greenhouse gas emissions and eutrophication. Soy has been described as the miracle plant (Mindell, 1995) and the environmental comparison between dairy and soy milk categorically confirms this. Per litre of milk, soy milk requires 14 times less land, 23 times less water, emits three times less greenhouse gas emissions and causes 10 times less eutrophication (pollution with excess nutrients) compared to cow's milk (Ritchie, 1992). We are often asked about which milk is the most sustainable. It is difficult to give a definite answer to such questions (Marinova and Bogueva, 2020). What is clear is that cow's and any dairy milk is environmentally the worst option. The best approach is for people to diversity their choices as reliance on only one species for food potentially locks us in dependence and overexploitation of that plant or animal.

Gen Z are fully aware about the environmental benefits of non-dairy milks and have seriously cut their consumption of cow's milk. Some even claim dairy milk is "evil" and "gross" stressing that it is only for the young calves (Tatum, 2020). While dairy continues to be a significant part of the Western diet in countries, such as Australia, UK or USA, Gen Z is changing the landscape setting example and demanding more environmentally friendly options (Putch, 2024). Gen Z feels compelled to do something about climate change and opting out from consuming cow's milk seems like a straightforward action to take. This environmental awareness is not abstract for Gen Z. It translates into everyday decisions, as illustrated in the words of the Gen Zer:

> *"Mate, when you actually stop and look at the stats — how many litres of water go into producing just one glass of cow's milk, or the hectares of land being used... it's kinda hard to unsee. It's not just about being lactose-intolerant or trendy; it's about doing something that feels right. Switching to oat or soy milk isn't just a vibe... it's smart, sustainable, and honestly, the least we can do for the planet. It's a small change with a big ripple effect."* (Male, born 1998)

Gen Z are aware that they are contributing towards cutting emissions, particularly methane, with every glass of plant-based instead of dairy milk. The livestock-based dairy industry is being severely disrupted by Gen Z's opinions, attitudes and most importantly rejection of its cow's milk because of other better options. While there are still young consumers who continue to use livestock-based dairy products, the number of adopters of plant-based options is so high that they are already shaping a dairy-free future when it comes to milk.

Animal Cruelty

Avoidance as well as reduction in the consumption of livestock-based products is highly influenced by consumer attitudes towards animals and animal rights. Even people on an omnivore diet are more inclined to cut meat, eggs and dairy in response to concerns about animal welfare (Dhont and Ioannidou, 2024). Aversion to animal cruelty is a strong reason for shifting towards plant-based milks. This awareness hits home for many Gen Z consumers, who find the treatment of dairy cows and their offspring emotionally confronting. A lot of Gen Zers are genuinely shaken by what goes on behind the scenes in the dairy industry; seeing or learning about how cows and their calves are treated can be pretty heartbreaking. It is that emotional wake-up call that pushes people to rethink what they are consuming:

> *"I used to drink dairy without thinking, but when I learned about what really happens to the calves, I couldn't keep doing it. It just felt wrong."* (Female, born 2004)

For Gen Z, milk is not just a drink. It is a symbol of ethical compromise they no longer want to make. Such an emotional response is a major factor in their food transition.

This is a powerful motivator for consumers who follow a vegan diet (Janssen *et al.*, 2016) as well as for vegetarians (Dhont and Ioannidou, 2024). The popularity of veganism and vegetarianism seems to be higher among Gen Z. For example, in Australia — a country with one of the world's highest per capita meat and cow's milk consumption, Gen Z has the highest prevalence of vegan and vegetarians among all generations currently alive, namely 10% in 2023 compared to, e.g. 3% for the Baby Boomers (Statista, 2025). The global trends are even stronger in 2023 with 27% of Gen Z across

31 countries reporting that they eat plant-based foods "mostly" or "all the time", which was a 5% increase on 2021 (Globe Scan, 2024). Irrespective of what diet people follow, the rise in sales of plant-based milks worldwide is an indicator of people's rejection of the exploitation of animals for food (Millstein, 2024), including in dairy factory farms where at least 270 million cows are raised, impregnated and separated from their calves to produce milk for human consumption (WWF, 2025). This ethical awareness goes beyond dietary labels and influences a much wider segment of Gen Z, as expressed by another Gen Zer:

> *"It's moved beyond just thinking about my own body or dietary needs. Once you learn about the realities of how animals are treated in the dairy industry — the constant pregnancies, the separation of calves from their mothers, the stress and discomfort they endure — it's hard to keep looking the other way. That kind of suffering for something as simple as milk or cheese just doesn't sit right with me anymore. I'd rather find an alternative that aligns with compassion and leaves a lighter footprint."* (Female, born 2002)

Such view shows that rejecting animal exploitation is not limited to strict vegans or vegetarians, but is becoming mainstream among younger generations. The dairy industry creates images of happy cows grazing freely in green fields whilst the majority of the milk is produced in intensive production factory farms where the cows are generally constrained and milked several times per day with electronic mechanised suction pump equipment. For Gen Z to justify claims to be an ethical consumer (Walters, 2021), it needs to navigate this industry deceit.

Gen Z's concerns about animal welfare manifest in other purchasing behaviour, such as avoiding buying products from brands that engage in animal cruelty (Statista, 2024) or prioritising cosmetic companies committed to animal-free testing (Bonifacio *et al.*, 2024). These young people are also concerned about live exports, cage eggs, factory farms and slaughter house cruelty as one girl from Australia explains: "If I do have to eat another animal, I want to ensure it had a good life and was treated decently and had a cruelty free childhood and pain free death" (year13, 2022, para. 12). In Spain, where bullfights continue to be a traditional pastime, Gen Z is opposing all animal-based tourist attractions (Carles *et al.*, 2023). They are also active in organisations, such as People for the Ethical Treatment of Animals (PETA) that describes them as the PETA Generation (PETA, 2025). It is estimated that 73% of Gen Z consider themselves animal rights activists (PETA, n.d.) and this translates in their food choices. Because of its use of social media and digital technologies, Gen Z can be wrongly stereotyped as obsessed only with itself; in reality, it is a generation that thinks about global issues, such as climate change, biodiversity loss and soil fertility, and is constantly assessing its role in relation to the world and all its inhabitants (Esaki-Smith, 2024).

Trends

It is clear that the plant-based milks are the alternatives to livestock products that are proving the most popular among consumers and Gen Z is in many ways responsible for this trend. In the USA and Canada, more than half of Gen Z approve these alternative milks (Innova Market Insights, 2024). In these two countries, the consumption of plant-based milks represents 40% of all — dairy and plant-based — milk consumption and Gen Z are the strongest contributors (Innova Market Insights, 2024). For many in Gen Z, the adoption of plant-based milks is not just about following trends but reflects layered motivations tied to taste, health and sustainability:

> *"Oat milk's just part of the rhythm of my day now — morning coffee, smoothies, even baking. I didn't jump on it because everyone else was doing it, I stuck with it because it actually works for me. It's smooth, doesn't give me that bloated feeling dairy used to, and the environmental side of it is a no-brainer. Knowing it's gentler on the planet just makes it feel like the smarter choice, not just a trendy one."* (Female, born 2001)

Hence, the appeal of plant-based milk is multidimensional, combining personal well-being with ethical and environmental priorities, which helps drive consistent consumption over time.

Another US market report shows that only 8% of American Gen Zers are buying conventional cow's milk (Nwanne, 2023). Irrespective as to what the exact figures are, what is clear is that these young people are drastically changing the Western diet. The mainstream Western diet is heavily reliant on dairy products, more so than, for example, traditional South-East Asian or African diets (Ritchie *et al.*, 2023). A possible contributing factor for this is the increased focus on health and wellness since the COVID-19 pandemic (NIQ, 2024).

Globally, plant-based and other sustainable ingredients are one of the six megatrends shaping the future of food, together with the resilient farm, food as medicine, the smart supply chain, digital retail and food automation (DigitalFoodLab, 2025). The digitally native Gen Z are making use not only of the new dairy-free products but are also using online shopping and other smart distribution channels to obtain the foodstuff they want. There is a lot of excitement around biomass fermentation and precision fermentation which aim at producing replacements for livestock-based products, including milks. By comparison with the plant-based meat analogues where the market is slow and there is a level of disillusion, milk alternatives are going very strong, wholeheartedly supported by Gen Z. As the DigitalFoodLab (2025) report states, a food revolution is taking shape with numerous start-ups, serious investments and emerging technologies which advance solutions to the global environmental, health and other social challenges. This sense of discovery and participation in a broader movement is endorsed in the way Gen Z talks about its role in shaping food systems:

> *"Scrolling through socials and seeing all these cool new milk options and plant-based cheeses... cashew, pea protein, even potato... it's honestly exciting. Trying them out isn't just about swapping one ingredient for another; it feels like we're part of something bigger. We're building a new food culture, one that's more inclusive, more sustainable and totally creative. It's not just about consumption anymore... it's about exploration, identity and impact."* (Male, born 2003)

Rather than seeing these changes as passive trends, Gen Z perceives itself as a co-architect of a new food future — curious, connected and motivated to leave behind outdated norms. Previously reluctant to see attractive opportunities in agriculture and food production, even venture capitalists are now heavily involved in the new ways of making, distributing and preparing food (Casamitjana, 2025). The global value of plant-based dairy products is forecast to grow with a compound annual growth rate (CAGR) of 4.1% during 2024–2029 exceeding the historically high 2.3% during 2019–2024 (Chehtman, 2024). Its global value is estimated at US$ 22.53 billion in 2025 and almost double by 2034 (Precedence Research, 2025). The alternative milks industry however is not without its challenges, driven mainly by perceptions about its products being ultra-processed and more expensive compared to cow's milk (Eastlake, 2024).

Using the latest technological advances, artificial intelligence (AI) and sales monitoring, the plant-based milk sector is striving to be more efficient in production, expanding its product portfolio, providing cutting-edge design and environmental packaging and younger consumers are attracted becoming key drivers for its growth (Research and Markets, 2025). Coffee shops around the world are responding to the growing interest and demand for plant-based milk options that have high standards of flavour, taste, texture, foamability and allow baristas to produce attractive "latte art" (Stanley-Foreman, 2025). Improved transparency about the methods and ingredients used allows young people to find answers about the processed nature of these products.

Gen Z is embracing the new options of plant-based milks as well as yogurts, creamers, ice creams and cheese in an unprecedented way. In fact, this is the first time in the history of food production in the West when one young and upcoming generation is making drastic changes in dietary preferences when it comes to milk and dairy products. These young people are disrupting well-established traditions because of their value systems combined with experiences of novel tastes and new responsibilities for the well-being of the planet and their own health.

Packaging and Ingredients

The lower environmental footprint of plant-based milks can be significantly affected by packaging. Which materials are used, what the design and the information on the packaging are, all impact Gen Z's perceptions of the alternative milks. According to Braberman (2024), for packaging to be successful for Gen Z, it needs to be visually appealing with mood-boosting colours as well as to support mental well-being and provide comfort against the challenges this generation faces. As one male Gen Z explains:

> *"We are exploiting the earth resources like there is no tomorrow. Plant-based will perhaps solve the need for additional food."* (Male, born 1999)

Gen Z embrace imperfection in their strive to be authentic; they are socially and environmentally aware, highly informed with digital sources on their fingerprints, but they also want transparency about the included food ingredients and the methods of production used (Bogueva and Marinova, 2024). When Gen Z is opting for non-animal milks, they expect products that are trendy and reflect their consciousness. They also need hope and reassurance that there are solutions to the planetary problems created by previous generations. In this day and age, when there are geopolitical tensions and digital overload, Gen Z are looking for packaging that lifts the spirit and brings joy. They are looking for designs to be environmentally honest and simplistic that provide promises and guarantees the plant-based milk for which they pay a premium, is worth it. A study of Portuguese Gen Z consumers (Gomes *et al.*, 2023) confirms that perceived green quality and green benefits increased their willingness to pay more for environmentally better products, including food, and this greenness should be made explicit through the packaging itself. This sentiment is reiterated by many young consumers who demand visual honesty and traceability in the packaging itself, not just vague branding or vague 'green' labels:

> *"If the packaging... the box looks cheap and low-effort or like it's just pretending to be eco-friendly... fake-green, I'm out. I'm*

> *not buying it. I want to feel confident that what I'm picking up actually reflects what the brand stands for... transparency, real sustainability and quality. Show me what's inside, how it's made and where it comes from. If it feels thoughtful, I'm in. If it feels like greenwashing or like they're cutting corners, that's a pass. I'm trying to make choices that work for me and don't mess with the planet. I like to feel I'm making a decent choice for the planet and myself."* (Female, born 2002)

Packaging acts as both a practical and emotional interface for Gen Z, shaping trust, perceived value and the willingness to pay more for eco-friendly alternatives.

Overpackaging and the use of non-recyclable materials is a non-desirable territory for Gen Z. Instead, eco-friendly, sustainable options that encourage shoppers to reuse and/or properly recycle the packaging materials are essential for this generation. For Gen Z, the alignment of packaging with ethical and environmental standards is inseparable from brand loyalty and consumer trust:

> *"I'm not just drinking oat milk... I'm voting with my wallet. If it's got dodgy ingredients or non-recyclable packaging, it doesn't matter how cool it looks."* (Male, born 1999)

This opinion highlights that design and materials are not just aesthetics for Gen Z. They are a reflection of brand integrity, and anything less can undermine product credibility, no matter the nutritional content. As Joel (2023) explains, they are focussed on transparent marketing information with minimised packaging elements with high standards of authenticity which distinguishes one brand from another. This certainly applies to the myriad of alternative non-animal milk options where packaging helps navigate the complexity of consumer choices. An appropriate bold, bright, simple sustainable packaging that shows the alternative milks in a confidently positive manner, can create trust and a connection with the Gen Z consumer.

The information on the packaging should also satisfy the desire of Gen Z to know more about the milk alternatives. Gen Z consumers are interested in the type of manufacturing processes used to fabricate the plant-based milk substitutes as well as their composition, including extracted oil bodies, emulsifiers and ingredients (McClements, 2020). Appearance, texture, sensory properties, physical and chemical stability as well as shelf-life are similarly important. Bioavailability of the digested nutrients from these alternative milks is another issue that affects their use (McClements, 2020). As plant-based milks are nutritionally different from cow's milk, they are often fortified with a range of key nutrients, such as calcium, iodine, vitamins A, B1, B2, B12 and D. Among the hundreds of plant-based milks sold in Australia and Singapore, approximately two-thirds are fortified (Brooker *et al.*, 2023). As there are currently no standards for plant-based beverages intended to be used as alternatives to cow's milk (Drewnowski *et al.*, 2021), there are vast differences in the nutritional profiles of the products available to consumers. Their protein content also varies. Overall, Gen Z expects these products, their packaging and the way they are being supplied to the consumer, to guarantee them to be safe, nutritional and tasty as well as have environmental, ethical and health benefits for which they are prepared to pay a premium price (Ramsing *et al.*, 2023).

Synthetic Milk

Synthetic milk, and down the track products made from it, are a different kind of disruption to the livestock-based dairy industry. Produced through precision fermentation, it does not require cows or other animals, but instead is cultured from cells (Bojovic, 2022). The fermentation process, an extension of those used for cheese and yogurt, starts with yeast to produce the same whey proteins and caseins as found in cow's milk (CSIRO, 2021). It is similar to the processes used to produce insulin for diabetics since the 1970s and rennet in cheesemaking since the 1990s (Whitttaker, 2024).

Synthetic milk has the same taste and qualities, such as texture and frothiness, as cow's milk without its environmental footprint and animal exploitation. One estimate shows that synthetic milk requires 1,740 times less land to produce the same amount of protein as the most efficient soy plant (Linder, 2019). If renewable energy is used in the manufacturing process, synthetic milk can revolutionise food production (Monbiot, 2022). It is also free from lactose and cholesterol friendly with minerals, fats, flavours and sugars added if needed to improve its properties (CSIRO, 2021). The new technology is attracting investors in Australia and all over the world (Whittaker, 2024). As a Gen Zer said:

> *"Synthetic milk feels like the future. AI and biotech are finally making food that's smart and sustainable. I'm here for it, especially if it helps fight climate change."* (Female, born 1999)

Reflecting the curiosity and optimism of digitally native Gen Z, many young consumers see synthetic milk as part of a broader technological revolution in sustainable food innovation and production. Synthetic milk, produced through techniques such as precision fermentation, is expected to eventually replace livestock-based milk and to be used in a wide range of products from ice cream to cheese. However, despite the promises, Gen Z remains cautious in its preparedness to add these foods to its diet, especially when concerns about health, safety and transparency linger. A survey of Gen Z youth in Türkiye shows that they do not find synthetic milk neither healthy nor natural (Aydemir *et al.*, 2023), underscoring a global hesitancy toward unfamiliar food technologies. Gen Z remains cautious about integrating these foods into its diet:

> *"I get why city kids are into synthetic milk... it's cool and eco-friendly. But where I'm from, people still trust the farm and real cows more than lab stuff."* (Male, born 2004)

A similar sentiment is echoed across diverse Gen Z communities, where scepticism about biotechnology often intersects with deeper concerns about long-term health impacts and corporate transparency. As one Gen Zer explained:

> *"Synthetic milk... I mean, it sounds awesome... like some futuristic milk out of a sci-fi film. Lab-grown, ultra sustainable, no cows in sight? I'm intrigued. But I also have questions... What's really in it? How's it made? Is it safe for my body over time? I'd definitely give it a shot, but I won't be the first in line. Let the brave souls*

> *test it out first... I'll follow once it feels legit and backed by solid info."* (Male, born 2000)

These perspectives express how Gen Z's willingness to experiment is tempered by critical thinking and a demand for clear, science-based communication. They are not just passive consumers; they are active participants in shaping the future of food.

While synthetic milk and dairy products are still in their early stages and not yet widely available on the market, Gen Z is already driving interest and leading the demand for precision fermentation, especially in the USA where 40% of these young consumers are willing to try the products with this new food technology (Advanced Biotech, 2023). They explain that the first driving force for them to adopt synthetic dairy would be taste while being better for the environment is the second-biggest reason (Advanced Biotech, 2023). Such comments show that Gen Z's openness to synthetic dairy is tightly linked to ethical production, environmental impact and most importantly, trust in the information provided. As one Gen Zer pointed out:

> *"If it tastes like the real thing and doesn't hurt animals or the planet, then I'm open to it. Just who makes it don't need to lie to us about how it's made.... it's gotta come with honesty. No fancy marketing spin or vague claims. Just show us how it's made, what's in it, and what it really stands for. Because if brands think they can get away with greenwashing or hiding behind labels, they've missed the point. We want real change, not a polished illusion."* (Male, born 2005)

This reinforces the need for full transparency, particularly around production processes and nutritional safety, if synthetic milk is to succeed with Gen Z consumers. However, real preferences for synthetic milk and dairy are yet to be revealed. This is a very different situation for Gen Zers whose attitudes towards plant-based milks are clear and explicit. Many of them see plant-based milks as a status symbol with one young American explaining: "The pictures that I see of oat milk are always in this minimalist style, it just makes you feel like whoever is drinking that has their life together or something...It's like, definitely a status symbol, like kind of a flex to drink oat milk" (Adams, 2021, para. 19).

Gender and Non-Dairy Alternatives

Are the attitudes towards plant-based and synthetic milks influenced by gender the way they are when it comes to meat alternatives (Bogueva *et al.*, 2022)? Is Gen Z associating the concept of masculinity with the milk young men and women drink? Our Australian study does not show any gender-based differences neither for Gen Z or other age groups (Bogueva and Marinova, 2025). It seems that the opaque white fluid is equally acceptable irrespective of its plant-based or synthetic origin. A 2022 cross-generational study in the UK also shows equal acceptance of plant-based milks at 47% for women and 46% for men with 52% of both genders rejecting them as part of their diets (Statista, 2022). Among the 14 different varieties of plant-based milks, women were more inclined to include almond, oat and coconut milk in their diet while men slightly preferred hazelnut, rice, cashew, walnut, quinoa, hemp and spelt milk with the preferences for soy, potato, pea and chickpea milk being exactly the same. In the USA, 24% of women of all ages consumed plant-based milks compared to 22% of men in 2023 (Peekage, 2023), but these milks were the preferred option for 27% of people aged 20–30 (compared to 29% of them opting for dairy milk).

No correlation between gender or age and non-dairy cheese alternatives was found in a study by Zollman Thomas and Bryant (2021) covering participants from Brazil, Germany, India, UK and USA, although Brazilian women were more inclined to purchase such food products. Similarly, there was no difference between men and women in their willingness to try plant-based milks in a UK study that analysed the images and words used in their packaging (Baptista and Schifferstein, 2023). However, a study of young people from Türkiye (Aydemir *et al.*, 2023), a country with a predominantly Muslim population, shows that men are slightly more inclined than women to consume animal-free milk, 34% for men compared to 30% for women.

It appears that gender is a relatively insignificant factor when it comes to milk. As one Gen Zer highlighted:

> *"To be honest, I've never thought milk had anything to do with masculinity. I drink oat because I like how it tastes, and it fits my lifestyle — it's not some statement about being tough or traditional."* (Female, born 2003)

There are only a few studies which delve deeper to explain the reasons behind the choice of milk. For Gen Z, plant-based milks are particularly trendy and promoted through social media (Yehl, 2024). They are seen as the new-wave characteristic of their generation which distinguishes it from their parents and grandparents. This attitude seems to be common across Gen Z genders. A cross-generational analysis of gender differences in the reasons behind the choice of plant-based milks shows that people simply enjoy the taste of different drinks and foods, but women are more likely than men to object against the ill treatment of animals with forceful impregnation and separation of the young from their mothers (Wunsch, 2024).

While ethical concerns tend to feature more strongly in women's motivations, many young men are also engaging thoughtfully, not just with nutrition but with production ethics and sustainability:

> *"I'm not switching milks because it's trendy — I'm doing it because it makes sense. If it's better for animals, the planet, and still gets the job done in my protein smoothie, then why wouldn't I?"* (Male, born 2002)

The shared opinion reflects a shift among Gen Z men toward more mindful consumption, where practical needs like nutrition coexist with deeper ethical concerns. It suggests that choosing plant-based milk is not about trends or identity politics, it is increasingly seen as a smart and responsible decision. This emerging attitude helps dissolve traditional gender boundaries around food choices and reinforces a broader cultural acceptance of sustainability-driven habits.

What's in the Name

We are reluctant to engage in lengthy debates as to what non-dairy milk and other products should be called. The dairy industry feels threatened by these new products and is trying to protect its market as well as linguistic territory. It argues that milk describes only the opaque white fluid secreted by the mammary glands of female mammals, mainly cows, for the nourishment of their young (Brey and Ankeny, 2019). A question that arises is why human mammals should be drinking the milk produced by the mammary glands of another species. The answer to this question is probably in the fact that humans are opportunistic eaters who essentially can eat almost anything that is not a health threat. Linguistically, the word "milk" is not a monopoly of the dairy industry. Both science and history show that plant-derived liquids have been called "milk" for thousand years, including in the Christian and Islamic world, China and throughout Asia (Brey and Ankeny, 2019).

Notwithstanding this, Food Standards Australia and New Zealand (2016) has opted for the two terms "plant-based milk alternatives" and "plant-based beverages" while the European Court of Justice ruled out in 2017 that soy products should not be marketed as "milk" or "cheese" (Palmer, 2019). In 2023, however the US Food and Drug Administration issued a guidance (to be finalised in 2025) that plant-based milk alternatives may be named as "beverage", "drink" or "milk" (FDA, 2023). This was in response to the common usage of the term "milk" for plant-based alternatives to the extent that it has firmly entered the consumers' vocabulary. Moreover, with the increased availability of plant-based milk alternatives, consumers are unlikely to become confused and can figure out the difference for themselves without the need to be protected. Consumers however need to know the type of plants used to produce the milk because of potential allergies, taste preferences, nutritional, environmental and other considerations.

Irrespective as to where legislation and the courts will take the debate about the appropriate naming of these foods, Gen Z consumers are simply referring to plant-based milk alternatives as soy, oat, almond etc. milk. This has become common usage and it is unlikely to change, particularly as young people are now accustomed to checking the contents of the foods they eat as well as to looking for other relevant information (Cha *et al.*, 2014; Joseph and Tan, 2023).

Importantly, Gen Z is buying less cow's milk than the average consumer (Yehl, 2024) and in the UK, 49% of these young people admit that they feel ashamed and embarrassed to order dairy in public, particularly in front of their peers (Rabb, 2022). One of them explains: "When I'm with other people, they don't see drinking [cow's] milk as normal" (Yehl, 2024, para. 9). Some describe Gen Z as the "Not Milk generation" (Seariac, 2023) while in fact, they are the alternative milk generation.

Conclusion

The overall view of Gen Z regarding dairy milk can be summarised as: "milk is bad for the cow, milk is bad for you and milk is bad for the planet" (Tatum, 2020). Despite most of Gen Zers having this knowledge, not all of them are prepared to give up cow's milk completely. Those who are adhering to their high environmental, health and animal welfare values are part of a transition that is surely disrupting the livestock sector. The trends show that: firstly, out of all plant-based alternatives to livestock-based foods, milk is the most popular animal-free option; and secondly, Gen Z is the trend-setting generation. Whether the choice is soy, oat, almond, coconut or any other non-dairy milk or products, the consumption of these foods is contributing towards reducing human environmental footprint by freeing up land, cutting down greenhouse gas emissions, decreasing the pressure on soil fertility and allowing better biodiversity prospects. The health benefits of plant-based milk options are probably more pronounced for older generations who have higher rates of non-communicable diseases, including obesity and overweight. However, in recent years we have witnessed these conditions affecting much younger people and what Gen Z is doing is not just a fancy trend in opposition to their parents, but a shift to protect themselves as well as the planet today and for tomorrow.

In the future, new dairy replacement options will emerge as precision fermentation and other breakthrough technological processes develop further, become commercially viable and widely available to consumers. The milk alternatives industry has witnessed a lot of interest from private investors but not so much government encouragement. Gen Z is prepared to pay higher prices for products that they perceive as being more sustainable but it is also time to reconsider where government subsidies should be going. These young people are shaping a future in which there will be less exploitation and presence of livestock, more choice and hopefully, better health outcomes for all.

References

Adams, J. (2021). How oat milk became Generation Z's status symbol. https://ascjcapstone.com/terms/spring-2021/jadams57/

Advanced Biotech. (2023). Gen Z is leading the demand for precision fermentation. https://adv-bio.com/gen-z-is-leading-the-demand-for-precision-fermentation/

Allergy & Anaphylaxis Australia. (2025). Soy. https://allergyfacts.org.au/allergy/soy

Amyoony, J., Moss, R., Dabas, T., Gorman, M., Ritchie, C., LeBlanc, J., & McSweeney, M. B. (2023). An investigation into consumer perception of the aftertaste of plant-based dairy alternatives using a word association task. *Applied Food Research*, 3(2), 100320. https://doi.org/10.1016/j.afres.2023.100320.

Australian Macadamias. (n.d.). An Australian legend: History and industry. https://www.australian-macadamias.org/an-australian-legend/history-and-industry

Aydar, E. F., Tutuncu, S., & Ozcelik, B. (2020). Plant-based milk substitutes: Bioactive compounds, conventional and novel processes, bioavailability studies, and health effects. *Journal of Functional Foods*, 70, 103975. https://doi.org/10.1016/j.jff.2020.103975.

Aydemir, M. E., Okan, Y. T., & Takim, K. (2023). Generation Z consume animal-free milk? A Türkiye experience. *Food and Health*, 9(3), 254–261. https://doi.org/10.3153/FH23023

Baptista, I. Y. F. & Schifferstein, H. N. J. (2023). Milk, mylk or drink: Do packaging cues affect consumers' understanding of plant-based products? *Food Quality and Preference*, 108, 104885. https://doi.org/10.1016/j.foodqual.2023.104885

Brey, H. & Ankeny, R. A. (2019). Crying over plant-based milk: Neither science nor history favours a dairy monopoly. The Conversation. https://theconversation.com/crying-over-plant-based-milk-neither-science-nor-history-favours-a-dairy-monopoly-123852

Bogueva D. & Marinova D. (2025). Will Australians eat alternative proteins? *Foods*, 14(9), 1526. https://doi.org/10.3390/foods14091526

Bogueva, D. & Marinova, D. (2024). Generation Z and novel plant-based food alternatives. In Bertella, G., Santini, C. (eds) *Plant-Based Food Consumption: Products, Consumers and Strategies*, pp. 105–129. Elsevier.

Bogueva, D., Marinova, D., & Bryant, C. (2022). Meat me halfway: Sydney meat-loving men's restaurant experience with alternative plant-based proteins. *Sustainability*, 14(3), 1290. https://doi.org/10.3390/su14031290

Bojovic, M. (2022). Not like udder milk: 'synthetic' dairy milk made without cows may be coming to a supermarket near you. The Conversation. https://theconversation.com/not-like-udder-milk-synthetic-dairy-milk-made-without-cows-may-be-coming-to-a-supermarket-near-you-189046

Bonifacio, R. A., Baui, A. F., Uy, M. Y. T., Bandojo, A. J., & Etrata, A. (2024). Generation Z's purchasing intention of cruelty-free cosmetic products: The moderating role of environmental and animal welfare concerns. *International Business Education Journal*, 17(1), 118–132. https://doi.org/10.37134/ibej.Vol17.1.10.2024

Braberman, H. (2024). Gen Z's packaging revolution: 7 trends reshaping design. Packaging World. https://www.packworld.com/trends/package-design/article/22915168/gen-zs-packaging-revolution-7-trends-reshaping-design

Brooker, P. G., Anastasiou, K., Smith, B. P. C., Tan, R., Cleanthous, X., & Riley, M. D. (2023). Nutrient composition of milk and plant-based milk alternatives: A cross-sectional study of products sold in Australia and Singapore. *Food Research International*, 173(Part 2), 113475. https://doi.org/10.1016/j.foodres.2023.113475

Casamitjana, J. (2025). Report predicts growth and innovation for plant-based food in 2025. Vegan FTA. https://veganfta.com/2025/02/04/report-predicts-growth-and-innovation-for-plant-based-food-in-2025/

Cha, E., Kim, K. H., Lerner, H. M., Dawkins, C. R., Bello, M. K., Umpierrez, G., & Dunbar, S. B. (2014). Health literacy, self-efficacy, food label use, and diet in young adults. *American Journal of Health Behavior*, 38(3), 331–339. https://doi.org/10.5993/AJHB.38.3.2

Chehtman, A. (2024). The key trends shaping dairy products and alternatives into 2025. Euromonitor International. https://www.euromonitor.com/article/the-key-trends-shaping-dairy-products-and-alternatives-into-2025

Chen, L. (2023). Soy milk. Encyclopedia of Human Nutrition, Volume 2, 4th ed. https://www.sciencedirect.com/topics/agricultural-and-biological-sciences/soya-milk

Commonwealth Scientific and Industrial Research Organisation (CSIRO). (2021). Animal-free dairy. https://www.csiro.au/en/research/production/food/eden-brew

Craig, W. J. & Fresán, U. (2021). International analysis of the nutritional content and a review of health benefits of non-dairy plant-based beverages. *Nutrients*, 13(3), 842. https://doi.org/10.3390/nu13030842

Dairy Australia. (2022). How does milk compare to plant-based beverages? https://cdn-prod.dairyaustralia.com.au/-/media/dairy/files/health-resources/dairy_and_plant-basedbeverages.pdf

Dairy Australia. (2024). Is it true that most Asian people are lactose intolerant? https://www.dairy.com.au/dairy-matters/you-ask-we-answer/yawa-29---is-it-true-that-most-asian-people-are-lactose-intolerant

DigitalFoodLab. (2025). Foodtech trends in 2025. https://www.digitalfoodlab.com/foodtech-trends-2025/

Dhont, K. & Ioannidou, M. (2024). Health, environmental, and animal rights motives among omnivores, vegetarians, and vegans and the associations with meat, dairy, and egg commitment. *Food Quality and Preference*, 118, 105196. https://doi.org/10.1016/j.foodqual.2024.105196

Drewnowski, A., Henry, C. J., & Dwyer, J. T. (2021). Proposed nutrient standards for plant-based beverages intended as milk alternatives. *Frontiers in Nutrition*, 8. https://doi.org/10.3389/fnut.2021.761442

Dynan, N. (2020). Reducing the personal and health impacts on those suffering lactose intolerance. The Good Nutrition Co. https://member.dietitiansaustralia.org.au/Common/Uploaded%20files/DAA/Resource_Library/2020/Lactose_Intolerance_Ndynan.pdf

Eastlake, D. (2024). What's next for the plant-based trend? Food Navigator Europe. https://www.foodnavigator.com/Article/2024/12/10/whats-next-for-the-plant-based-trend/

Esaki-Smith, A. (2024). What Generation Z is really like, according to students. Forbes. https://www.forbes.com/sites/annaesakismith/2024/08/21/what-generation-z-is-really-like-according-to-students/

Food and Drug Administration (FDA). (2023). Labeling of plant-based milk alternatives and voluntary nutrient statements: Guidance for industry. Draft guidance. https://www.fda.gov/media/165420/download

Food Standards Australia and New Zealand. (2016). Plant-based milk alternatives. https://www.foodstandards.gov.au/consumer/nutrition/milkaltern

Globe Scan. (2024). Plant-based Consumption across Generations. https://globescan.com/2024/06/05/insight-of-the-week-plant-based-consumption-across-generations/

Gomes, S., Lopes, J. M., & Nogueira, S. (2023). Willingness to pay more for green products: A critical challenge for Gen Z. *Journal of Cleaner Production*, 390, 136092. https://doi.org/10.1016/j.jclepro.2023.136092

Janssen, M., Busch, C., Rödiger, M., & Hamm, U. (2016). Motives of consumers following a vegan diet and their attitudes towards animal agriculture. *Appetite*, 105, 643–651. https://doi.org/10.1016/j.appet.2016.06.039

Joel, R. (2023). How Gen Z shoppers are forging the future landscape of packaging. Packaging Gateway. https://www.packaging-gateway.com/comment/how-gen-z-shoppers-are-forging-the-future-landscape-of-packaging/?cf-view&cf-closed

Joseph, J. & Tan, S. T. (2023). Food label reading habits and their correlations with food choice motives among young consumers during the transition to the endemic phase of COVID-19. *Measurement: Food*, 12, 100115. https://doi.org/10.1016/j.meafoo.2023.100115.

Halabi, N., Hristova, V., & Vlaev, I. (2024). Milking the alternatives: Understanding coffee consumers' preferences for non-dairy milk. *Behavioural Sciences*, 14(7), 569. https://doi.org/10.3390/bs140705692024

Innova Market Insights. (2024). Plant-based milk trends in the US and Canada. https://www.innovamarketinsights.com/trends/plant-based-milk-trends

Insights Symrise (2024). Top Gen Z food and beverage preferences. https://in-sight.symrise.com/article/top-gen-z-food-preferences

Lawrence, K. (2022). Study reveals that Gen Z is saying no to dairy products. Power of positivity. https://www.powerofpositivity.com/dairy-products-gen-z/

Linder, T. (2019). Edible microorganisms — An overlooked technology option to counteract agricultural expansion. *Frontiers Sustainable Food Systems*, 3. https://doi.org/10.3389/fsufs.2019.00032

Lutze, T. (2024). Australian supermarket vegan milk taste test: One soy is gold-standard, all almond milks are watery. The Guardian. https://www.theguardian.com/food/2024/jan/09/australian-supermarket-vegan-milk-taste-test-best-soy-almond-oat-milk-australia

Marinova, D. & Bogueva, D. (2019). Planetary health and reduction in meat consumption, *Sustainable Earth*, 2, 3. https://doi.org/10.1186/s42055-019-0010-0

Marinova, D. & Bogueva, D. (2020). Which 'milk' is best for the environment? We compared dairy, nut, soy, hemp and grain milks. The Conversation. https://theconversation.com/which-milk-is-best-for-the-environment-we-compared-dairy-nut-soy-hemp-and-grain-milks-147660

Martínez-Padilla, E., Faber, I., Petersen, I. L., & Vargas-Bello-Pérez, E. (2023). Perceptions toward plant-based milk alternatives among young adult consumers and non-consumers in Denmark: An exploratory study. *Foods*, 12(2), 385. https://doi.org/10.3390/foods12020385

McClements, D. J. (2020). Development of next-generation nutritionally fortified plant-based milk substitutes: Structural design principles. *Foods*, 9(4), 421. https://doi.org/10.3390/foods9040421

MedlinePlus. (2023). Lactose intolerance. https://medlineplus.gov/genetics/condition/lactose-intolerance/#frequency

Métraux, J. (2025). Plant Milk Is Better for Us and the Climate. So Why Do We Subsidize Dairy? https://www.motherjones.com/food/2023/10/milking-it-plant-milk-starbucks-oat-soy-dairy-subsidy-climate/

Millstein, S. (2024). Is Veganism Really Growing? Using Data to Track the Trend. https://sentientmedia.org/increase-in-veganism/

Mindell, E. (1995). *Soy Miracle.* Fireside, New York.

Monbiot, G. (2022). Embrace what may be the most important green technology ever. It could save us all. The Guardian. https://www.theguardian.com/commentisfree/2022/nov/24/green-technology-precision-fermentation-farming

NIQ. (2024). Beyond plant-based: Trends reshaping alternative milk. Nielson Consumer. https://nielseniq.com/global/en/insights/education/2024/alternative-milk-trends/

Nwanne, E. (2023). Gen V(egan)? Gen Z Shifts Away from Meat and Dairy. Brightfield Group. https://blog.brightfieldgroup.com/gen-z-shifts-away-from-meat-and-dairy

Otegui Carles, A., Fraiz Brea, J. A., & Araújo, N. (2023). The ethical attitudes of generation z in spain toward animal-based tourism attractions. *Tourism*, 71(2), 285–300. https://doi.org/10.37741/t.71.2.4

Palmer, S. (2019). Milk wars: Farmers take on dairy alternatives. https://www.nswfarmers.org.au/NSWFA/NSWFA/Posts/The_Farmer/Trade/Milk_wars_farmers_take_on_dairy_alternatives

Paul, A. A., Kumar, S., Kumar, V., & Sharma, R. (2020). Milk analog: Plant based alternatives to conventional milk, production, potential and health concerns. *Critical Reviews in Food Science and Nutrition*, 60(18), 3005–323. https://doi.org/10.1080/10408398.2019.1674243

Peekage. (2023). Peekage CPG insights: The reign of dairy milk challenged by plant-based alternatives! https://peekage.com/blog/peekage-cpg-insights-the-reign-of-dairy-milk-challenged-by-plant-based-alternatives

People for the Ethical Treatment of Animals (PETA). (2025). Generation Z: The PETA Generation. https://prime.peta.org/news/generation-z-the-peta-generation/

People for the Ethical Treatment of Animals (PETA). (n.d.). Vegan Offerings Are Smart Business. https://www.peta.org/issues/animals-used-for-food/animals-used-food-factsheets/vegan-offerings-are-smart-business/

Putch, K. (2024). Consumers changing how they approach dairy consumption. Dairy Processing. https://www.dairyprocessing.com/articles/2823-consumers-changing-how-they-approach-dairy-consumption

Precedence Research. (2025). Plant-based Milk Market Size, Share, and Trends 2025 to 2034. https://www.precedenceresearch.com/plant-based-milk-market

Rabb, M. (2022). Is it embarrassing to order dairy milk at a coffee shop? Gen Z thinks so. The Beet. https://thebeet.com/is-it-embarrassing-to-order-dairy-milk-at-a-coffee-shop-gen-z-thinks-so/

Ramsing, R., Santo, R., Kim, B. F., Altema-Johnson, D., Wooden, A., Chang, K. B., Semba, R. D., & Love, D. C. (2023). Dairy and plant-based milks: Implications for nutrition and planetary health. *Current Environmental Health Reports*, 10, 291–302. https://doi.org/10.1007/s40572-023-00400-z

Raptou, E., Tsiami, A., Negro, G., Ghuriani, V., Baweja, P., Smaoui, S., & Varzakas, T. (2024). Gen Z's willingness to adopt plant-based diets: Empirical evidence from Greece, India, and the UK. *Foods*, 13, 2076. https://doi.org/10.3390/foods13132076

Research and Markets. (2025). 2025 Plant-based Milk Market Report — Industry Size, Competition, Trends and Growth Opportunities by Region — Forecast by Types and Applications (2024–2032). https://www.researchandmarkets.com/reports/5801228/plant-based-milk-market-report-industry-size

Ritchie, H. (1992). Dairy vs. plant-based milk: what are the environmental impacts? https://ourworldindata.org/environmental-impact-milks

Ritchie, H., Rosado, P., & Roser, M. (2023). Meat and Dairy Production. Our World in Data. https://ourworldindata.org/meat-production

Seariac, H. (2023). From the 'Got Milk?' generation to the 'Not Milk' generation. https://www.deseret.com/2023/4/6/23672564/gen-z-dairy-free-milk-alternatives/

Sethi, S., Tyagi, S. K., & Anurag, R. K. (2016). Plant-based milk alternatives an emerging segment of functional beverages: A review. *Journal of Food Science and Technology*, 53(9), 3408–3423. https://doi.org/10.1007/s13197-016-2328-3

Stanley-Foreman, Z. (2025). How plant milks are pushing for further innovation in specialty coffee. Perfect Daily Grind. https://perfectdailygrind.com/2025/01/plant-milks-innovation-in-specialty-coffee/

Statista. (2022). Which, if any, of the following do you include in your diet nowadays? https://www.statista.com/statistics/1388118/consumer-preference-of-different-types-of-plant-based-milks-gender/

Statista. (2024). Share of Gen Z do not buy from brands that participate in animal abuse in the U.S. in 2022, by select situations. https://www.statista.com/statistics/1353456/us-avoid-animal-abuse-brands-gen-z

Statista. (2025). Types of diets followed by consumers in Australia as at October 2023, by generation. https://www.statista.com/statistics/1418604/australia-diets-types-followed-by-generation

Tatum, M. (2020). Creative challenge: can milk be futureproofed? The Grocer. https://www.thegrocer.co.uk/analysis-and-features/creative-challenge-can-milk-be-futureproofed/650463.article

Vegconomist. (2023). Poll: Majority of Gen Z is in Favor of a Vegan Diet, With 7-in-10 Planning to Stay Vegan Over Next Five Years. https://vegconomist.com/society/majority-gen-z-favor-vegan-diet/

Walters, P. (2021). Are Generation Z ethical consumers? In: Stylos, N., Rahimi, R., Okumus, B., & Williams, S. (eds) *Generation Z Marketing and Management in Tourism and Hospitality*. Palgrave Macmillan, Cham. https://doi.org/10.1007/978-3-030-70695-1_12

Welsh, J. A., Braun, H., Brown, N., Um, C., Ehret, K., Figueroa, J., & Boyd Barr, D. (2019). Production-related contaminants (pesticides, antibiotics and hormones) in organic and conventionally produced milk samples sold in the USA. *Public Health Nutrition*, 22(16), 2972–2980. https://doi.org/10.1017/S136898001900106X

Whitttaker, M. (2024). The Australian dairy farmers trying to make cows obsolete. Forbes Australia. https://www.forbes.com.au/covers/entrepreneurs/precision-fermentation-fake-milk-making-cows-obsolete

Worldwide Fund for Nature (WWF). (2025). Sustainable Agriculture: Dairy. https://www.worldwildlife.org/industries/dairy

Wunsch, N.-G. (2024). Consumers' reasons for drinking plant-based milk in the United Kingdom 2022, by gender. Statista. https://www.statista.com/statistics/1388294/consumers-reasons-for-drinking-plant-based-milk-in-the-united-kingdom/

year13. (2022). Gen Z's Animal Cruelty Awakening. https://www.youthsense.com.au/youth-insights/gen-zs-animal-cruelty-awakening/

Yehl, C. (2024). The "not milk" generation: How Gen Z prompted milk's rebrand. https://theblackandwhite.net/77677/feature/the-not-milk-generation-how-gen-z-prompted-milks-rebrand

Zechmann-Khreis, M. (2024). Lactose intolerance: Worldwide distribution. https://www.nahrungsmittel-intoleranz.com/en/lactose-intolerance-worldwide-distribution/

ZeroSmart (2020). What Kind of Milk is Most Sustainable? https://www.zerosmart.co.uk/post/what-kind-of-milk-is-most-sustainable

Zollman Thomas, O. & Bryant, C. (2021). Don't have a cow, man: Consumer acceptance of animal-free dairy products in five countries. *Frontiers in Sustainable Food Systems*, 5. https://doi.org/10.3389/fsufs.2021.678491

Zuberbier, T. (2016). Cow's Milk Allergy. https://www.ecarf.org/en/information-portal/allergies-overview/cows-milk-allergy

Zulkosky, C. (2023). Gen Z, veganism, and the future. The Food Institute. https://foodinstitute.com/focus/gen-z-vegans-and-the-future/

Chapter 9

Precision Fermentation: The Science behind Gen Z's Food Choices

"You cannot escape the responsibility of tomorrow by evading it today."

— Abraham Lincoln

"Science is not only a disciple of reason, but, also, one of romance and passion."

— Stephen Hawking

From climate change to food insecurity, Gen Z is not just living in turbulent times of urgent global challenges; they are being called to respond to them. Abraham Lincoln's timeless warning reminds us that this generation cannot afford to delay solutions. Its everyday food choices, more than a matter of taste, have become an act of responsibility. In this context, precision fermentation emerges as a scientific marvel and a symbol of how young consumers are attempting to meet the moral and ecological responsibilities of tomorrow, today.

Abraham Lincoln's words ring truer than ever for this generation, who increasingly understands that the choices these young people make — what they eat, wear and support, shape the world they will inherit. For many, food has become more than sustenance; it is a form of activism and active actions towards change. In this landscape, precision fermentation stands out as both a symbol and a solution, a new way to produce the foods we love without the environmental or ethical cost of traditional agriculture.

At first glance, this technology might sound clinical, i.e. using genetically programmed microbes to produce proteins like whey or casein, without animals. For Gen Z, however, it is anything but cold or distant. As Stephen Hawking beautifully observed, science is driven not just by logic, but by romance and passion. That is exactly how many young people approach innovations such as precision fermentation — not just with curiosity, but with an emotional investment in the kind of future it could help create. This is a generation that brings feelings into food technology. It cares deeply about climate as well as about identity, equity and trust. These young people are not passively consuming whatever shows up on shelves, they are questioning, researching, sharing and deciding what aligns with their values. For many, precision fermentation is exciting because it is smart or sustainable, but also because it aligns with their desire for bold, ethical and intelligent change.

In many ways, Gen Z is the ideal audience and co-creator of precision fermentation's unfolding story. This technology, which uses genetically programmed microorganisms including bacteria, yeast or fungi to produce specific functional ingredients and proteins such as casein, whey or egg whites without the use of animals, holds immense promise for sustainable food systems. Yet, its success will depend not only on technical efficacy, but on cultural acceptance, emotional resonance and narrative clarity.

In this chapter, we explore the evolving relationship between Gen Z and precision fermentation, unpacking its excitement as well as scepticism, and the social forces shaping both. We examine how Gen Z's curious nature and hunger for transparency, sustainability and innovation intersect with powerful emotional and psychological drivers. After briefly explaining what precision fermentation is, we look at how Gen Z, while hopeful, faces immense pressure to choose wisely, consume responsibly and lead transformation of a system this generation did not create. The young people are not just reacting to the current food system, their knowledge, expectations and psychological responses are rewriting it.

A Generation Curious by Nature

We have always been curious about how Gen Z perceives precision fermentation. Known for their digital fluency and environmental activism, Gen Z has become a driving force behind a new food culture — one that is both experimental and deeply values-driven. The rise of precision fermentation, a technology that uses microorganisms to produce ingredients which mimic traditional animal-based products, has not only captured scientific and commercial attention but has also resonated with this generation's ideals.

This chapter unpacks how Gen Z is engaging with precision fermentation, exploring the science, psychology and sociology behind their food choices. With 84% of Gen Z in the USA having tried at least one innovative food trend in the last year (Rand, 2025), their curiosity is unquestionable. The trends that attract their attention include plant-based alternatives, functional and fermented foods as well as mocktails and non-alcoholic beverages. Are these young people's curiosity strong enough to establish a persisting willingness to consume lab-made proteins? What lies beneath their interest in precision fermentation and what barriers remain in their journey toward adopting this transformative technology?

Before we tease out how Gen Z is shaping this particular transformation in the food production processes, we need to understand what precision fermentation actually entails. Why is this technology hailed as a breakthrough in sustainable food production? How does it work and why is it resonating so strongly with this generation in particular? In the next section, we outline the foundations of precision fermentation, setting the scene for a deeper analysis of its cultural and generational impacts.

What Is Precision Fermentation and Why It Matters

Although there is not yet a widely accepted international definition of what this technology is (Sturme *et al.*, 2025), precision fermentation is not a buzzword — it is a biotechnological advancement that enables the production of animal-free proteins, fats and enzymes by programming microorganisms to produce specific ingredients in a novel way (Bogueva and Danova, 2024). Microorganisms, such as yeast and fungi, are engineered to produce substances that have the same taste, texture and nutrition as the original animal-based products (CSIRO, n.d.). Fermentation used for millennia for food and beverages such as yogurt, kefir, vinegar, miso, sauerkraut, kimchi, bread and beer, is now employed to produce microbial protein or single-cell protein (Sturme *et al.*, 2025).

The precision fermentation process offers an appealing alternative to animal husbandry, especially in terms of environmental impact and animal welfare. This technology has been used widely since the 1980s to produce many high-value pharmaceuticals, including vaccines and insulin for the treatment of diabetes, vitamins for fortified foods, such as B2, B12, C and D2, many flavours, aromas and sweeteners as well as enzymes, e.g. the replacement of the animal-derived rennet with the fermentation-produced chymosin used in cheese (CSIRO, n.d.; Hinds, 2020). The costs associated with precision fermentation have also fallen drastically making the technology affordable, encouraging and accelerating its widespread use.

A report by Fortune Business Insights (2025) projects that the global precision fermentation market will surge from $3.03 billion in 2024 to $57.01 billion by 2032, with Europe accounting for nearly half of the market share. This projection points to a rapid growth and production maturation, but it also implies the emergence of a few dominant players shaping the industry and the market's direction. Whether Gen Z will be the driving force behind this expansion, however, remains an open question. We heard clear uncertainty and doubt from one Gen Zer:

> *"It sounds like something out of a sci-fi flick. Like, are we just swapping paddocks for petri dishes or another brewer? I get that it's better for the environment, but there's still a weirdness to it. Not sure if I'm ready to be knocking back lab-made milk or cheese every day, just feels a bit off... too synthetic... you know?"* (Male, born 1999)

This view reflects a broader sentiment within Gen Z, staying open-minded but cautious. While many in this generation are climate-aware and curious about future food solutions, emotional comfort and cultural familiarity still play a big role in their choices. The idea of replacing something as every-day and emotionally embedded as milk with a lab-engineered version can trigger hesitation, especially if it feels too far removed from nature or tradition. Such an indirect resistance is not necessarily a rejection. It is scepticism shaped by identity, trust and a desire for transparency. As we accent it across all chapters of this book, many Gen Zers want innovation that aligns with their values but does not compromise authenticity or comfort. If precision-fermented products feel "too synthetic" or disconnected from the idea of real food, adoption may stall.

It is easy to assume that Gen Z would naturally embrace precision fermentation; after all, they have grown up as digital natives, educated about climate change from an early age and are often open to bold innovations that promise a better future. The reality however may be more complex as it hinges on properly communicating and building consumer trust.

Many Gen Z consumers are aware of precision fermentation and its role in food innovation. A study found that 40% of US adults are open to trying foods made through precision fermentation, with Gen Z and Millennials leading the way (French, 2023). The Hartman Group's (2023) research suggests that younger consumers are generally more interested in food tech and are early adopters of new products. Some Gen Z consumers may not fully understand the scientific process behind precision fermentation, but they are enthusiastic about its potential benefits (Green Queen, 2025).

Beyond technological innovation, affordability and ethics, storytelling, sensory familiarity and clear communication will be key for the adoption of precision-fermentation products. Brands that can demystify the process, link it to emotional and cultural experiences and position it as a meaningful evolution, not a radical replacement, may stand a better chance of winning over this demographic cohort. The insight from the words of Gen Z ties back to the earlier point that this generation's relationship with food is about more than nutritional function. It is about how food fits into the identity, values and emotions of Gen Z. How much they know and what they can find about the foodstuff produced through precision fermentation will influence consumption. To win young consumers over, precision fermentation must not only promise a better future, it has to feel right in the present.

Gen Z's Awareness and Knowledge about Precision Fermentation

For Gen Z, curiosity about new foods is tempered by knowledge gaps and this frames their relationship with precision fermentation. Although without the full knowledge about the underlying science and technology, the young people are intrigued by the idea of animal-free dairy or meat; however, they may be unaware of the genetic engineering involved or the regulatory and cost hurdles (Green Queen, 2025; Sawano, 2024; Sturme *et al.*, 2025).

Gen Z's attitudes regarding precision fermentation are framed against a background of using food choices as a form of activism. A 2022 survey by Deloitte (Paoletti, 2022; Pacas, 2023) found that nearly half (i.e. 45%) of UK Gen Z consumers have stopped buying products from brands that do not align with their sustainability values. This generation is more likely than any other to embrace food technologies that assure ethical sourcing, climate resilience and transparency, but they also need to deliver on these promises.

Sawano (2024) and the report by Sartorius (2023) highlight the complexity of the precision fermentation process combined with the economic and political structures influencing its scalability and accessibility. Gen Z values transparency, yet ironically, the young people may not be getting the full picture from marketing campaigns that focus on novelty and ethics more than science. This highlights an important gap and food education must evolve to match the pace of innovation. Without clear communication from food technology companies and regulators, even this highly curious generation may struggle to distinguish between hype and substance.

Why Does Precision Fermentation Resonate or Not with Gen Z?

At first glance, precision fermentation may seem like a perfect fit for Gen Z — a generation raised on climate consciousness, digital fluency and a hunger for innovation. It speaks the language of sustainability, ethics and scientific ingenuity; these are values that seem to guide Gen Z's choices more than tradition ever could. As a technology, precision fermentation is sleek, future-forward and promises a gentler footprint on the planet. However, the situation is not that simple.

Through our conversations with Gen Z, what emerges is not blind enthusiasm but a thoughtful negotiation. For some, precision fermentation ignites a spark — a sense that we are finally transcending outdated systems and designing food with intention, intelligence and integrity. It aligns with the vision for a better world, one that does not sacrifice sentient animals, health or the earth for the sake of convenience and taste. For others, the promise of this technology does not land quite so effortlessly. There is hesitation, rooted in questions about cost, transparency, unnaturalness and control. In a world oversaturated with curated truths and corporate spin, some young consumers wonder: Who truly benefits from this new science? Is the technology completely safe? Can something made in a lab ever feel nourishing to the soul, not just the body?

Thus, the appeal of precision fermentation is not universal, nor is it static. It is filtered through the lens of lived experience, emotional instinct and collective memory. Gen Z does not just assess food by its label or source code; these young consumers measure it against their identity, their hopes and their fears. They want more than functionality, they want meaning. In this way, precision fermentation stands as both an invitation and a test. It asks not only whether we can reimagine the building blocks of our diet, but whether this reimagining feels right, ethically, emotionally and existentially, for a generation that sees food as far more than fuel.

Environmental impact is one of the key values in focus that plays a major role in Gen Z's acceptance of precision fermentation. At its core, this generation wants to know that its dietary choices matter, that the food on the plate does not come at the expense of the planet. Products created through precision fermentation are claimed to generally use less water, land and energy, and emit fewer greenhouse gases (French, 2023; Pacas, 2023). They are offering a fairer, nutritious and environmentally responsible alternative, making it a key player in shaping the future of food systems

(Ajayeoba and Ijabadeniyi, 2025). This aligns with Gen Z's strong climate consciousness and desire to support transformative, not performative, sustainability efforts (Pacas, 2023). These attributes appeal to a generation raised from an early age amid climate disruption and taught that the future hinges on present-day decisions. For many, embracing new food technologies becomes a form of ecological stewardship. Nevertheless, values alone do not make a product successful; it also requires the flavour and taste to be right.

Taste and texture are critical in determining Gen Z's willingness to adopt precision fermentation products. While they care deeply about sustainability, their taste buds still lead the conversation. In fact, many Gen Z consumers admit to prioritising flavour over environmental benefit when making food choices. Research showed that they are more likely to embrace new foods that deliver sensory satisfaction alongside sustainability (Advanced Biotech, 2023). For a lot of young Australians, precision fermentation might win points for ethics and sustainability, but if it does not hit the mark on taste, it is a hard pass. They are open to innovation and the science behind animal-free dairy and meat feels smart, scalable and aligned with their desire to fix what is broken in the food system:

> *"I reckon precision fermentation is the way forward if we want proper change. We can't keep smashing the planet just to get a cheese toastie, y'know?"* (Male, born 2004)

Gen Z tends to think about the bigger picture. However, even with all good promise, young people are not naive. They have grown up in a world of greenwashing, dodgy labels and corporate spin. Their bar for trust is high:

> *"At the end of the day, if it doesn't taste right, no one's going to stick with it. We're not just going to eat something 'cause it's good for the planet. It's gotta taste like the real deal, otherwise it's just another fad. We had enough of fake food and shady practices with deliberately mislabelled food."* (Female, born 2005)

This pessimism is not groundless. Gen Z has witnessed many cases with false claims — from "honey laundering", where nearly half the honey sold in Australian supermarkets was found to be filled in with sugar syrup but still sold as "100% pure" (Commonwealth Fraud Prevention Centre, 2021), to the free-range egg and meat scandal, where "free-range" labels masked factory-farmed conditions (Ranadheera, 2024). This generation has seen how marketing can deceive, and how ethics can be faked for profit. It wants better. For this young cohort, it is not a choice between flavour and values, it is about demanding both. Research backs this mindset showing that taste and texture remain top priorities in Gen Z's food decisions, no matter how planet-friendly the process is (Advanced Biotech, 2023). If precision fermentation is to win Gen Z's loyalty, it must deliver honestly, transparently and deliciously, or risk being left on the shelf with yesterday's false promises.

What excites Gen Z about precision fermentation is certainly its potential to deliver both: indulgence without guilt, pleasure without compromise (Advanced Biotech, 2023). This however needs to be delivered delicately and with balance. If the foods do not taste good, even the most ethical origin will not be enough. Despite sustainability aspirations, Gen Z wants to eat food that satisfies not just the mind and conscience, but also the senses (Advanced Biotech, 2023).

> *"Honestly, what pulls me in is the idea that you shouldn't have to pick between doing the right thing and enjoying what you eat. I don't want to feel guilty every time I grab a snack or put cheese on my pasta. If this tech can give us that sweet spot — tastes legit but doesn't cost the Earth or harm animals — then that's the kind of future I want to be part of. But it's got to be real. No greenwashing, no fake 'ethical' labels. We've seen too much of that already. We back brands that walk the talk, and if they mess that up, we'll move on. Our food choices aren't just meals — they're a reflection of who we are and what we stand for."* (Male, born 2003)

The ethical compass guiding this generation extends far beyond environmentalism. Ethical food choices are central to how Gen Z navigates the marketplace. The young people are drawn to brands that align with their broader worldview, those that champion animal welfare, climate responsibility and social justice (Green Queen, 2025). Their purchasing behaviour is an extension of their identity, and they expect transparency, integrity and action from the companies, brands and products they support. This alignment of values and curiosity is fuelling what was once seen as niche, it has become a mainstream cultural movement. With projections of 132 million US consumers using precision fermentation products by 2027 (Green Queen, 2025), this trend is no longer a niche. It is accelerating into the mainstream, potentially pushed forward by a generation that wants more from its food, more meaning, more intention and more innovation.

Despite these positives, not all Gen Z consumers are on board with precision fermentation. A UK Food Standards Agency review (FSA, 2025) found that while 52–68% of all consumers were willing to try precision fermentation dairy or egg products, only 17–35% were willing to regularly purchase or incorporate them into their diets. Concerns include perceived unnaturalness, safety and regulation, affordability, access and labelling confusion. Some Gen Z consumers are wary of genetically modified organisms and lab-grown foods, viewing them as "unnatural" or overly-processed (FSA, 2025). Consumers are unsure how to categorise precision-fermented foods: are they vegan, vegetarian or something else entirely? This ambiguity can create barriers to adoption. There is also uncertainty about the long-term health effects and regulatory oversight of precision fermentation products, especially given their novelty. These products are often priced at a premium, which may alienate younger consumers with limited disposable income (FSA, 2025). Addressing concerns around safety, transparency and affordability will be essential for broader acceptance. As the technology matures, an ongoing dialogue between producers, regulators and consumers will shape its future for the Gen Z's food landscape.

Importantly, companies are beginning to understand that Gen Z's engagement with sustainability is not merely performative. It is not a trend to be commodified, but a transformative lens through which they view the world. As All G Foods notes, this mindset gives brands permission to innovate boldly, to break from the past and co-create a future that aligns with the values and dreams of their most influential consumer base (Pacas, 2023).

If we look at this trend as a dance between science and society, Gen Z is not just content to be spectators, and it is not just following the rhythm. Gen Z is choreographing the next move. Precision fermentation, with all its promise and its challenges, offers young people a medium to explore with an evolving narrative, where responsibility meets experience, and where food becomes a statement of both identity and intent.

What Gen Z Does Not Know and Why It Matters

Gen Z gravitates toward hands-on, interactive learning — they want to experience information, not just absorb it (Chardonnens, 2025). When it comes to precision fermentation, even the appetite for innovation and engagement has not fully closed the knowledge gap needed to embrace new food technologies. Despite the enthusiasm for sustainability and food innovation, Gen Z is not immune to confusion, misinformation and blind spots in its understanding of precision fermentation. The range of knowledge gaps (see Table 9.1) may impact the young cohort's ability to make informed decisions, advocate effectively for what is right for them and drive meaningful change. Key areas where knowledge gaps persist are in relation to the technology itself, how it is regulated and the costs of the new foodstuffs as well as how it will be adopted, what disruptions are likely in traditional farming and how the associated food market will look.

The disconnect between Gen Z's values and their knowledge about precision fermentation could hinder broader acceptance if not addressed through engagement, education and participatory innovation. Universities and research institutions should integrate educational initiatives such as biotechnology and food science into the curriculum to improve scientific literacy among younger generations (Huss, 2023). Companies must provide clear, accessible information about precision fermentation, including its benefits, challenges and ethical considerations (Knychala *et al.*, 2024). Gen Z's activism could play a role in shaping regulatory frameworks, ensuring that precision fermentation aligns with sustainability goals while protecting small-scale farmers (EU CAP Network, 2024). Addressing cost and scale challenges through investment in infrastructure and

Table 9.1. Knowledge gaps related to precision fermentation.

Knowledge Gap	Why It Matters
Scientific and Technological Process	Gen Z often supports R&D outcomes related to food without fully understanding the methods used. Many are unaware of how microorganisms are engineered to produce proteins and other food ingredients through precision fermentation (Sawano, 2024). This lack of understanding can lead to misconceptions about safety and efficacy as well as to ethical concerns surrounding this new biotechnology method.
Regulatory Landscape	Strict regulations should govern precision fermentation, particularly regarding labelling, genetically modified organisms (GMOs) and consumer acceptance. Gen Z may underestimate how long approval processes can delay product rollout (Sawano, 2024). The European Food Safety Authority (EFSA) has highlighted challenges in novel food applications, causing uncertainty for innovators (FFE, 2023). In the US, the Food and Drug Administration (FDA) and Department of Agriculture (USDA) have complex approval pathways that can take years (KPMG, 2025). Although some precision fermentation products have been formally approved in Australia and New Zealand, the general perception is that the existing common regulatory framework in these countries is adequate to manage any new products (Gumulya *et al.*, 2025), which could raise some issues of concern and ambiguity among young people.
Cost and Scale	While Gen Z prioritises sustainability, a few grasp that for now precision fermentation processes remain costly and difficult to scale, making products not yet widely accessible (Sartorius, 2023). The technology is still in the early stages, with high production costs limiting accessibility. Reports indicate that scaling up fermentation facilities requires significant investment and cost reductions depend on advances in strain engineering and bioprocess optimisation (Watson, 2023; Lever, 2023).

Table 9.1. (*Continued*)

Knowledge Gap	Why It Matters
Adoption Delays	Despite its potential, precision fermentation faces market adoption challenges due to economic uncertainty and investment risks. Gen Z may not fully grasp the financial hurdles companies face in securing funding and scaling production (Sartorius, 2023). A survey found that while 40% of Americans are open to precision fermentation foods, regulatory and economic barriers slow adoption (French, 2023).
Disruption to Traditional Farming	Precision fermentation could disrupt traditional farming practices, including dairy and livestock industries. Concerns about small-scale producers' livelihoods are rarely considered (Sawano, 2024). Studies show that younger generations are less likely to return to family farms, raising concerns about rural economies but also offering new opportunities (Michigan News, 2023; Krymowski, 2021).
Market Consolidation	The possibility of corporate monopolies in biotech foods may conflict with Gen Z's values (Kühl *et al.*, 2024). While the young generation favours ethical consumption, it may not realise that despite many new players, the major food conglomerates are interested in dominating the precision fermentation space (Knychala *et al.*, 2024). Consolidation in the biotech food markets could limit competition and innovation.

Source: Authors.

technological advancements will be crucial for widespread adoption (Watson, 2023).

By bridging these knowledge gaps, Gen Z can become more-informed advocates for sustainable food innovation, ensuring that precision fermentation develops in a way that aligns with their values and priorities. Precision fermentation should also fit consumers' expectations and particularly Gen Z's profile for foods that offer healthier choices coupled with environmental and animal welfare benefits.

Understanding Gen Z's Psychological Profile

Understanding Gen Z's psychological profile reveals valuable insights into their interest and engagement with precision fermentation. This generation is not monolithic in its response to emerging food technologies. While many are excited by the innovation, others remain sceptical or uneasy. Their food choices are deeply intertwined with values, emotions and social influences, making precision fermentation either a compelling solution or a source of concern.

One major factor shaping Gen Z's choices is health consciousness. Many of these young consumers actively seek nutrient-rich foods that support and promote both physical and mental well-being (Food Insight, 2022). They actively seek products that enhance gut health, cognitive function and overall vitality (Jakubowska *et al.*, 2024). Precision fermentation enables the production of bioavailable nutrients and functional ingredients, and could align well with a health-conscious mindset. Research suggests that Gen Z is particularly drawn to alternative proteins and fermented foods due to the perceived benefits for digestion and immunity (FSA, 2025).

As one Gen Z individual with whom we had the opportunity to engage in a conversation shared:

> *"I love the idea of sustainability, but I don't want to eat something if I don't fully understand what's in it. I worry it's too processed, which automatically for me is equal to unhealthy, even if it's meant to be better for the planet."* (Female, born 2000)

Sustainability, however, remains a cornerstone of Gen Z's mindset. These young people often make intentional food decisions that reflect strong environmental values (Halicka *et al.*, 2025; Jakubowska *et al.*, 2024). Precision fermentation offers a low-impact alternative to traditional animal agriculture, producing dairy and protein without the ethical concerns related to factory farming (FSA, 2025). Empirical evidence indicates that Gen Z is more willing to adopt lab-grown and precision-fermented foods if they are marketed as environmentally friendly (French, 2023). This moral compass is central to how Gen Z views new food technologies. Nevertheless, this can clash with the needs for transparency, control and naturalness, which if not met, have psychological consequences of resentment and rejection. Another Gen Zer put it this way:

> *"If it's made in a lab, how can I trust it? I feel like we're being turned into experiments, and no one's asking if we're okay with that. Precision or not, it's not clear what exactly this new innovation will serve... our health, ... humanity's needs.... or someone's interests. I don't want to be sceptical, but before embracing something we need to know all the pitfalls it holds."* (Male, born 1997)

Other representatives of Gen Z cohort see food innovation, including precision fermentation, as part of the solution. Another young representative of this generation offers a different perspective:

> *"If precision fermentation means fewer animals suffer and we use fewer resources, I'm all for it. But it has to be transparent. I want to know who's behind it and how it works."* (Female, born 2005)

These seemingly different views highlight the generation's value-based curiosity. Young people want food that aligns with their ethics but also supports their emotional and physical needs; they do not want to be fooled by something that may create trouble for them and humanity in the long run.

Social media also play a pivotal role in shaping Gen Z's food preferences. Viral trends, influencer endorsements and digital storytelling significantly impact young people's purchasing decisions (Jakubowska *et al.*, 2024). One Gen Z consumer with whom we spoke, shared:

> *"I first heard about precision fermentation from a vegan chef I follow on TikTok. If a creator I trust explains how it's made and why it's better for the planet, I'll usually give it a go. But it's not just about the food — it's the story behind it. If I vibe with the brand and what they stand for, even if it's a bit weird or new, I'm way more likely to try it."* (Female, born 2000)

This sentiment highlights how platforms, such as TikTok and YouTube, and influencer-driven food trends act as gatekeepers to what is "in" and what is trusted (BritainThinks, 2019). Novel foods like precision-fermented dairy and other proteins can gain traction when framed as wellness-oriented or environmentally conscious, especially when endorsed by relatable social media voices.

Emotional eating is another defining characteristic of this group. Food serves as a tool for emotional regulation for many Gen Z consumers. Many associate foods with stress management, often turning to meals that enhance mood and energy (Food Insight, 2022). Young consumers seek foods that improve mood, boost energy levels and provide comfort during stressful times (Jakubowska *et al.*, 2024). Fermented foods with functional properties, such as probiotic benefits for gut health or mood regulation, appeal to this emotional and health-centred approach. Precision fermentation can enhance these benefits by producing customised bioactive compounds that support emotional health. A Gen Zer explains:

> *"Some weeks I'm just trying to get through the stress of uni. If food can help my gut or mood, I'm in. I've had anxiety since high school, and eating well is part of how I manage it. So, if precision fermentation can make that healthier, support my body, and still taste good, even if it's lab-made... why not?"* (Male, born 2001)

This reflection captures a generation willing to embrace innovation when it aligns with its emotional well-being and daily realities. Precision fermentation, with its ability to deliver tailored, bioactive benefits, resonates deeply, provided it also meets Gen Z's standards of taste, trust and meaning.

Despite young people's idealism, Gen Z is also pragmatic. Convenience, affordability and accessibility heavily influence what they consume. Ethical aspirations are frequently moderated by budget constraints, making price sensitivity a key adoption barrier for any food, including the one made by precision fermentation (Halicka *et al.*, 2025). The below perspective of one Gen Zer captures a core tension for many in his generation — their desire to support sustainable innovations like precision fermentation is real, but so are the constraints of everyday budgets:

> *"I reckon it's awesome what precision fermentation stands for... good for the planet, no animals involved, all that. But if it's seven bucks for a bottle of precision-fermented milk, I'm still gonna grab the cheap one. Wish I could always go for the ethical option, but yeah, price matters. Maybe if more of us bought it, the cost would come down and it'd be more of a normal thing, hey?"* (Male, born 1999)

For this generation, ethical ideals do not exist in a vacuum. They are filtered through financial realities, especially for students and young adults navigating casual jobs, study loads and the rising cost of living. While Gen Z often leads in values-based consumption, its choices are shaped by a pragmatic awareness of what is realistically within reach. This highlights a critical insight that affordability is not just a marketing strategy, it is an enabler of values-based consumption. Studies show that cost-effective production and mainstream availability are critical factors in Gen Z's willingness to embrace new food technologies (French, 2023). Brands that offer affordable, convenient and ethically sourced precision-fermented products are more likely to gain trust, loyalty and succeed in this market, not just because they align with Gen Z's nutritional or environmental standards and values, but because they fit into its life and financial constraints.

What emerges is a generation that is simultaneously cautious and courageous, one that views food not just as sustenance, but as identity, ideology and innovation. The psychological makeup of Gen Z reveals why precision fermentation must not only serve its purpose, taste good and be ethically sound, it must also feel right. It must speak to Gen Z's minds and hearts. Most importantly, it also needs to deliver the quality products it promises.

Balancing Promise and Perception: Gen Z Concerns about Precision Fermentation

While Gen Z is generally receptive to food innovation and sustainability-oriented technologies such as precision fermentation, their openness is tempered by a number of nuanced concerns and underlying tensions. Despite being heralded as a promising technology for the future of food, offering solutions for climate resilience, reduced resource consumption and animal welfare, precision fermentation remains subject to scepticism among Gen Z consumers, particularly due to issues surrounding safety, naturalness and corporate ethics.

A central worry is related to the perceived health and safety risks of consuming foods derived from precision fermentation. Since these products are relatively novel and often involve synthetic biology, some Gen Z consumers express unease about the long-term implications of their consumption (FSA, 2025). This reflects a broader generational anxiety around "unknowns" in food technology, especially when regulatory frameworks or longitudinal health studies are still in their infancy. As one Gen Zer commented:

> *"Look, I get that it sounds innovative and all, cutting-edge, even. But at the end of the day, I don't want to be the guinea pig for some lab-made food we barely understand yet and know the long-term effects of. It's all well and good to say it's sustainable, but if it messes with your health down the line, that's a hard 'no' from me. Let someone else be the crash test dummy first, thanks."* (Male, born 2005)

This sentiment illustrates a cautious optimism — curiosity tinged with a desire for more robust safety assurances. Additionally, the perception that precision-fermented foods are "unnatural" poses a substantial psychological barrier. Gen Z tends to favour food that is minimally processed, organic or described as "clean", and they are often highly attuned to ingredient transparency and sourcing. In this context, precision fermentation, despite its sustainability merits, can be viewed as overly-engineered or synthetic, too artificial, evoking a sense of detachment from traditional food systems (Kühl *et al.*, 2024). The excerpt below reflects an emotional and symbolic connection to food that challenges purely scientific definitions of health or sustainability:

> *"I get that it's sustainable, but if it's made in a lab, I just can't trust it the same way as something grown in the ground. And don't*

get me wrong — I reckon it's clever tech. But if it's whipped up in a lab, I just can't wrap my head around trusting it the same way I trust something that's come out of the soil. Give me something that's grown under the sun, not under a microscope, you know?" (Female, born 2002)

This unease is compounded by the fact that some precision fermentation techniques utilise genetically modified microorganisms (GMOs), which remain a polarising topic. Even though GMOs are scientifically validated as safe in many contexts, the term itself continues to carry negative connotations for a significant subset of Gen Z, especially those aligned with non-GMO, plant-based or organic food preferences (Kühl *et al.*, 2024).

Economic and ethical dimensions further complicate the narrative. While precision fermentation has the potential to decentralise production and reduce environmental impact, some Gen Z individuals worry that it could instead reinforce existing food system inequalities. There is a growing concern that powerful corporations might monopolise this emerging technology, marginalising small-scale farmers and aggravating economic disparities in the agricultural sector (Kühl *et al.*, 2024). For a generation that often values equity, social justice and grassroots movements, the prospect of corporate control raises red flags about who benefits from innovation and who might be left behind.

Another layer of resistance stems from a lack of awareness and understanding. While Gen Z is known for being digitally savvy and engaged with innovation, this does not automatically translate into deep familiarity with novel food technologies. As Green Queen (2025) notes, despite their curiosity, many young consumers remain unsure about what precision fermentation involves, how it differs from traditional fermentation and why it might be preferable in terms of sustainability or nutrition. This knowledge gap contributes to hesitancy, suggesting that mere availability of such products is not enough. Effective communication, education and labelling transparency are crucial to build trust.

Comparison with Older Generations

Understanding how different generations respond to food innovation helps explain the unique position Gen Z occupies in shaping the future of food. While each cohort brings its own values and preferences to the table, it is the contrast between tradition and transformation that makes these generational divides very revealing.

Older generations, particularly Baby Boomers and Generation X, tend to favour traditional diets rooted in familiarity and long-standing health beliefs. According to the UK Food Standards Agency (FSA, 2020), these older cohorts prioritise food quality, safety and origin, often placing trust in conventional farming and recognisable ingredients. Sustainability matters, but it is often interpreted through the lens of personal health and food security rather than systemic transformation. A 2024 study by Makowska *et al.* (2024) found that Baby Boomers were the most likely to check expiration dates and buy only what they needed, reflecting a frugal and responsible consumption ethic. They also placed the highest importance on ecological certifications but were less open to unfamiliar food technologies. This aligns with broader findings that older consumers are more risk-averse and less likely to adopt novel food innovations such as lab-grown or precision-fermented products (Zeren *et al.*, 2023).

Millennials (or Generation Y) share Gen Z's concern for sustainability but are less digitally immersed in their food decisions. While they helped popularise organic, plant-based and locally sourced foods, their choices are often shaped by lifestyle trends and wellness culture rather than activism. Marinova and Bogueva (2022) note that Millennials are more likely to adopt sustainable diets for personal health benefits than for environmental or ethical reasons. They are also more sceptical of lab-grown or genetically engineered foods, preferring "natural" alternatives, even when those come with a higher environmental cost. A study found that Millennials are more likely than Gen Z to share food experiences online, but their motivations are often rooted in aesthetics and personal branding rather than systemic change (Poyoi *et al.*, 2024).

Although these generational differences can be extended beyond precision fermentation to all novel foods, what is clear is that Gen Z has a new approach to its dietary choices. These young people are cautiously adventurous and inquisitive as to what food can do for them personally as well as for the planet. In fact, they are a cultural catalyst when it comes to acceptance or rejections of new foods.

Gen Z as a Cultural Catalyst

Gen Z is rewriting the rules. Raised in a digital-first world and shaped by climate anxiety, social justice movements, global crises and a health pandemic, this generation approaches food as a tool for change. These young people are more experimental, increasingly value-driven and curiously open to food technologies like precision fermentation and cultivated meat (Halicka *et al.*, 2025). However, unlike the generations before them, Gen Z's food choices are deeply intertwined with identity and activism. Young people expect brands to be transparent, inclusive and accountable. According to Ajayeoba and Ijabadeniyi (2025), Gen Z sees precision fermentation not just as a scientific innovation, but as a solution to systemic issues related to food insecurity, climate change and animal welfare.

As highlighted by Knychala *et al.* (2024), Gen Z is particularly drawn to precision fermentation's ability to replicate the taste and texture of animal products without the ethical or environmental baggage. Platforms like GenV (2024) are helping demystify these technologies, showing how precision fermentation can produce milk, meat and eggs without animals — an idea that resonates with Gen Z's demand for innovation with purpose and animal welfare. Gen Z's food choices reflect their values, digital engagement and evolving health priorities, making them a key driver of food innovation. As Makowska *et al.* (2024) argue, Gen Z's openness to change is not just about novelty. It is about aligning consumption with conscience.

In sum, although Gen Z exhibits a forward-thinking orientation that aligns well with the goals of precision fermentation, its concerns cannot be overlooked or dismissed as technophobia. These apprehensions reveal broader themes in Gen Z's consumer behaviour: a desire for authenticity, scepticism toward corporate motives, a demand for health and environmental accountability, and an expectation for clear, digestible information. Addressing these concerns through inclusive dialogue, rigorous safety standards, transparent labelling and equitable innovation pathways could significantly improve the generational acceptance and integration of precision fermentation in future food systems.

References

Advanced Biotech. (2023). Gen Z is Leading the Demand for Precision Fermentation. https://adv-bio.com/gen-z-is-leading-the-demand-for-precision-fermentation/

Ajayeoba, T. A. & Ijabadeniyi, O. A. (2025). Transforming Food for the Future: Precision Fermentation as a Key to Sustainability, Nutrition, and Health. *SSRN*. http://dx.doi.org/10.2139/ssrn.5119686

Bogueva, D. & Danova, S. (2024). Comparing precision fermentation and traditional fermentations: Consumer views. In Bogueva, D. (ed). *Consumers Perceptions and Food* (pp. 563–588). Springer Nature, Singapore.

BritainThinks (2019). Future Consumer: Food and Generation Z. Food Standard Agency. https://www.food.gov.uk/sites/default/files/media/document/fsa191102bt-gen-z_rea_full-report_081019-final.pdf

Chardonnens, S. (2025). Adapting educational practices for Generation Z: Integrating metacognitive strategies and artificial intelligence. *Frontiers in Education*, 10, 1504726. https://doi.org/10.3389/feduc.2025.1504726

Commonwealth Fraud Prevention Centre. (2021). Eyes bigger than their stomach: Deceit in the Australian food market, https://www.counterfraud.gov.au/news/general-news/eyes-bigger-their-stomach-deceit-australian-food-market

Commonwealth Scientific and Industrial Research Organisation (CSIRO). (n.d.). Reimagining food using fermentation. https://www.csiro.au/en/about/challenges-missions/future-protein-mission/novel-protein-production-systems/precision-fermentation

EU's Common Agricultural Policy (EU CAP) Network. (2024). Thematic Group on Gen Z: Leading Generational Renewal in Farming. TG Generational Renewal: Background Paper. https://eu-cap-network.ec.europa.eu/sites/default/files/2024-10/eu-cap-network-tg-genren-background-paper.pdf

Food Fermentation Europe (FFE). (2023). Precision Fermentation: Practices, challenges and suggestions for the novel food framework. EFSA ad-hoc stakeholder meeting, Parma. https://www.efsa.europa.eu/sites/default/files/2023-11/4-food-fermentation-europe-presentation.pdf

Food Insight (2022). 2022 Food and Health Survey Spotlight: Generation Z. https://foodinsight.org/spotlight-generation-z/

Food Standards Agency (FSA). (2020). The Future Consumer — Food and Generation Z. Research Report. https://www.food.gov.uk/sites/default/files/media/document/generation-z-full-report-final.pdf

Food Standards Agency (FSA). (2025). A rapid evidence review on consumer responses to precision fermentation. *FSA Research and Evidence*, 1–55. https://science.food.gov.uk/article/136898

Fortune Business Insights. (2025). Food Processing & Processed Food: Precision Fermentation Market. https://www.fortunebusinessinsights.com/precision-fermentation-market-109824

French, R. (2023). Millennials, Gen Z pave the way for foods made with precision fermentation. https://www.supplysidefbj.com/market-trends-analysis/millennials-gen-z-pave-the-way-for-foods-made-with-precision-fermentation

GenV. (2024). Meat, Milk, and Eggs from Precision Fermentation. https://genv.org/meat-milk-and-eggs-from-precision-fermentation/

Green Queen (2025). 77% of Consumers Will Try Precision Fermentation Once They Understand the Benefits, Survey Finds. https://www.greenqueen.com.hk/precision-fermentation-survey-hartman-cargill-perfect-day/

Gumulya, Y., Peng, H., Ebert, B. E., Johnson, H., van der Pols, J. C., Bansal, N.... Turner, M. S. (2025). Advancing Australia's food future: Opportunities and challenges in precision fermentation. *Future Foods*, 11, 100630. https://doi.org/10.1016/j.fufo.2025.100630

Halicka, E., Kaczorowska, J., Rejman, K., & Plichta, M. (2025). Investigating the consumer choices of Gen Z: A sustainable food system perspective. *Nutrients*, 17(3), 591. https://doi.org/10.3390/nu17030591

Hartman Group (2023). Fermenting the Future: The Growing Opportunity for Products Made with Precision Fermentation. https://www.hartman-group.com/documents/2112785198/precision-fermentation

Hinds, T. (2020). Precision fermentation is nothing new and it's here to stay. https://www.rethinkx.com/blog/precision-fermentation-is-nothing-new-and-its-here-to-stay

Huss, J. A. (2023). Gen Z students are filling our online classrooms: Do our teaching methods need a reboot? *InSight: A Journal of Scholarly Teaching*, 18, 101–112. https://files.eric.ed.gov/fulltext/EJ1407503.pdf

Jakubowska, D., Dąbrowska, A. Z., Pachołek, B., & Sady, S. (2024). Behavioural intention to purchase sustainable food: Generation Z's perspective. *Sustainability*, 16(17), 7284. https://doi.org/10.3390/su16177284

Knychala, M. M., Boing, L. A., Ienczak, J. L., Trichez, D., & Stambuk, B. U. (2024). Precision fermentation as an alternative to animal protein, a review. *Fermentation*, 10(6), 315. https://doi.org/10.3390/fermentation10060315

KPMG (2025). Ten Key Regulatory Challenges of 2025, Rolling through the Shift. https://kpmg.com/kpmg-us/content/dam/kpmg/pdf/2024/ten-key-regulatory-challenges-of-2025.pdf

Krymowski, J. (2021). Generation Z is the next driving force in food and ag. Ag Daily. https://www.agdaily.com/insights/generation-z-is-the-next-driving-force-in-food-and-ag

Kühl, S., Schäfer, A., Kircher, C., & Mehlhose, C. (2024). Beyond the cow: Consumer perceptions and information impact on acceptance of precision fermentation-produced cheese in Germany. *Future Foods*, 10, 100411. https://doi.org/10.1016/j.fufo.2024.100411

Lever (2023). Lever VC Insights: Scale-up Costs of Precision Fermentation. https://www.levervc.com/wp-content/uploads/2023/07/Lever-VC-Insights-Scale-up-Costs-of-Precision-Fermentation.pdf

Makowska, M., Boguszewski, R., & Hrehorowicz, A. (2024). Generational differences in food choices and consumer behaviours in the context of sustainable development. *Foods*, 13(4), 521. https://doi.org/10.3390/foods13040521

Marinova, D. & Bogueva, D. (2022). *Food in a Planetary Emergency.* Springer, Singapore.

Michigan News (2023). Are you ready for Gen Z farming. https://www.michiganfarmnews.com/are-you-ready-for-gen-z-farmers-recent-study-shows-perception-gaps-between-generations

Pacas, J. (2023). Inspiring a Lifestyle Without Limits for Gen Z: Fermenting with Precision. All G Food. https://www.allg.com/blog/inspiring-a-lifestyle-without-limits-for-gen-z-fermenting-with-precision

Paoletti, J. (2022). Gen Z And Environmental Issues: How to Earn Young Consumers' Trust. Forbes. https://www.forbes.com/councils/forbescommunicationscouncil/2022/06/01/gen-z-and-environmental-issues-how-to-earn-young-consumers-trust/

Poyoi, P., Gassiot-Melian, A., & Coromina, L. (2024). Generation Z and Millennials' food-sharing behaviour: A cross-generational analysis of motivations, satisfaction and behavioural intention. *British Food Journal*, 126(13), 207–225. https://doi.org/10.1108/BFJ-10-2023-0899

Ranadheera, S. (2024). From 'honey laundering' to fake free-range: food fraud costs billions. Pursuit, University of Melbourne. https://pursuit.unimelb.edu.au/articles/from-honey-laundering-to-fake-free-range-food-fraud-costs-billions

Rand, S. (2025). Gen Z food trends: What today's young US consumers want. Attest. https://www.askattest.com/blog/research/gen-z-food-trends

Sartorius. (2023). Is Precision Fermentation the Future of Food? https://www.sartorius.com/en/knowledge/science-snippets/precision-fermentation-the-future-is-almost-now-1464778

Sawano, T. (2024). Precision fermentation the next-generation food production technology — industry collaboration accelerating market development. Mitsui & Co. Global Strategic Studies Institute Monthly Report https://www.mitsui.com/mgssi/en/report/detail/__icsFiles/afieldfile/2024/04/15/2402t_sawano_e_1.pdf

Sturme, M., van der Berg, J. P., & Kleter, G. (2025). Precision fermentation — With a focus on food safety. FAO, Rome. https://doi.org/10.4060/cd4448en

Watson, E. (2023). Scale can transform the unit economics of precision fermentation, but it's 'not the most powerful lever,' says Synonym. AFN. https://agfundernews.com/scale-can-transform-the-unit-economics-of-precision-fermentation-but-its-not-the-most-powerful-lever-says-synonym

Zeren, H. Z., Bali, E. B., & Demir, H. (2023). Nutritional habits comparison of the baby boomer, X, Y, and Z generations located at a private college in Muğla, Türkiye. *Czech Journal of Food Science*, 41(6), 436–445. https://doi.org/10.17221/105/2023-CJFS

Chapter 10

Personalised Nutrition: Gen Z's Search for Tailored Wellness

"The food you eat can either be the safest and most powerful form of medicine or the slowest form of poison."

— Ann Wigmore

These words by Ann Wigmore, a Lithuanian-American holistic health practitioner, naturopath and raw food advocate, serve as a powerful reminder that food is more than just sustenance. It has to be approached with caution and understanding. In fact, it is a fundamental driver and a key determinant of long-term health and well-being. Observing and having conversations with Gen Z, it has become clear that these young people do not just eat for survival. They eat with intention. Their eating is with a purpose, viewing food as an essential tool for optimising their health and enhancing their daily lives. Every dietary choice they make is deliberate, reflecting not only their personal health needs but also their broader values. Each of them carves out their own unique path to their relationship with food and well-being.

With half of Gen Z reporting mental health problems (see Chapter 2), it is not surprising that personal well-being is high on the agenda for these young people. Although many follow particular diets ranging from plant-rich to low-carb (see Rand, 2025 for the dietary choices of US Gen Z), this generational cohort expects solutions that are personalised to its needs, expectations and values to be able to achieve optimal health. To gain a deeper understanding of how Gen Z approaches nutrition and wellness, we spoke with a few young individuals from Sydney, Australia and these conversations are captured in this chapter.

We start with defining the meaning of personalised nutrition before explaining Gen Z's holistic approach to wellness. The role of technology, including omics, in personalising nutrition is then discussed. Other important approaches we cover relate to adaptogens, superfoods and functional foods. Finally, we comment on the challenges and future of personalised nutrition.

Introducing Personalised Nutrition

Personalised nutrition is an approach that creates customised dietary recommendations for individuals by considering factors such as genetics, eating patterns, health status and other personal traits. Gibney *et al.* (2016, p. 237) define it as a strategy that "assists individuals in achieving a lasting dietary behaviour change that is beneficial for health". Personalised nutrition is believed to hold the potential to transform the way nutrition-related health issues are managed, with the possibility of preventing or even resolving them through precise, individualised dietary interventions (Ferrario *et al.*, 2021). Unlike broad, one-size-fits-all nutritional guidelines, this approach adapts to each person's unique needs and preferences, offering tailored guidance for better health outcomes based on understanding food, genes and individual physical traits.

When testing the benefits of personalised nutrition, the Food4Me Study funded by the European Union, the largest randomised controlled trial on this topic, found that tailored dietary advice led to healthier eating and lifestyle changes compared to a one-size-fits-all approach (Celis-Morales *et al.*, 2017). The study, which involved over 1,600 adults across seven European countries, showed that personalisation based on current diet analysis was most effective, while adding genetic or phenotypic data provided no extra benefit (Celis-Morales *et al.*, 2015a). Conducted as an internet-based intervention, Food4Me demonstrated the potential for scalable, cost-effective solutions to improve public health through personalised nutrition (Celis-Morales *et al.*, 2015b; Livingstone *et al.*, 2016).

A systematic literature review of 11 studies of personalised nutrition from USA, Canada and Europe (Jinnette *et al.*, 2021) shows that dietary intake is improved with personalised advice. The levels of personalisation vary from advice based on general lifestyle, preferences, culture and dietary patterns, to the use of observable characteristics (phenotypes) and biomarkers (e.g. blood pressure and cholesterol) which can be measured, to omics data (metabolomics, genetics, transcriptomics, proteomics and metagenomics) and a systems approach combining all of the above (Qina, 2025). There are different ways of delivering the personalised nutrition services with some of the most common ones being the use of devices, including wearables, coaching and intake of supplements, recipes, meal plans and meal kits, diagnostic tests and telehealth (Qina, 2025).

Why Personalised Nutrition Matters to Gen Z

Gen Z is a special cohort that is redefining wellness with a strong emphasis on personal health and nutrition. Raised in a digital world, this generation values individuality, technology-driven solutions and proactive health management. Unlike previous generations, these young people seek tailored dietary experiences that align with their values, including sustainability, mental well-being, ethnic and ethical food choices. Increased awareness of any food-related problems, such as food intolerances, allergies and gut health, and the connection between diet and mental well-being make personalised nutrition a powerful tool for optimising Gen Z's health based on its unique needs, preferences and lifestyles.

A section of Gen Z, estimated at 17% (Qina, 2025), is already using personalised nutrition, the main reason being looking for ways to reach their health goals. They also seek foods, drinks and alternative medicines to help them deal with stress and maintain good brain, eye and gut health (Daily, 2025). This generally comes at a cost but Gen Z seems to be well-aware of the benefits related to using nutrition for good health.

Gen Z's Holistic Approach to Wellness

Gen Z is considered the most anxious generation (Haidt, 2024). Haidt (2024) explains this with the extensive use of digital devices and soaring screen time which have replaced free play and physical contacts, including with nature. The tech-savvy Gen Z however has grown up with having access to information at its fingertips. Unlike previous generations, these young people are increasingly health literate and view health as one of the most important aspects of life.

This generation approaches health holistically, recognising the interconnection between fitness, nutrition and well-being. They prioritise comprehensive wellness, integrating physical, mental and emotional health. Their approach to nutrition is deeply connected to their overall wellness, focusing on balanced diets that support active lifestyles and mental clarity. As shared by one Gen Z individual, wellness is considered incredibly important and personalised nutrition has a role to play:

> *"You know, it's crazy how wellness has become such a big deal for our generation. It's not just about looking good anymore; it's about feeling good — physically, mentally and emotionally. This is like being happy and healthy. This is why we need a personalised food... nutrients"* (Male, born 1997)

Gen Z is growing up in an era of high stress, depression, burnout, the COVID-19 pandemic and chronic health conditions. These issues plague people of all ages due to internal and external stressors such as economic problems, personal or social issues and work-life balance (Koutsimani *et al.*, 2019). Climate change and the geopolitical tensions put further strain on this generation. Such an environment makes Gen Z more aware of the need to prioritise their own health to avoid serious consequences later in life. Mental health has become a significant part of mainstream wellness for Gen Z. They now engage in various forms of therapy and personalised approaches, using apps to track progress, and incorporating tailored nutrition and supplements to boost brain health. This generation values finding what specifically works for them individually, reflecting a proactive and individualised attitude towards overall wellness. As one of the Gen Z we spoke with explained:

> *"I love that mental health is becoming a big part of mainstream wellness now... a big part of what foods are good for you. It used to be something more emerging, but now it's really a priority for so many people, especially young people our age. It's not just about 'hitting the gym' anymore. People are openly talking about the different types of therapy they do to stay on top of their personal wellbeing. Whether it's talk therapy, art therapy or even just journaling, or the foodie thing.... And it's not just that. People are using multiple apps to track their progress, eating foods that work for them, drinking things like matcha and healthy smoothies, and some even take supplements to boost brain health preventatively... as they have relatives with dementia or Parkinson or whatever scary autoimmune thing... It's not just food, nutrition. Personalisation is huge these days. Everyone's figuring out what works for them specifically, which is awesome."* (Female, born 2005)

This young cohort is also dealing with negative emotions, such as stress, anxiety and sadness, often provoked by concerns imposed on them by previous generations, including climate change uncertainties (Bardales Salguero *et al.*, 2024). Such negative emotions can lead to other consequences. They can include unhealthy eating habits, such as emotional and binge eating, as well as consuming calorie-dense, nutrient-poor foods (Abdoli *et al.*, 2023). One of the Gen Zers explains:

> *"Yes! Stress and mood on top of food are a huge deal. And honestly, meditation is massive right now.... for a reason. With everything happening in the world, stress levels are through the roof. Me personally... mindfulness apps have helped me a lot, but sometimes even that doesn't feel like enough. For some people I know, quick breathing exercise or guided meditation can completely reset their day, but not everyone finds that helpful. I often comfort myself with food... You know what I mean... just like with food.... there's no one-size-fits-all solution. We all need different things to manage stress and feel our best. I don't want to say it, but... I often eat randomly and uncontrollably for comfort... Many of my friends or people I know are doing the same."* (Female, born 2000)

Despite their higher levels of understanding of healthy living and the interest in personalised nutrition, emotional, comforting and binge eating is quite popular among Gen Z, predominantly due to their personalities and lifestyles (Priporas *et al.*, 2022; Singh and Gupta, 2024). The eating behaviours of Gen Z are influenced by their unique characteristics, and various studies highlight that stress impacts eating behaviour through both physiological and biological responses (Rogers, 1999). Gen Z, in particular, tends to favour highly palatable foods and snacks, leading to calorie overconsumption driven by hedonic mechanisms (Newman *et al.*, 2007). The links between stress and eating behaviour are well-documented in psychology, with eating frequently used as a coping strategy for emotions (Wardle *et al.*, 2000). Emotional eating often serves as a coping mechanism, and personality traits like perfectionism can mediate the relationship between stress and eating behaviour. Moreover, research indicates that chronic stress and an unhealthy diet play a major role in the early development of lifestyle diseases, such as diabetes, obesity and cardiovascular disorders (Huston, 2022; Schweren *et al.*, 2021). Gen Z is aware of the pressing need to change habits to thrive in the future. As explained by one Gen Z representative:

> *"We've seen firsthand the consequences of neglecting health with our grandparents and now parents. Stress, burnout... chronic health conditions, poor diet are everywhere. We kind of know that if we don't prioritise our health now, we could end up like our parents and grandparents... we definitely could face serious consequences later — both physically and mentally."* (Male, born 2002)

Unlike previous generations, Gen Z is more inclined toward preventive health measures as a strategy for disease avoidance. This perspective is supported by research showing that Gen Z is taking proactive steps to improve their well-being. In a study in Canada and USA, 37% of Gen Z reported getting more sleep, 35% increased their exercise and 27% adjusted their diet for better nutrition (Innova Market Insights, 2023). Moreover, Gen Z is more likely than other generations to use food and beverages as tools for managing health conditions (26%) and weight loss (25%) (Innova Market Insights,

2023). Gen Z's prioritisation of health is not just about immediate well-being but also about long-term preventive measures. As one Gen Zer explained:

> *"... I think we've just realised that everything is connected. Sleep affects our mental health, diet affects our energy, and fitness impacts our mood. We're looking at wellness holistically, something that hit different (e.g. meaning instead of being the same, is better), something that previous generations, our parents and their parents never really thought of. It's sad in a way that it's happening. It's sad seeing our parents and grandparents neglecting their health... and you know we will be the ones to take the burden of looking after them. It's really sick to see how they are killing themselves... still eating big steaks, bacon and sausages, the same way they raised us. I've started pushing my family to be mindful of their food choices. Personalised nutrition is something that will become more and more important for our generation. I can already see it happening."* (Male, born 2000)

According to a survey of 5,000 respondents from China, UK and USA, Gen Z particularly views the future of health and wellness as centred on maintaining overall well-being (McKinsey and Company, 2024b). The belief in tailored nutritional plans is underscored by their potential to lead to tangible improvements in health outcomes. Gen Z's approach to wellness goes beyond physical health to involve mental and emotional well-being, highlighting their comprehensive view of food and health. Young people are purchasing more wellness services and products than older generations (McKinsey and Company, 2024b). They however can feel overwhelmed by the information and dietary recommendations available online:

> *"Food is a very complicated thing. You can't even imagine how complicated it is.... You know.... very often I'm so confused about what to eat. You could definitely keep scrolling through endless diet trends online. It's giving me brain-rot* [author note: meaning mental exhaustion because of too much screen time and information overload]. *One day, it's about high protein, the next it's about... bulking... then about... plant-based.... There are so many messages out there. How am I supposed to know what's*

> *actually good for me? What is good for our health now and also for our future health...*" (Male, born 1998)

Gen Z's commitment to holistic wellness is evident, not just for themselves but also for their entire family. By actively encouraging their parents and relatives to adopt healthier dietary habits, Gen Z recognises the importance of influencing their loved ones to make better food choices. This is especially crucial when facing challenges related to weight and physical health:

> *"I want to influence my mum and dad to change the way they eat... as they constantly struggle with their weight, look and related physical inability to move in a way I want them to be. That's exactly where personalised nutrition comes into play. Everyone's dietary needs are different. Everyone's body is different, so why should we all follow the same generic dietary advice? It's about tailoring what we eat to what actually works for us and only for us — our bodies, our goals and even our lifestyles."* (Female, born 2003)

Because everyone's body and dietary needs are unique, this young cohort advocates for personalised nutrition plans tailored to individual needs, goals and lifestyles. This perspective acknowledges the limitations of generic dietary advice and promotes a more customised approach to health and wellness. As another person from this generation suggests:

> *"If we can fine-tune our nutrition to fit our specific needs, we might actually start seeing real progress — whether that's in performance, recovery or just feeling better overall. It's not some magic fix for everything, but honestly, it could be a game-changer for a lot of people."* (Female, born 2001)

Gen Z and Striking the Balance with Personalised Nutrition

Gen Z is at the forefront of wellness trends, embracing highly personalised approaches to nutrition, health optimisation and lifestyle choices (McKinsey and Company, 2024b). Their focus has shifted towards tailored dietary solutions, integrating technology and data-driven insights to enhance personal well-being. However, Gen Z's approach to nutrition is marked by curiosity and caution as one of them explains:

> *"You know, supplements have somehow become a key component of my friends' and my personalised nutrition tactics. Have you noticed how pharmaceutical products that were once prescription-based are now shifting into everyday wellness sections in pharmacies? You can buy them almost anywhere now. I'm not sure if that's entirely good... It does open a niche for tailored wellness solutions, but too much freedom with supplements and self-medication might lead to new problems... can bring other troubles."* (Female, born 2000)

According to this young individual, the deregulation of pharmaceutical products and their increasing availability in mainstream wellness spaces offer the potential for personalised solutions. However, this person's cautionary note about the risks of excessive accessibility is both timely and important. Without proper oversight, the growing ease of obtaining supplements and over-the-counter pharmaceuticals may inadvertently encourage self-medication, leading to unforeseen health complications. This observation stresses a delicate balance in the realm of personalised nutrition. On one hand, the wide availability of supplements empowers individuals to customise their health strategies, addressing their unique needs; on the other hand, this freedom brings risks, particularly when individuals lack the knowledge or professional guidance to use such products responsibly. The shift of pharmaceutical products from prescription-only to the everyday wellness shelves reflects a burgeoning demand for personalised wellness solutions but also highlights the need for robust education and informed decision-making.

Personalised nutrition, by its very nature, thrives on tailored approaches to meet individual requirements. Supplements often serve as the foundation of these strategies, helping to bridge nutritional gaps or address specific health concerns. However, in the absence of professional consultation (e.g. with dietitians, healthcare providers or proponents of emerging technology-based solutions) misuse can result in unintended side effects or adverse interactions with other treatments. The importance of integrating expert guidance into personalised nutrition cannot be overstated. By addressing the risks associated with unrestricted access, we can promote a more thoughtful and responsible approach. This ensures that the benefits of personalised nutrition are fully realised without compromising health or safety. Striking this balance seems to be vital for Gen Z in fostering an era of wellness that is as accessible as it is informed:

> *"Totally! Things like sleep aids, gut health probiotic, and nootropics... you name it... are becoming more accessible. It's like today's medical advancements are tomorrow's over-the-counter essentials. I get what you mean about potential risks, though. It's great to have options, but we don't need to use them unless they genuinely serve a purpose. If our nutrition is truly personalised and meets our actual needs, we won't need all these extra supplements and medications."* (Male, born 1996)

This statement offers a thoughtful take on the balance between the accessibility of modern wellness products and the principles of responsible consumption. By recognising the growing opportunity of access to items like sleep aids, probiotics and nootropics, this Gen Zer acknowledges how advancements in medicine and wellness have opened doors to new, convenient options for enhancing health. However, there is still a caution about relying on these solutions without a genuine need, which reflects a deeper understanding of wellness that prioritises personalised approaches over generalised, one-size-fits-all remedies.

This young person's point gives emphasis to a key principle of personalised nutrition: if tailored properly to meet an individual's specific needs, there should be less dependency on external supplements and medications. This perspective is a reminder that while accessibility is beneficial, overuse or misuse can detract from achieving holistic well-being. It also promotes the idea that personalisation, when thoughtfully executed, can ensure that health strategies are both effective and viable. Ultimately, this perspective aligns with a growing sentiment among Gen Zers in the wellness space that the future of health lies not in having endless options available but in understanding which ones truly matter for each unique individual. It strikes a compelling balance between embracing innovation and encouraging critical decision-making about health practices. Gen Z's needs are real and they want them to be met by the latest innovations and technological solutions:

> *"And the problem is that the demand is real. We don't just want to follow trends; we want solutions that match our fast-paced lives. Subscription-based food services are booming — customised meal plans, adaptogenic drinks and functional foods designed for energy, focus or relaxation. They're packed with so many ingredients, some of which we might not even need. But honestly... I think it's cool. It brings a tech-driven edge to food, and that's exciting."* (Female, born 1998)

Gen Z is redefining wellness with a hyper-personalised approach to nutrition, but not without some healthy scepticism. While they embrace the convenience of supplements and functional foods, they also question the growing ease of access to products that were once prescription-only. There is appreciation for the convenience and innovation in nutrition solutions, but concerns arise about potential risks linked to self-medication and unnecessary consumption. Is this newfound freedom empowering or does it open the door to overuse and unintended health risks?

Demand for tailored nutrition is undeniable. From subscription-based meal plans to adaptogenic drinks promising energy and focus, food is becoming more tech-driven than ever. The excitement is real, but so is the awareness — just because something is marketed as essential does not mean it actually is. Gen Z is not just following trends; they are looking for a balance and seeking solutions that fit their fast-paced lives, endorsing innovation with a mindful approach to what young people truly need.

Digital Influence and Technology Shaping Gen Z's Personalised Nutrition

Gen Z's relationship with food is deeply interconnected with their digital habits. While food has always been an essential part of cultural identity, for Gen Z it has become a symbol of empowerment. They view their food choices as an extension of their broader identity and commitment to a healthier, more sustainable future. Like everything in their life, their understanding of nutrition is also shaped by the accessibility of digital information and open online discussions, enabling them to define well-being in a highly personalised and inclusive manner (BetterLYF, 2024). They actively seek out brands, food and ingredients that align with these values, making food a tool for both personal and social transformation. This generation prioritises self-actualisation and individuality, moving away from fixed societal expectations. They recognise systemic health issues and advocate for wellness solutions that foster diversity, accessibility and genuine connections.

As pointed out earlier, what sets Gen Z apart from other generations is their first digital approach to discovering food trends. Social media platforms, particularly Instagram and TikTok, have become essential tools for staying on top of the latest food innovations, including personalised nutrition trends. The connection between Gen Z and these platforms is more than just trend-following. It is about learning, experimenting and connecting with like-minded individuals who share similar values. Social media have created a dynamic community of food enthusiasts who prioritise wellness, sustainability and authenticity — values at the core of Gen Z's food choices. One representative of this generational cohort shared:

> *"Discovering wellness-focused foods such as personalised nutrition and other superfoods was not just about jumping on trends but about finding products and brands that resonate with their generation's values and lifestyles."* (Female, born 2003).

This is an important distinction as Gen Z does not simply follow the latest fad but instead looks for integrating these products into their lives in meaningful ways. By following food influencers and other peers, they gain access to valuable information about wellness trends, from adaptogens to superfoods, and can make informed decisions that align with their health goals. From the many conversations we had with Gen Zs we can infer that this generation's food choices are driven by a combination of personal well-being, community and social influence. Their awareness of what they eat has been elevated by the accessibility of information and resources through social media.

For example, research underscores Gen Z's heightened engagement with food trends and nutrition compared to previous generations through the social media. A study conducted in the United States and Canada found that nearly half (47%) of adult Gen Z follow food trends on social media, significantly higher than the 31% of other generations (Innova Market Insights, 2023). Additionally, 42% of Gen Z actively share their opinions about food online, far surpassing the 30% of older generations who do the same (Innova Market Insights, 2023).

Facilitated by technology and by actively seeking out foods that align with their health, sustainability and ethical values, Gen Z is leading the charge for a new kind of food culture, one that prioritises not just taste but purpose and responsibility. Technology plays a pivotal role in Gen Z's nutritional habits, with 49% of adult Gen Z seeking dietary advice through digital platforms and 21% using technology to track their health (Innova Market Insights, 2023). From nutrition apps to wearable devices, these tools provide real-time data and insights, enabling informed dietary choices. Notably, 45% of Gen Z consumers prefer food and beverage products that incorporate a digital or online component (Innova Market Insights, 2023). As one Gen Z individual highlighted:

> *"It's important because we all talk about things that are familiar to us but are still kind of taboo, like body image and mental health. When it comes to food and nutrition, it's no longer about one-size-fits-all diets. It's about finding what actually works for you — the foods that make you feel good and support your lifestyle. And honestly, technology makes it so easy to track everything now. I use three different apps just to monitor my food intake, nutrient quality, hydration and sleep. AI and biometrics can now tailor food recommendations based on your genetics, lifestyle and even your mood. It's wild how personalised nutrition is getting!"* (Male, born 2000)

Social media platforms and apps offer features such as calorie counting, meal planning and nutrient analysis, making it easier to maintain a healthy diet. For centuries, it has been widely recognised that people have varying reactions to different foods, and what is a beneficial nutrient for one person could be harmful or provoke guilt for another (Levenstein, 2012). The shift from one-size-fits-all diets to personalised nutrition is crucial to Gen Z. The integration of artificial intelligence (AI), biometrics and digital tracking makes health management more precise and holistic. The ability to tailor nutrition based on genetics, lifestyle and mood underscores how advanced personalisation has become. One Gen Zer compared personalised nutrition to fitness needs:

> *"Exactly! My fitness app is bussin* [slang for surprisingly good]. *It gives me real-time insights into my workouts and recovery. I honestly don't know what I'd do without it. There are so many personalised programs now, whether you're into particular strength training, HIIT* [High-Intensity Interval Training], *or even meditation, which I do sometimes with my friends. But when it comes to food, it's a whole different story. In the fitness world, it's always about protein — like: 'Eat more protein, it's good for you!' But good for what exactly? Just to keep bulking? That's totally not gucci* [slang for good, fancy and exceptional]. *AI can tell me what to eat based on my workouts, sure, but what about my mood, stress level, or other needs? Like you said, there's no such thing as one-size-fits-all, and food should be personalised just like workouts are."* (Male, born 1999).

Technology is revolutionising personalised nutrition by integrating multiple health metrics through apps and devices (see Figure 10.1 which is an AI-created image). This holistic approach moves beyond generic dietary advice to tailored solutions that support individual physical, mental and emotional well-being. The digital revolution in food consumption among Gen Z is not only about immediate trends but also about long-term shifts in food values that could have lasting impacts on the food industry. As Gen Z continues to embrace wellness-focused foods and share their experiences online, the influence of social media will only grow stronger, further shaping the future of food in ways that reflect the values of this dynamic generation.

Figure 10.1. Gen Z and personalised nutrition (ChatGPT-created image) https://chatgpt.com/c/684004f5-efb4-8004-a4c0-bbe0e80de96c.

The Rise of Omics and Precision Nutrition

The evolution of omics technologies is transforming nutrition science (Brennan and de Roos, 2021). They include techniques and tools, such as: metabolomics — a combination of analytical technologies for the detection and quantification of small molecules (metabolites) with the application of tools for data mining and analysis (Bio21 Institute, n.d.), transcriptomics — the study of all ribonucleic acid (RNA) molecules in a cell to investigate genes and their link to diseases, such as cancer (National Cancer Institute, n.d.), proteomics — the study of proteins and their cellular activities

(Al-Amrani *et al.*, 2021) and metagenomics — the study of a particular community of microorganisms (e.g. microbes) found in a specific place, such as the human skin or in the soil (National Human Genome Research Institute, 2025). These cutting-edge tools drive the rise of precision nutrition, personalising dietary recommendations based on an individual's genetic, metabolic and microbiome profile. As research continues to uncover how different people respond to the same foods, personalised nutrition is poised to revolutionise health outcomes and Gen Z is genuinely interested in these opportunities.

In the meantime, Gen Z is paying attention to its food intake. A European study reported that more than half of the Gen Z participants (52%) were tracking their food consumption in some way (EIT Food, 2021). Among them, 36% focused on counting calories, while 24% monitored their macronutrient intake (EIT Food, 2021). The COVID-19 pandemic contributed to this health awareness and interest in a proactive approach to diet management, emphasising personalisation over generic health guidelines.

Compared to older generations, Gen Z is investing more in wellness-related products and services that align with their unique health goals, such as nutrition apps, DNA-based diet plans, precision supplementation and functional foods designed for specific physiological needs (BetterLYF, 2024). Their spending patterns extend to personalised meal subscriptions, food tracking technologies and holistic health services that cater to individual lifestyles (McKinsey and Company, 2024b).

Taste for Adaptogens, Superfoods and Functional Foods

Gen Z consumers say they are looking for foods and beverages that support their immune health. They are deeply invested in making mindful food choices. These young people are redefining well-being with their obsession for wellness-focused products, such as adaptogens — herbs and mushrooms that can boost the body's natural energy, and superfoods — nutrient-rich foods for minimal calories (Rodriguez, 2024). At the same time, they are digitally savvy, often discovering food trends via social media and food influencers, leading to a mix of traditional and modern influences in their eating habits.

Gen Z's approach to food reflects a merging of the traditional and the modern. Adaptogens, the natural solution to the modern world's stress crisis, are not a recent discovery but rather an age-old secret that has supported generations across different cultures (Exon Publications, 2025). These natural substances, deeply embedded in the traditions of Ayurvedic, Chinese and other folk medicine, have been revitalised and celebrated for their exceptional capacity to boost resilience, vitality and overall well-being. As one Gen Z explained:

> *"I think all the wellness practices like adaptogens and superfoods are grounded in ancient traditions, but now we are rediscovering them. Social media allows us to integrate them into contemporary lifestyles."* (Female, born 1998)

This fusion of old and new is indicative of how Gen Z is redefining wellness. It is not about choosing between one or the other but about creating a balanced, integrated approach to living well to improve health, reduce stress and boost energy levels.

Gen Z's relationship with food is deeply interconnected with their digital habits. They actively seek out brands and ingredients that align with their values, making food a tool for both personal and social transformation. Adaptogens such as Maca, Ashwagandha and Reishi mushrooms are particularly popular among Gen Z, who turn to them as "mood-boosters" with stress-relieving properties and ability to maintain a balanced, energised lifestyle (Rodriguez, 2024). Superfoods such as Açai are popular staples because of their high nutrient density and rich antioxidant content. They are highly sought-after by individuals who aim to enhance their physical and mental well-being.

Gen Z demonstrates a distinct focus on the functionality of food, viewing it primarily as a source of fuel and energy. Half of Gen Z western respondents strongly agree that food is essential because it provides the energy they need (Forkes-Rees, 2024). Compared to older generations, Gen Zers are more concerned about the protein content in their meals (+6%) and less focused on factors such as salt (−11%), sugar (−9%) and fat content (−11%), which are traditionally associated with health (Forkes-Rees, 2024). This indicates a shift in priorities toward functional benefits over conventional health concerns.

Interestingly, Gen Z's views do not always align with scientific evidence. For example, when it comes to red meat, these young people see more its benefits, e.g. a good source of omega-3 fatty acids and protein, and are less inclined (−7%) to consume it in moderation (Forkes-Rees, 2024). Similarly, Gen Z sees dairy as a good functional food and a source of calcium, despite 65% of the global human population having a reduced ability to digest lactose after infancy (MedlinePlus, 2023). Furthermore, unlike animal milks plant-based beverages generally do not have saturated fat that raises the low-density lipoprotein (LDL) or bad cholesterol levels, and have about the same amount of calcium (Huber, 2018).

Health plays a significant role in shaping the food choices of many Gen Z consumers and they need to be properly informed. These younger individuals are nearly twice as likely as the general adult population to prioritise healthy eating, with 7% claiming they always eat healthily and rarely indulge in unhealthy foods, compared to just 4% of adults overall (Forkes-Rees, 2024).

All evidence underscores a broader trend among Gen Z with a strong emphasis on wellness, driven by a desire for long-term health, environmental sustainability and ethical consumption. Unlike older generations, who may adopt healthy eating habits later in life, Gen Z is integrating these choices into their lifestyles earlier, reflecting a proactive approach to well-being. This focus on health is likely fuelled by Gen Z's digital savviness, which exposes them to an abundance of health-related content and trends on social media. Platforms like TikTok and Instagram have normalised discussions about nutrition, fitness and mindfulness, making health a visible and aspirational goal. Furthermore, this generation's inclination to question traditional food systems and seek transparency from brands about their products further aligns with their preference for healthier, more natural and ethically sourced options.

Offering products that align with Gen Z's values, such as functional foods, plant-based options, and transparent labelling, will be key to capturing this influential market segment. Understanding Gen Z's health-conscious mindset is not just about meeting current demand but anticipating the broader shifts they are driving in food culture.

Business Response to Gen Z's Wellness Expectations

Unlike previous generations, Gen Z is taking greater control of their health, demanding science-backed, effective solutions from food and wellness companies. Their expectations extend beyond basic functionality. They seek authenticity, sustainability and digital engagement with the products and services they use. Gen Z's health priorities also vary by gender, emphasising the need for targeted and functional offerings that cater to their diverse needs. For this young cohort, wellness is a serious matter (McKinsey and Company, 2024c).

Recognising the priorities of this new generation, businesses should be looking at adapting their strategies to engage Gen Z in meaningful ways. Nutrition and wellness brands are curating appealing social media content that aligns with Gen Z's digital consumption habits. Hotels and travel companies are integrating AI-driven wellness solutions and nature-based experiences, such as forest bathing, plant-based cooking courses and mindfulness retreats to appeal to Gen Z's holistic approach to health. Workplace well-being programs now include gym memberships, mental health resources and flexible wellness initiatives to attract and retain young talent.

Ultimately, for Gen Z, wellness is not a passing trend. It is a lifestyle that is deeply intertwined with their values, social interactions and career choices. Businesses and health professionals that embrace this shift will be better positioned to connect with and support this generation's personalised approach to well-being. Gen Z places a significantly higher priority on fitness and wellness than previous generations. Research indicates that 56% of young consumers in the US consider fitness a "very high priority," compared to just 40% of the general population (McKinsey and Company, 2024a). Their focus on fitness extends beyond physical health, incorporating preventive measures to achieve long-term well-being, including aging-related concerns.

The digital era has granted Gen Z unprecedented access to health-related content, particularly through social media platforms, such as TikTok and Instagram. Living through the COVID-19 pandemic further reinforced the importance of maintaining good health, influencing Gen Z's fitness habits and their preference for wellness-oriented environments (Bogueva and Marinova, 2024). Today, gyms, fitness studios and group classes are not just places to exercise, they serve as social hubs where like-minded individuals connect over shared wellness goals.

A Gen Zer explained the importance of choice and individuality:

> *"Services that are out there are cool to have. Same with variety of products that are addressing personal nutritional needs. It's cool, but more of a side thing for me. I like to know that they are out there, but I also like to keep it simple — a good simple meal, steak, burger, pizza, some veggies, some fruit... this is just what I like. That's my version of personalised nutrition. I mean, no one has time to meal prep every single day, but we still want to eat healthy. These services aim to take out the guesswork and make sure we're getting what we need without overthinking it. Not sure if I like this. I like to have a choice."* (Male, born 2000)

Understanding Gen Z's evolving expectations, businesses should be reshaping their wellness strategies by providing tailored content and leveraging technology for convenience and customisation, ensuring scientific credibility and proven efficacy in their wellness products. Businesses that align with this holistic approach will be better positioned to engage and resonate with this generation.

According to Francis and Hoefel (2018), businesses need to recognise three key implications for this generation: viewing consumption as access rather than ownership, including subscribing to meal delivery services; seeing consumption as a way to express personal identity, and treating consumption as an issue of ethical importance.

By embracing technology and inclusiveness, social justice and responsibility, Gen Z is redefining the future of wellness. Their expectations are reshaping the food and wellness industries, pushing brands toward greater transparency, innovation and digital engagement. Sustainability, ethical production and social responsibility are key considerations in their purchasing decisions. Gen Z demands not just convenience and personalisation, but also scientific credibility and proven efficacy (McKinsey and Company, 2024b). Those businesses that align with Gen Z's value-driven, holistic approach to health will thrive in this rapidly evolving landscape.

Navigating the Future: Challenges and Opportunities in Personalised Nutrition

The future of personalised nutrition will focus on increasingly individualised approaches, including highly specific dietary recommendations based on an individual's unique genetic profile, gut microbiome, lifestyle habits and detailed biometric data. These recommendations will be supported by advanced technologies, such as AI algorithms, and delivered through easy-to-use apps to track nutrient intake and identify nutritional gaps (Matusheski *et al.*, 2021). There may also be potential for 3D food printing to create customised meals with precise nutrient compositions and correct food additives, all aimed at optimising health based on individual needs and preventing diseases (Abedini *et al.*, 2024; Eswaran *et al.*, 2023).

Furthermore, the application of genetics and omics biomarkers through advances in nutrigenomics may allow for diet recommendations based on genetic predispositions, such as how an individual metabolises certain nutrients or their risk for specific diet-related conditions (Mullins *et al.*, 2020; Keijer *et al.*, 2023). The robust integration of bioinformatics and genomics is seen as critical for advancing personalised nutrition (Singar *et al.*, 2024). Advances in nutrigenomics now allow for diet recommendations based on genetic predispositions, helping individuals optimise their nutrient intake and mitigate the risk of diet-related conditions.

Gen Z, known for its data-driven and health-conscious mindsets, is likely to show strong interest in genetics, biomarkers and bio-individual wellness. Unlike previous generations, which relied on general dietary guidelines, Gen Z seeks personalised and science-backed approaches to optimise its health. This generation is not just consuming health information, they are actively using technology to shape their well-being.

One of the most significant transformations in nutrition is the rise of food tracking apps, a proven favourite among Gen Z. These tools are revolutionising personalised nutrition by offering intuitive, real-time feedback. With a simple barcode scan, users can instantly determine whether a product aligns with their dietary goals. A Gen Zer explains:

> *"I feel like the next big thing will be more around our personal health. The future is all about bio-individual wellness.... Fascinating things like genetic testing to create a personalised diet and fitness plans tailored to our unique bodies. It's not just about what's generally healthy but what's best for you specifically. It could be ultra-cool things to experience."* (Male, born 2004)

Wearable technology, AI-driven meal planning tools and smart apps are making personalised nutrition more accessible than ever before. These innovations analyse real-time health data, offering tailored guidance that aligns with an individual's nutritional needs, fitness goals and metabolic responses.

Going beyond calorie counting and macronutrient tracking, microbiome analysis is emerging as the next frontier in personalised nutrition. By examining gut bacteria, AI-powered apps can generate highly specific dietary recommendations that optimise digestion, nutrient absorption and even mental well-being. Gen Z's dreams and expectations are high:

> *"Imagine a world where your nutrition app syncs with your biome, microbiome analysis, to be exact, and gives you precise meal recommendations for optimal gut health. It's like sci-fi. No more guesswork, just science-backed, personalised nutrition. And that means no more one-size-fits-all diets. Instead of generic calorie tracking, we'll have AI-driven recommendations tailored to our*

> *body's unique needs. It's a complete game-changer."* (Female, born 2003)

This level of personalisation is shifting the way nutrition and health industries operate, as companies invest in precision-based wellness solutions to meet Gen Z's expectations for customisation and efficacy.

Unlike previous generations that often focused on reactive healthcare, Gen Z is proactively addressing health concerns before they arise. Their approach is preventive, tech-driven and deeply personalised:

> *"That's the whole vibe! We're not just waiting for health problems to pop up... myself and the people around me, my friends... we're actually preventing them. Our generation is making wellness way more personal, proactive and sustainable if you wish. We're not just following trends; we're shaping the future of health. And honestly? This is just the beginning."* (Female, born 2002)

This shift toward proactive and preventive health management fuels the demand for scientifically validated, technology-enhanced solutions that help individuals optimise their well-being before they experience health issues.

While personalised nutrition offers numerous benefits, it also raises concerns about accessibility, affordability and ethical considerations. These are of concern as precision-based health solutions, such as genetic testing and microbiome analysis, often come at a high cost, limiting accessibility. The increasing reliance on AI and personal health data raises concerns about privacy, consent and data security. There are also potential issues with inclusivity. Not all consumers have equal access to cutting-edge technology. Gen Z is at the forefront of advocating for more inclusive and affordable solutions; their voices need to be loud here too.

Despite these challenges, Gen Z is driving innovation in personalised nutrition, pushing for scalable, ethical and science-backed approaches. As this generation continues to embrace technology and advocate for transparency, with the right efforts the future of personalised health and wellness will become increasingly democratised, data-driven and accessible to all.

Concluding Words

Gen Z's strong inclination toward tailored solutions, ones that align with their individual health goals, genetic predispositions and preference for immersive experiences, makes Wigmore's insight particularly relevant. This generation approaches food with intention and strategy, making choices that reflect both their personal well-being and broader societal values. Their demand for authenticity, sustainability and seamless digital engagement underscores a fundamental shift in how food and wellness are perceived.

As the food industry evolves, it is essential to recognise that Gen Z prioritises sustainability, ethics and transparency, ensuring that their food choices align with both their health goals and moral principles. While there will always be individual exceptions, for most, food is more than sustenance. It is a statement of identity, a tool for self-optimisation and a means of shaping a healthier future.

Going back to Wigmore's perspective, it really captures this mindset perfectly. For Gen Z, food is not an afterthought, but rather an intentional and deeply personal choice. As personalised nutrition continues to advance, its potential to transform health outcomes, prevent chronic diseases and empower individuals will redefine how society approaches nutrition, wellness and self-care in the modern world.

References

Abdoli, M., Scotto Rosato, M., Cipriano, A., Napolano, R., Cotrufo, P., Barberis, N., & Cella, S. (2023). Affect, body, and eating habits in children: A systematic review. *Nutrients*, 15(15), 3343. https://doi.org/10.3390/nu15153343

Abedini, A., Sohrabvandi, S., Sadighara, P., Hosseini, H., Farhoodi, M., Assadpour, E.... Jafari, S. M. (2024). Personalised nutrition with 3D-printed foods: A systematic review on the impact of different additives. *Advances in Colloid and Interface Science*, 328, 103181. https://doi.org/10.1016/j.cis.2024.103181

Al-Amrani, S., Al-Jabri, Z., Al-Zaabi, A., Alshekaili, J., & Al-Khabori, M. (2021). Proteomics: Concepts and applications in human medicine. *World Journal of Biological Chemistry*, 12(5), 57–69. https://doi.org/10.4331/wjbc.v12.i5.57

Bardales Salguero, R., Bogueva, D., & Marinova, D. (2024). Australia's university Generation Z and its concerns about climate change. *Sustainable Earth Reviews*, 7, 8. https://doi.org/10.1186/s42055-024-00075-w

BetterLYF (2024). Wellness and Self-Care Trends for Gen Z. https://www.betterlyf.com/articles/mental-health/self-care-trends-for-gen-z

Bio21 Institute. (n.d.). Metabolomics Australia. The University or Melbourne. https://www.bio21.unimelb.edu.au/metabolomics-australia

Bogueva, D. & Marinova, D. (2024). Health, nutrition and exercise. In Seemiller, C. & Grace, M. (eds) *Gen Z Around the World: Understanding the Global Cohort Culture of Generation Z* (pp. 79–89). Emerald Publishing, Leeds, UK.

Brennan, L. & de Roos, B. (2021). Nutrigenomics: Lessons learned and future perspectives. *American Journal of Clinical Nutrition*, 113(3), 503–516. https://doi.org/10.1093/ajcn/nqaa366

Celis-Morales, C., Livingstone, K. M., Marsaux, C. F., Macready, A. L., Fallaize, R., O'Donovan, C. B .... Mathers, J. C. (2017). Effect of personalised nutrition on health-related behaviour change: Evidence from the Food4Me European randomised controlled trial. *International Journal of Epidemiology*, 46(2), 578–588. https://doi.org/10.1093/ije/dyw186

Celis-Morales, C., Livingstone, K. M., Marsaux, C. F., Forster, H., O'Donovan, C. B., Woolhead, C., ... Mathers, J. C. (2015a). Design and baseline characteristics of the Food4Me study: A web-based randomised controlled trial of personalised nutrition in seven European countries. *Genes & Nutrition*, 10(1), 450. https://doi.org/10.1007/s12263-014-0450-2

Celis-Morales, C., Livingstone, K. M., Woolhead, C., Forster, H., O'Donovan, C. B., Macready, A. L., ... Mathers, J. C. (2015b). How reliable is internet-based self-reported identity, socio-demographic and obesity measures in European adults? *Genes & Nutrition*, 10(5), 28. https://doi.org/10.1007/s12263-015-0476-0

Daily, R. (2025). 2025 Wellness trends: Gen Z turns to mood-boosting foods and beverages. https://www.foodnavigator-usa.com/Article/2025/01/24/gen-z-shoppers-improve-moods-with-foods

EIT Food. (2021). Our Food, Our Food System: What Generation Z wants from a healthy food system. European Union. https://www.eitfood.eu/media/news-pdf/Our_Food%2C_Our_Food_System_-_EIT_Food_report_.pdf

Eswaran, H., Ponnuswamy, R. D., & Kannapan, R. P. (2023). Perspective approaches of 3D printed stuffs for personalised nutrition: A comprehensive review. *Annals of 3D Printed Medicine*, 12, 100125. https://doi.org/10.1016/j.stlm.2023.100125

Exon Publications (2025). Adaptogens: Examples, Supplements, Drops, Health Benefits, and Side Effects. Brisbane, Australia. https://doi.org/10.36255/adaptogens-examples-supplements-drops-health-benefits-side-effects

Ferrario, P. G., Watzl, B., Møller, G., & Ritz, C. (2021). What is the promise of personalised nutrition? *Journal of Nutritional Science*, 10, e23. https://doi.org10.1017/jns.2021.13.

Forkes-Rees, C. (2024). Gen Z: A Generation of Health-Conscious Consumers. UK Agriculture and Horticulture Development Board (AHDB). https://ahdb.org.uk/news/consumer-insight-gen-z-a-generation-of-health-conscious-consumers

Francis, T., & Hoefel, F. (2018). 'True Gen': Generation Z and its implications for companies. McKinsey. https://www.mckinsey.com/industries/consumer-packaged-goods/our-insights/true-gen-generation-z-and-its-implications-for-companies

Gibney, M., Walsh, M., & Goosens, J. (2016). Personalized nutrition: Paving the way to better population health. In Kraemer, K., Cordaro, J. B., Fanzo, J., Gibney, M., Kennedy, E., Labrique, A., Steffen, J., & Eggersdorfer, M. (eds) *Good Nutrition: Perspectives for the 21st Century* (pp. 235–248). Karger Publishers. https://doi.org/10.1159/000452389

Haidt, J. (2024). *The Anxious Generation: How the Great Rewiring of Childhood is Causing an Epidemic of Mental Illness.* Alen Lane, London.

Huber, J. (2018). Christopher Gardner busts myths about milk. Stanford Medicine. https://med.stanford.edu/news/all-news/2018/08/christopher-gardner-busts-myths-about-milk.html

Huston P. (2022). A sedentary and unhealthy lifestyle fuels chronic disease progression by changing interstitial cell behaviour: A network analysis. *Frontiers in Physiology*, 13, 904107. https://doi.org/10.3389/fphys.2022.904107.

Innova Market Insights (2023). Zooming in on Generation Z: Exploring Lifestyles and Attitudes in the US and Canada, August. https://www.innovamarketinsights.com/reports/zooming-in-on-gen-z/

Jinnette, R., Narita, A., Manning, B., McNaughton, S. A., Mathers, J. C., & Livingstone, K. M. (2021). Does personalised nutrition advice improve dietary intake in healthy adults? A systematic review of randomized controlled trials. *Advances in Nutrition*, 12(3), 657–669. https://doi.org/10.1093/advances/nmaa144

Keijer, J., Escoté, X., Galmés, S., Palou-March, A., Serra, F., Aldubayan, M. A. ... del Bas, J. M. (2023). Omics biomarkers and an approach for their practical implementation to delineate health status for personalised nutrition strategies. *Critical Reviews in Food Science and Nutrition*, 64(23), 8279–8307. https://doi.org/10.1080/10408398.2023.2198605.

Koutsimani, P., Montgomery, A., & Georganta, K. (2019). The relationship between burnout, depression, and anxiety: A systematic review and meta-analysis. *Frontiers in Psychology*, 10, 84. https://doi.org/10.3389/fpsyg.2019.00284

Levenstein, H. (2012). *Fear of Food: A History of Why We Worry about What We Eat*. University of Chicago Press.

Livingstone, K. M., Celis-Morales, C., Navas-Carretero, S., San-Cristobal, R., Macready, A. L., Fallaize, R.... Mathers, J. C. (2016). Effect of an Internet-based, personalized nutrition randomised trial on dietary changes associated with the Mediterranean diet: The Food4Me Study. *American Journal of Clinical Nutrition*, 104(2), 288–297. https://doi.org/10.3945/ajcn.115.129049

Matusheski, N. V., Caffrey, A., Christensen, L., Mezgec, S., Surendran, S., Hjorth, M.F.... Péter, S. (2021). Diets, nutrients, genes and the microbiome: Recent advances in personalised nutrition. *British Journal of Nutrition*, 126, 1489–1497. https://doi.org/10.1017/S0007114521000374

McKinsey & Company. (2024a). Mind the gap. Curated reads for Gen Z and their Z-curious colleagues. https://www.mckinsey.com/~/media/mckinsey/email/genz/2024/01/2024-01-23d.html

McKinsey & Company. (2024b). The trends defining the $1.8 trillion global wellness market in 2024. https://www.mckinsey.com/industries/consumer-packaged-goods/our-insights/the-trends-defining-the-1-point-8-trillion-dollar-global-wellness-market-in-2024

McKinsey & Company. (2024c). What is the future of wellness? https://www.mckinsey.com/featured-insights/mckinsey-explainers/what-is-the-future-of-wellness

MedlinePlus. (2023). Lactose intolerance. https://medlineplus.gov/genetics/condition/lactose-intolerance/#frequency

Mullins, V. A., Bresette, W., Johnstone, L., Hallmark, B., & Chilton, F. H. (2020). Genomics in personalised nutrition: Can you "eat for your genes"? *Nutrients*, 12(10), 3118. https://doi.org/10.3390/nu12103118

National Cancer Institute. (n.d.). Transcriptomics. https://www.cancer.gov/publications/dictionaries/cancer-terms/def/transcriptomics

National Human Genome Research Institute. (2025). Metagenomics. https://www.genome.gov/genetics-glossary/Metagenomics

Newman, E., O'Connor, D. B., & Conner, M. (2007). Daily hassles and eating behaviour: The role of cortisol reactivity status. *Psychoneuroendocrinology*, 32(2), 125–132. https://doi.org/10.1016/j.psyneuen.2006.11.006

Priporas, C.-V., Vellore-Nagarajan, D., & Kamenidou, I. (E). (2022). Stressful eating indulgence by generation Z: A cognitive conceptual framework of new age consumers' obesity. *European Journal of Marketing*, 56(11), 2978–3006. https://doi.org/10.1108/EJM-06-2021-0386

Qina. (2025). What is personalized nutrition? https://www.qina.tech/blog/health-digital-personalized-nutrition-background#h40slvzc6derbayll7nwf13fvjllpa

Rand, S. (2025). Gen Z food trends: What today's young US consumers want. https://www.askattest.com/blog/research/gen-z-food-trends

Rodriguez, I. (2024). Ashwagandha, Maca & Reishi: How Adaptogens Became Gen Z's Favorite Mood Boosters. https://www.wokewaves.com/posts/gen-z-adaptogens-smoothies-coffee

Rogers, P. J. (1999). Eating habits and appetite control: A psychobiological perspective. *The Proceedings of the Nutrition Society*, 58(1), 59–67. https://doi.org/10.1079/pns19990009

Schweren, L. J. S., Larsson, H., Vinke, P. C., Li, L., Kvalvik, L. G., Arias-Vasquez, A., Haavik, J., & Hartman, C. A. (2021). Diet quality, stress and common mental health problems: A cohort study of 121,008 adults. *Clinical Nutrition*, 40(3), 901–906. https://doi.org/10.1016/j.clnu.2020.06.016

Singar, S., Nagpal, R., Arjmandi, B. H., & Akhavan, N. S. (2024). Personalised nutrition: Tailoring dietary recommendations through genetic insights. *Nutrients*, 16(16), 2673. https://doi.org/10.3390/ nu16162673

Singh, S. & Gupta, A. (2024). Exploring psychological and emotional factors: The interplay of emotions and cognitive behavioural aspects on binge eating in Gen Z. In: Jyoti Kukreja, J., Geetha M., Chandra Poonia, R., & Verma, R. R. (eds) *Neuroscientific Insights and Therapeutic Approaches to Eating Disorders* (pp. 229–237). IGI Global. https://doi.org/10.4018/979-8-3693-3230-6.ch015

Wardle, J., Steptoe, A., Oliver, G., & Lipsey, Z. (2000). Stress, dietary restraint and food intake. *Journal of Psychosomatic Research*, 48(2), 195–202. https://doi.org/10.1016/s0022-3999(00)00076-3

Chapter 11

Transparency Is Key: Gen Z and Food Safety Standards

"Honesty and transparency make you vulnerable. Be honest and transparent anyway."

— Mother Teresa

"A lack of transparency results in distrust and a deep sense of insecurity."

— Dalai Lama

Introduction

Transparency has become a defining expectation for Gen Z, particularly when it comes to food. Unlike previous generations, Gen Z views food not just as nourishment but as an ethical and environmental choice. The quotes from Mother Teresa and the Dalai Lama highlight two key perspectives — the courage required to be transparent and the consequences of failing to do so. The demand for honest labelling, traceability, regulated safety standards and ethical sourcing reflects Gen Z's commitment to sustainability and trust.

This chapter explores how Gen Z's insistence on transparency is reshaping the food industry, forcing brands to prioritise accountability in ways never seen before. We examine clear labelling, digital verification, corporate responsibility, activism and how a lack of transparency can erode trust. The perspectives of Mother Teresa and the Dalai Lama provide a valuable lens, one that reinforces the courage required for honesty and the consequences of failing to be open, a lens that perfectly outlines Gen Z's attitude.

The power of transparency is about building trust, not breeding distrust. Mother Teresa's reflection on the vulnerability that comes with honesty captures the essence of Gen Z's expectations from the food industry. This generation is willing to reward transparency, even when it reveals imperfection, because it signals authenticity. Brands that openly share their sourcing struggles, processing methods or sustainability limitations are often met with empathy and trust, rather than backlash.

In contrast, the Dalai Lama's warning about the consequences of opacity illustrates Gen Z's deep-rooted scepticism toward corporations, institutions and even the contributions of previous generations. Transparency lapses, be it in labelling, labour practices or sustainability claims, trigger rapid distrust. A common definition of social trust is when somebody is given the benefit of the doubt (Rahn and Transue, 1998). This is no longer the case with Gen Z, given the multitude of problems and challenges they are inheriting — from global warming to widespread pollution with plastics and forever chemicals to loss of biodiversity and sea level rise. With the occurrence of misinformation scandals and greenwashing precedents, Gen Z's threshold for ambiguity is low; its trust must be earned through consistent, verifiable action.

Gen Z's Influence: Data-Driven Awareness and Ethical Consumption

For Gen Z, consumption is not just a transaction. It is an act of informed decision-making shaped by digital literacy, ethical priorities and social activism. They approach food choices with investigative rigour, demanding accountability from brands and aligning their purchases with their values. Their ethical consumerism is guided by data, activism and social justice. According to McKeever *et al.* (2021), Gen Z are global citizens, digital accelerators, future proofers, climatarian consumers and equality activists and these characteristics are all related to demand for transparency. They are particularly manifested in Gen Z's approach to food consumption and production.

A 2024 YouGov study of US Gen Z adults highlights their commitment to sustainability, revealing that 70% of them support climate-smart agriculture and 55% are willing to pay more for environmentally sustainable products (Just, 2024). Additionally, 68% express the need for greater transparency from retailers regarding food sustainability (Mitchell, 2024). These findings affirm that for Gen Z, food safety extends beyond hygiene or contamination, it includes moral and environmental dimensions of ethics, sustainability and corporate integrity:

> *"When I buy food, I'm not just thinking about what I will be putting on my plate. I'm thinking about the people who picked it, the land it came from and whether the company actually cares about doing the right thing or just wants my money. If I can't find that information easily, I just don't trust it. This is not only with me, but all my friends are having the same issue."* (Female, born 2002)

This opinion reflects the holistic, value-driven mindset of many Gen Z consumers. It reinforces the idea that transparency is not only about product safety but about justice, sustainability and ethics. For Gen Z, trust is built on visible, verifiable actions, not brand promises. The inability to access such information erodes consumer confidence and drives disengagement (Khamitov *et al.*, 2024).

Transparency is the key to securing Gen Z's trust. Brands that openly disclose environmental impact, labour conditions and ingredient sourcing gain not just consumers, but passionate advocates. Conversely, companies that withhold information risk losing credibility, especially in an era where social media can amplify scrutiny and expose deceptive practices instantly.

Mother Teresa's wisdom: "Honesty and transparency make you vulnerable. Be honest and transparent anyway", rings especially true in the food industry. Transparency can reveal mistakes, yet Gen Z respects businesses that take accountability and commit to improvement. Admitting imperfections, rather than concealing them, fosters deeper trust and loyalty.

Conversely, the Dalai Lama's warning: "A lack of transparency results in distrust and a deep sense of insecurity', reflects the consequences of opacity. When brands fail to disclose sourcing practices or sustainability efforts, Gen Z grows sceptical. Food scandals, misleading labels and unethical labour practices have reinforced this young generation's distrust, pushing them toward companies that prioritise radical transparency. As one Gen Z expressed it:

> *"It's simple... If a company messes up and owns it, I respect that way more than one that pretends everything is perfect. Nobody's perfect and mistakes happen, but what matters is how you handle them. Hiding the truth only makes it worse and breaks trust. Hiding things is what makes me stop buying from them... I want real, not fake. There are so many fake things around us... filtered images, scripted apologies, carefully crafted marketing gimmicks. The least a brand can do is be honest, because once I feel like they're lying, I'm done... I am afraid that these days there are not much honesty and transparency..."* (Male, born 2001)

This view directly aligns with the argument that transparency, particularly when it reveals imperfection, fosters trust. Gen Z does not demand perfection but demands honesty. The remark illustrates why authenticity and vulnerability have become key strategies for brands seeking loyalty among this generation.

Unlike previous generations, Gen Z is hyper-aware of how food is sourced, produced and marketed. Their access to digital tools empowers them to research deeply, fact-check claims and hold companies accountable. This has led to key shifts in food safety standards, such as clearer labelling with simplified ingredient lists and ethical certifications, digital tracking of food origins and corporate accountability in ethical practices. A report by Packaging Technology Today (2024) emphasises Gen Z's growing demand for transparency in food sustainability, with many consumers willing to spend more to support brands reflecting their values (Robichaud and Yu, 2021). As one Gen Z explained:

> *"Food used to be just something we ate. Now, for me, it's about knowing where it came from, how it was made and whether I'm supporting something good or just convenient. I don't want to be*

> *part of a system that hides the truth. I want something that I believe is good."* (Female, born 2003)

This view highlights how Gen Z has reframed consumption as an ethical act. The perspective illustrates the shift from passive eating to active ethical engagement, reinforcing the fact that food transparency is no longer a niche preference, but a generational demand that influences purchasing behaviour and industry standards (Confetto *et al.*, 2023).

Transparency is no longer just about avoiding backlash. It has become a non-negotiable pillar of food safety and a competitive advantage. Gen Z is reshaping the food industry with its insistence on honesty and accountability. For these young people, transparency is not a marketing strategy. It is not just a response to consumer demand, but a precondition and the foundation for trust, safety and brand loyalty in the eyes of Gen Z. Consumer trust of this generation is predominantly built on their experiences (Czernyszewicz and Wiśniewska, 2023). Companies that fail to meet their expectations risk irrelevance, while those that embrace openness gain a loyal following in an era where informed consumerism defines success.

The Rise of Food Transparency

Gen Z's demand for openness and transparency is a direct response to past failures in the food system, including scandals involving misleading ingredient lists, mislabelling and unethical sourcing (Roodenburg *et al.*, 2023; Thompson, 2023; BritainThinks, 2019). As one Gen Z told us:

> *"I've seen too many brands claim to be 'ethical' or 'sustainable' without proving it. I don't trust vague statements anymore... I want to see where my food comes from, how it's made, and whether a company is actually doing good or just marketing it that way."* (Female, born 2002)

Having grown up in an era of heightened corporate scrutiny, Gen Z has developed a zero-tolerance policy for ambiguity, rejecting vague claims and demanding clear, verifiable information about the food they consume (Marinova and Bogueva, 2022). This scepticism is deeply rooted in their exposure to food safety incidents and corporate greenwashing, where companies have falsely marketed products as sustainable or ethical without substantive proof.

Studies show that Gen Z is more likely than previous generations to challenge misleading claims, using digital tools and social media to hold brands accountable (Kymäläinen *et al.*, 2021). Its insistence on transparency in food labelling, sourcing and corporate accountability has reshaped industry standards, pushing companies toward radical honesty to maintain consumer trust. Moreover, Gen Z's digital fluency has empowered them to investigate food origins, verify sustainability claims and expose deceptive practices. This shift has forced brands to adopt clearer labelling, disclose sourcing details and embrace third-party certifications to maintain credibility (Halicka *et al.*, 2025). As a result, food transparency is no longer a niche preference, it is a generational demand that actively influences purchasing behaviour and industry regulations.

According to Salam *et al.* (2024), Gen Z's consumer preferences are shaped by their digital fluency and heightened awareness of sustainability issues. These young people rely on social media, peer reviews and independent certifications to assess whether a brand's ethical claims are genuine. In the UK, the Food Standards Agency found that Gen Z is more likely than any previous generation to challenge vague claims, demand accountability and use digital platforms to hold companies to account (BritainThinks, 2019). Its digital fluency enhances the ability to investigate and share concerns publicly, making transparency a non-negotiable expectation (Marinova and Bogueva, 2022). Social media have made this easier, shifting food safety from a bureaucratic domain into a viral conversation, where brands are constantly under public scrutiny (BritainThinks, 2019).

Clear Labelling: Demanding Honesty on the Shelf

In our conversations with Gen Z, one question kept surfacing: why does this generation have such a low tolerance for uncertainty? What fuels their resistance to vague or ambiguous claims? The answer lies in a defining trait of true Gen Z consumers. They are inherently sceptical, especially when it comes to marketing jargon. Terms like "natural" or "healthy" do not impress them unless backed by solid, scientific evidence. Unlike previous generations who may have taken such labels without looking further, Gen Z demands proof. They expect full transparency from brands, with verifiable details about ingredients, sourcing and production methods. For them, trust is built on clarity, not catchy slogans. As one Gen Z explained it:

> *"Honestly, I just don't buy into labels like 'clean', 'natural' or 'wholesome' anymore... what do they even mean?... Brands like to throw around fancy words like 'clean', "wholesome' or 'natural' without explaining what that actually means.... It kinda feels like they're just trying to sound good without doing much. What does "clean" even mean? No additives? Organic? Sustainably sourced?... Like, is it actually better for me, or is it just marketing?... So why should I be bothered to pay attention... I just move on. I'd rather spend extra time and dig a little deeper and make sure I'm getting something real rather than being fooled by a pretty package with fancy words that don't say anything... At the end of the day, I just want to know what I'm putting in my body without having to decode a marketing strategy. Transparency is everything."* (Female, born 2005)

These types of sentiments are far from uncommon. They reflect a broader generational shift toward informed and evidence-based consumerism. According to Kabaja *et al.* (2023), Gen Z's purchasing decisions are strongly influenced by environmental labelling, but only when it is perceived as trustworthy, credible and backed by evidence. In other words, it is not the presence of a label that matters, but the substance behind it. Studies show that over 60% of Gen Z consumers report stronger concerns about food sourcing and ingredient transparency compared to five years earlier (Just, 2024). This shift is no coincidence. Raised in a hyper-connected world, their digital fluency enables them to fact-check claims instantly, making misleading labels a serious reputational risk for brands. As they grow up, their ability to make independent judgements also increases.

The rise of clean labelling, where ingredient lists are short, readable and informative, has quickly become a new industry standard. However, Gen Z expects more. It wants brands to go beyond what is regulatory requirements, voluntarily disclosing allergens, sourcing details and the full story behind nutritional integrity. It is not about ticking regulatory boxes; it is about earning trust through radical transparency. As one Gen Z elaborates, these expectations are extremely important:

> *"You know, when I'm grabbing snacks or picking out groceries, I really appreciate when a brand is straight-up about what's in its food, where it comes from, and why they chose certain ingredients. It just makes me feel like I can actually trust what I'm eating.... and it is important for me and my friends as we look after our bodies...for us is important to stay fit... You know, if a company is telling me their eggs are from local farms or their coffee beans are ethically sourced, I'm way more likely to buy it. Being honest about food goes a long way in getting me to support a brand. If the opposite... get suspicious really fast."* (Male, born 1999)

Gen Z's scepticism is not baseless. It is grounded in a demand for authenticity. Research by Mitchell and Topic (2023) confirms that Gen Z links transparent ingredient information, voluntarily disclosing allergens, sourcing practices, nutritional integrity and environmental packaging directly to brand trust and purchase intent. Brands that meet this demand by embracing full transparency tend to secure the rewards of stronger customer loyalty. Those that rely on pseudo-ethical language or hide behind vague health halos often face the opposite — swift backlash on social media and a sharp erosion of consumer trust.

The demands for honesty and information go beyond the text on the labels. Gen Z uses numerous channels to develop understanding about the products with traditional marketing, family and friends being only the starting point. Social media with YouTube, TikTok and online influencers also shape the attitudes of these young people. Moreover, studies indicate that Gen Z consumers actively engage with digital tools such as QR codes and mobile apps to trace ingredients in real time, reinforcing their expectation for radical transparency (Kabaja *et al.*, 2023). Brands that offer full disclosure on sourcing and production methods, especially environmental responsibility have seen increased engagement and trust (Dragolea *et al.*, 2023). Conversely, companies that fail to substantiate their sustainability claims often face public scrutiny and market share losses when social media exposes inconsistencies (McKeever *et al.*, 2021).

Gen Z's expectations reflect a deeper cultural shift toward empowered, evidence-driven consumerism. They are not merely reacting to trends, they are actively shaping them. For brands, this means the age of vague claims is over. Transparency is not a nice-to-have, it is the price of admission.

Traceability from Farm to Fork in Real Time

Gen Z is reshaping consumer expectations by demanding full transparency in food supply chains. Unlike previous generations, they do not settle for indefinite assurances. They want real-time access to sourcing, production and distribution data. A socio-psychological study in Poland shows that the interplay between the components of the theory of planned behaviour (personal attitude, subjective norms and perceived behavioural control) combined with health concerns, consumer knowledge and trust influence sustainable food purchasing intentions (Jakubowska *et al.*, 2024). The trust component builds on detailed traceability of the information related to sourcing, production and labelling reinforcing knowledge, which ultimately drives purchase decisions. This demand stems from growing concerns about food safety, ethical sourcing and environmental impact. For Gen Z, traceability is a non-negotiable expectation (Czernyszewicz and Wiśniewska, 2024).

Gen Z wants to know the entire story behind the food they eat. This includes production practices, transportation and labour conditions. Technologies such as blockchain and artificial intelligence (AI) are emerging as a solution, revolutionising food traceability, offering secure, proof-tracking systems that allow consumers to verify food origins in real time, ensuring greater accountability in supply chains (Mitchell, 2024). Blockchain technology in particular can enhance transparency, traceability and data security in food supply chains, leading to better quality, safer produce and improved access (Ellahi *et al.*, 2023). It has been shown to improve food supply chain resilience, reduce food waste and enhance compliance with environmental, social and governance standards (Van Niekerk, 2024). Further strengthening the traceability efforts, AI-powered algorithms can analyse shipment delays, flag contamination risks and detect inefficiencies (Mahendra, 2024) and Gen Z is at ease with these applications.

In addition, sustainability certifications such as Fair Trade, B Corp, Rainforest Alliance, Climate Neutral and Organic, remain important trust markers for Gen Z shoppers and are no longer optional. These certifications are issued by third parties and although they vary in scope, procedures used and cost paid by the brand, they are a first step for the conscientious consumer in navigating the complexity of greenwashing. A study indicates that nearly two-thirds of US Gen Z consumers (61.3%) are willing to pay a premium for items that are sustainably produced or ethically sourced, reinforcing their commitment to responsible consumption (Rand, 2025).

Ethical Sourcing and Social Responsibility: Linking Food Safety to Values

Gen Z is redefining food safety by linking it directly to ethical sourcing and social responsibility. Unlike previous generations, Gen Z views food safety not just as a matter of hygiene and contamination prevention but as a broader issue encompassing fair labour practices, environmental sustainability and corporate accountability (BritainThinks, 2019). This shift reflects their growing awareness of how food production impacts human rights, climate change, nature's biodiversity and global supply chains. Sustainability and food safety ethical sourcing are closely tied to sustainability, and Gen Z sees environmental responsibility as a key component of food safety. A study in the USA shows that over 70% of Gen Z values sustainability in their food choices and they prefer brands that minimise carbon footprints, reduce food waste and use sustainable packaging (Rand, 2025).

Gen Z strongly associates food safety with social justice, recognising that unsafe working conditions and exploitative labour practices compromise the integrity of food production (BritainThinks, 2019; Blackstone *et al.*, 2023). They believe that ethical sourcing is not just about ensuring food quality but also about protecting workers' rights and promoting fair wages. This generation is particularly concerned about forced labour in agricultural supply chains, where poor working conditions can lead to contamination risks and unethical practices that compromise food safety. Ethical sourcing ensures that food is produced under humane conditions, reinforcing trust in brands.

There is a clear shift in expectations regarding sustainability and ethics. For Gen Z, food safety is not just about health, it is about ethics. They expect brands to uphold fair labour practices, reduce carbon footprints and support regenerative agriculture (Just, 2024). Over half of Gen Z consumers prefer companies that prioritise lower environmental impacts or locally sourced foods (Mitchell, 2024). Ethical sourcing is directly tied to trust, making it a key factor in purchasing decisions. For them, food safety includes moral assurance, products that are sustainably sourced, ethically produced and transparently labelled.

Recently, there has been a slight decline in consumers expecting companies and governments to take responsibility for environmental issues. This has been noted particularly in Europe (Innova Market Insights, 2024). While this shift is subtle, it may signal growing apathy or disillusion among consumers toward institutional efforts in sustainability. Gen Z, in particular, has the lowest expectations of corporate and governmental involvement, whereas Baby Boomers hold the highest expectations (Innova Market Insights, 2024). This generational divide reflects broader attitudes toward responsibility and action. One possible explanation for Gen Z's lower expectations is their preference for individual and community-driven solutions over institutional intervention. Unlike previous generations, Gen Z has grown up witnessing corporate greenwashing and slow governmental responses to climate change. As a result, they may be more sceptical about large-scale promises and instead focus on personal choices, grassroots activism and direct engagement with ethical brands (Deloitte, 2025). This shift aligns with their tendency to support businesses that demonstrate transparency and authenticity rather than relying on regulatory enforcement.

Most importantly, Gen Z's digital fluency allows them to access independent information and hold brands accountable without relying on government oversight. Social media platforms enable them to track corporate sustainability efforts, expose unethical practices and mobilise consumer pressure. This empowerment may reduce their reliance on traditional institutions to drive change (Evans-Reber, 2022). However, while Gen Z may have lower expectations of institutional involvement, this does not mean they are disengaged. Instead, they are reshaping the way sustainability is approached, favouring decentralised, consumer-led movements over top-down regulations (Wood, 2022). This trend suggests that companies and governments must adapt by fostering genuine engagement, transparency and collaboration with younger consumers rather than relying solely on policy-driven initiatives (Theocharis and Tsekouropoulos, 2025).

The Role of Government and Industry Regulations

Gen Z is increasingly critical of traditional government regulations, often viewing them as insufficient, reactive and inadequate in addressing the evolving landscape of food safety and sustainability. This cohort expects brands to go beyond basic legal compliance and instead adopt self-imposed ethical and sustainability standards. According to the UK Food Standards Agency, Gen Z's rising activism and vocal demands are beginning to influence policy debates, signalling a shift in how food governance must evolve to meet generational expectations (BritainThinks, 2019).

Regulatory frameworks, such as the European Union's General Food Law Regulation (EC, n.d.) and the US Food and Drug Administration (FDA) Food Code (US FDA, 2022), establish foundational standards for food safety, but Gen Z remains sceptical about their effectiveness. They often perceive these regulations as slow to respond to emergent issues like climate change, ethical sourcing, supply chain transparency and corporate accountability. As one Gen Z with whom we discussed the issues about government role in food regulations shared:

> *"Just following the law isn't enough anymore. If brands really care, they'll show us what they're doing, where their food comes from, how it's made, who's behind it. We expect more than just ticking boxes."* (Female, born 1998)

This sentiment reflects the broader Gen Z's expectation for radical transparency and proactive responsibility. For this generation, simply meeting regulatory compliance is the bare minimum, not the gold standard. They expect brands to lead with purpose and clarity, not hide behind minimum standards. This generation's deep digital fluency enables them to scrutinise supply chains and corporate claims, often using social media as a platform to voice and amplify concerns and demand accountability.

A study by the UK Food Standards Agency found that Gen Z's relationship with food safety is shaped by a broader mistrust of traditional institutions and governmental oversight (FSA, 2020b). Unlike older generations who may place confidence in state assurances, Gen Z gravitates toward third-party certifications, blockchain-enabled traceability and independent audits (FSA, 2020b). They value systems that can offer verifiable, real-time proof of ethical and safety practices. This shift has catalysed a rise in consumer-led movements, where Gen Z exerts pressure on brands to not only comply with laws but to lead in areas such as animal welfare, low-impact production and fair labour practices.

> *"I trust what I can verify, not just what a brand or government tells me. Honestly, some of these regulations feel outdated... they don't reflect what matters to us now, like climate impact, fair treatment of workers, or whether the ingredients are actually safe long-term. The same applies to novel ingredients, especially with the newly emerging alternatives these days.... If a company has nothing to hide, then why not make their whole supply chain visible? I want to see real-time data, not vague claims or a stamp that says it passed some inspection months ago. I'm not saying we don't need government rules... we do, but they have to evolve and actually force companies to be accountable. If a brand is cutting corners or*

> *greenwashing, I want regulations that call that out, not just let it slide. It shouldn't be up to consumers to do all the detective work. I don't want to be forced to be a whistle-blower."* (Female, born 1998)

This perspective reflects a growing frustration among Gen Z with outdated regulatory frameworks that fail to address the complexity and urgency of today's food system challenges. While they value transparency from companies, they also call for modernised government oversight that does not merely react but anticipates risks, enforces ethical standards and aligns with the values of a digitally empowered and socially conscious generation.

This generational push is not without impact. Regulatory bodies are beginning to recalibrate their approaches in response to Gen Z's concerns. The UK Food Standards Agency, for example, has acknowledged the importance of tailoring its communication and engagement strategies to resonate with younger audiences who consume information differently and demand more interactive, transparent messaging (FSA, 2020a). In parallel, the European Commission has underscored the need for greater openness in regulatory processes and clearer communication about food safety standards to rebuild public trust (EC, n.d.).

Gen Z's influence represents a broader cultural shift from passive consumer compliance to active civic participation in shaping food safety norms. Their demand for accountability and governance innovation is pushing both regulators and industry players to reimagine what it means to be truly responsible in the modern food system.

Consumer Activism and Gen Z's Influence

Gen Z's distinctive activism is reshaping the food industry. This young cohort uses social media as a watchdog to expose unethical practices, leading to boycotts and petitions (Mitchell, 2024). They rarely engage in activities, such as signing petitions or making donations for particular causes. Instead, they use their personal networks, following influencers, liking posts and commenting to support issues of interest to them (Lammers, 2025). Unlike previous generations, their dominant consumer activism is through online engagement and their energetic presence is felt from a young age. Exposed to many major challenges, such as climate change, inequality, economic distress, global pandemic and political uncertainties, combined with easy access to vast bodies of information, including with smart phones, Gen Z has also become the anxious generation (Haidt, 2024) which justifies its explicit stances against human and planetary injustice. As Carnegie (2022) explains, a lot of their activism is driven by frustration and they demand and insist on change. An example of this is flooding the American multinational coffee chain Starbucks with fake job applications in a protest against the company's decision to fire workers who join the trade unions (Carnegie, 2022).

Another example of consumer activism against unethical palm oil sourcing is the Rainforest Action Network's campaign targeting major US snack food brands. The organisation exposed how companies like PepsiCo, Nestlé and Kellogg's were sourcing palm oil grown in areas linked to deforestation and human rights violations. This led to widespread consumer backlash, petitions and boycotts, pressuring these brands to adopt more sustainable palm oil policies (Rainforest Action Network, 2013; WWF, 2013). A Gen Z-led campaign against unethical palm oil sourcing is *Gen Z for the Trees* (Z4T), a youth-driven initiative by the Rainforest Partnership. This group, co-founded in 2020 by the Gen Z's Roshan Khan, has actively investigated palm oil supply chains and pushed for corporate accountability (RSPO, 2023). Roshan Khan encourages other Gen Z members to promote and support sustainable palm oil. The group's mission is to reach global zero deforestation by 2030 by empowering youth to lead rainforest protection efforts through policy, research and advocacy (Rainforest Partnership, 2025). One of the group major efforts was the Palm Oil Project, researching supply chains and uncovering over 300 unlisted palm oil mills linked to deforestation (Khan and Ziah, 2021). As a Gen Zer explained:

> *"I am part of the Palm Oil watch project. Now when I found out that a major snack brand was sourcing palm oil from deforested areas, I stopped buying it immediately.... and so did most of my friends. We are followers of Roshan Khan's efforts to stop these*

Figure 11.1. Gen Z using social media for food campaigns (ChatGPT-created image).

Source: https://chatgpt.com/c/684d7196-7bb0-8004-97c3-24f0bbe58e04.

> *unsustainable practices. You know, we do things like posting in social media about it, but not just posting about it, but we tag the brand, share petitions and make sure people knew. It's not just cancel culture... it's accountability."* (Female, born 2003)

This illustrates how Gen Z consumers do not passively use information, but they mobilise it. Social media platforms become spaces for awareness as well as for action (see Figure 11.1). The *Gen Z for the Trees* activism is only one of many examples. Another impactful movement was the *All Eyes on Papua* campaign, which gained traction online by highlighting deforestation caused by palm oil plantations in Indonesia. This campaign mobilised Gen Z activists to support Indigenous tribes fighting against land destruction, demonstrating how digital activism can drive real-world change. Indicative of the impact of the campaign is the fact that for a week, the *All Eyes on Papua* image was shared nearly three million times on Instagram (France24, 2024). Digital boycotts, when widely shared, can have measurable economic and reputational impact, pressuring brands to change sourcing practices or publicly address ethical concerns. The collective, fast-moving nature of these campaigns makes them uniquely powerful tools for systemic change.

Evidence shows that around two-thirds (66%) of Gen Z's activism takes place online, where they effectively use social media platforms, including TikTok, Instagram, Snapchat, X (formerly Twitter), Reddit and Discord, for viral campaigns which amplify their voices and drive change (Rodriguez, 2024). While Gen Z leads in digital activism with most efforts happening online, about 34% of them also engage in real-world actions like rallies and protests (Rodrigues, 2024). These in-person activities complement their online presence, creating a powerful mix that strengthens their message and demonstrates genuine commitment to the causes they support (Rodriguez, 2024).

A notable example of Gen Z activism in Australia is the 2024 Grill'd fast-food worker strike in Melbourne. In October 2024, employees at the popular burger chain Grill'd staged a walkout to protest underpayment and exploitation (The Citizen, 2024). This was considered Australia's first-ever fast-food worker strike (Katanasho, 2024). Young workers, many of whom were Gen Z, accused the company of using traineeship programs to legally pay them below minimum wage, some earning as little as $14.90 per hour. The protest included flyers with quick response (QR) codes linking to boycott campaigns, social media activism and direct engagement with unions. The strike gained national attention, pressuring Grill'd to increase wages and reconsider its employment practices. This example highlights how Gen Z uses digital tools and direct action to demand corporate accountability. One of the Gen Z individuals we talked with reflected on this blended approach:

> *"Last year my friends in social media went to a protest outside a fast-food chain in Melbourne that was underpaying workers. Grill'd wasn't fairly treating their workers. They handed out flyers with QR*

codes linking to boycott campaigns and videos of employees speaking out. I was sharing things on Instagram. It's about connecting the dots... like the practices and who's affected." (Male, born 2001)

Similar activism campaigns were organised in the US against McDonald's and Chipotle where workers protested over poor working conditions and low wages (Alfonseca, 2021). In various American cities, fast-food employees and activists staged demonstrations, demanding fair pay and better labour conditions. Protesters have used flyers, QR codes and social media campaigns to spread awareness and encourage consumer boycotts (Alfonseca, 2021). Another example is the boycott of the burger fast food restaurants Wendy's in the UK, where activists have called out the chain for failing to sign onto the Fair Food Program, which ensures fair wages and safe working conditions for farmworkers. Protesters have used digital petitions and viral campaigns to pressure Wendy's to change its sourcing practices (Owens, 2023).

These examples reflect how Gen Z bridges online and offline activism, transforming digital outrage into physical presence. These efforts are not isolated gestures but part of a broader, coordinated movement demanding systemic accountability. Livestreams, QR codes and visual storytelling help merge digital and physical spaces, demonstrating how Gen Z turns everyday consumption choices into political acts. Gen Z's activism thrives on exposure. A single viral post can damage a brand's reputation, making corporate accountability more critical than ever. Brands must engage authentically with consumers, addressing concerns before they escalate into major crises.

Brands Adapting to Gen Z's Transparency Standards

Gen Z's demand for transparency has transformed the way brands communicate, operate and build consumer trust. Companies that successfully meet these expectations thrive, while those that fail face scrutiny and declining consumer loyalty. Brands that excel in transparency through authenticity, social responsibility and inclusive messaging, and openly share their values and sustainability efforts resonate with Gen Z. According to McKinsey, Gen Z prefers brands with a strong story or purpose (Storozhuk, 2024). Companies that align with Gen Z's social consciousness, particularly through sustainability initiatives, gain loyalty and advocacy. Moreover, businesses that embrace diversity and inclusivity in their messaging strengthen the relationship with Gen Z, a generation that values ethical engagement (Storozhuk, 2024).

However, brands that fail to clearly communicate their values and ethical practices risk losing consumer trust. Research from Edelman (2024) highlights how lack of clear communication leads to scepticism and disengagement. Companies that attempt to obscure their supply chains or fail to uphold their ethical commitments often experience public criticism and declining sales, fuelled by Gen Z's ability to mobilise social media campaigns against them (Edelman, 2024). Even when a company decides not to address an issue of concern to Gen Z, this is interpreted as taking a stance and the social media are the frontline for building trust (Edelman, 2024).

To stay ahead, brands must prioritise defining and communicating their values, ensuring that transparency is genuine rather than performative. Edelman (2024) suggests that businesses must actively engage in meaningful storytelling, making transparency not just about disclosure but about creating an authentic narrative that resonates with consumers. Successful companies commit to aligning their actions with their messaging, reinforcing their credibility and avoiding reputational damage (Storozhuk, 2024).

All these trends indicate that transparency is no longer optional — it is a competitive advantage that determines brand success. Several brands have successfully adapted to Gen Z's transparency demands, gaining loyalty and market share. Companies that fail to adapt face consumer-driven activism and market exclusion, while those that embrace transparency secure long-term loyalty and influence industry standards (Wilson, 2021). As food technology advances, transparency becomes more complex, presenting both opportunities and challenges for companies and brands.

Conclusion

Generational cohorts are shaped by the common experiences they go through during their formative years (Debevec *et al.*, 2013). The COVID-19 global pandemic was such a turbulent event which shaped Gen Z's views about health and transparency. This was intertwined with the stronger emerging evidence about the importance of sustainability, climate action and demand for social justice which continue to be essential involvements for Gen Z. They are all reflected in the way these young people relate to food.

Gen Z's unwavering demand for transparency is transforming the foundations of the food industry. Their insistence on clear labelling, traceability and ethical sourcing is more than a passing trend. It is a cultural shift redefining what trust and accountability look like in food systems. These values are not optional extras but essential criteria that determine which brands they support and which they reject.

For Gen Z, transparency is not a branding strategy, it is the baseline for ethical consumption. In a world of misinformation and greenwashing, this generation rewards honesty, even when it reveals imperfections. Companies that adopt a transparent, consumer-centric approach, acknowledging both their progress and shortcomings, will earn long-term loyalty and respect. In contrast, those that obscure their practices or fall short of ethical standards risk not just public backlash, but obsolescence.

As Gen Z continues to mature into the dominant consumer force, their values will shape the global food safety and sustainability agenda. Their digital fluency, activist mindset and global awareness are already influencing how regulations are written, how food is marketed and how safety is communicated. Transparency is no longer a nice-to-have, it is the competitive edge in an era of empowered, informed consumers. Ultimately, this generational shift is not just reshaping brands, but raising the bar for the entire food ecosystem, ushering in a future where integrity, openness and accountability define food desirability and consumer trust.

References

Alfonseca, K. (2021). Restaurant workers protest poor working conditions & low wages at McDonalds & Chipotle; incl. company comments. Business and Human Rights Resource Centre. https://www.business-humanrights.org/en/latest-news/restaurant-workers-protest-poor-working-conditions-low-wages-at-mcdonalds-chipotle-incl-company-comments/

Blackstone, N. T., Rodríguez-Huerta, E., Battaglia, K., Jackson, B., Jackson, E., Norris, C. B., & Sparks, J. L. D. (2023). Forced labour risk is pervasive in the US land-based food supply. *Nature Food*, 4, 596–606. https://doi.org/10.1038/s43016-023-00794-x

BritainThinks (2019). Future Consumer: Food and Generation Z. Food Standard Agency. https://www.food.gov.uk/sites/default/files/media/document/fsa191102bt-gen-z_rea_full-report_081019-final.pdf

Carnegie, M. (2022). Gen Z: How young people are changing activism. BBC. https://www.bbc.com/worklife/article/20220803-gen-z-how-young-people-are-changing-activism

Confetto, M. G., Covucci, C., Addeo, F., & Normando, M. (2023). Sustainability advocacy antecedents: How social media content influences sustainable behaviours among Generation Z. *Journal of Consumer Marketing*, 40(6), 758–774. https://doi.org/10.1108/JCM-11-2021-5038

Czernyszewicz, E. & Wiśniewska, M.Z. (2023). The determinants of trust in food safety from the perspective of Generation Z students. *Annales Universitatis Mariae Curie-Skłodowska, sectio H — Oeconomia*, 57(3), 57–79. http://dx.doi.org/10.17951/h.2023.57.3.57-79

Czernyszewicz, E. & Wiśniewska, M.Z. (2024). Credibility of food processing companies in terms of food safety: Perspective of Generation Z consumers. *Emerald Insight*, 32(2), 320–335. https://doi.org/10.1108/CEMJ-10-2022-0175.

Debevec, K., Schewe, C. D., Madden, T. J., & Diamond, W. D. (2013). Are today's millennials splintering into a new generational cohort? Maybe! *Journal of Consumer Behaviour*, 12(1), 20–31. https://doi.org/10.1002/cb.1400

Deloitte. (2025). 2025 Gen Z and Millennial Survey. Growth and the pursuit of money, meaning and well-being. https://www.deloitte.com/content/dam/assets-shared/docs/campaigns/2025/2025-genz-millennial-survey.pdf

Dragolea, L., Butnaru, G., Kot, S., Zamfir, C., Nuta, A., Nuţă, F.... Stefănică, M. (2023). Determining factors in shaping the sustainable behaviour of the generation Z consumer. *Frontiers in Environmental Science*, 11, 1096183. https://doi.org/10.3389/fenvs.2023.1096183

Edelman, A. (2024). Winning with Gen Z: Embracing intention and values for brand success. https://www.edelman.com/trust/2024/trust-barometer/special-report-brand/gen-z-embracing-intention-values-brand-success

Ellahi, R. M., Wood, L. C., & Bekhit, A. E.-D. A. (2023). Blockchain-based frameworks for food traceability: A systematic review. *Foods*, 12(16), 3026. https://doi.org/10.3390/foods12163026.

Evans-Reber, K. (2022). How to meet Gen Z's workplace expectations. Forbes. https://www.forbes.com/councils/forbeshumanresourcescouncil/2021/11/10/how-to-meet-gen-zs-workplace-expectations/

European Commission (EC). (n.d.). General Food Law. https://food.ec.europa.eu/horizontal-topics/general-food-law_en

Food Standards Agency (FSA). (2020a). Food and Generation Z. FSA. Generation Z Executive Summary. https://www.food.gov.uk/sites/default/files/media/document/fsa191102bt_executive-summary-140220-final.pdf

Food Standards Agency (FSA) (2020b). The Future Consumer — Food and Generation Z. https://cdn.prod.website-files.com/6578727b2437b2ef1c478c1b/65afbc460c86e87d60bafdaa_FSA-BT-Food-and-Gen-Z-Full-Report.pdf

France24. (2024). 'All Eyes on Papua' campaign generates interest in deforestation cases. https://www.france24.com/en/live-news/20240605-all-eyes-on-papua-campaign-generates-interest-in-deforestation-cases

Haidt, J. (2024). *The Anxious Generation: How the Great Rewiring of Childhood is Causing an Epidemic of Mental Illness.* Alen Lane, London.

Halicka, E., Kaczorowska, J., Rejman, K., & Plichta, M. (2025). Investigating the consumer choices of Gen Z: A sustainable food system perspective. *Nutrients*, 17(3), 591. https://doi.org/10.3390/nu17030591

Innova Market Insights. (2024). European Consumer Trends: Sustainability and Ethics. https://www.innovamarketinsights.com/trends/european-consumer-trends-sustainability/

Jakubowska, D., Dąbrowska, A. Z., Pachołek, B., & Sady, S. (2024). Behavioural intention to purchase sustainable food: Generation Z's perspective. *Sustainability*, 16(17), 7284. https://doi.org/10.3390/su16177284

Just, B. (2024). Study: Gen Z wants more transparency in food sustainability, quality. Food Business News. https://www.foodbusinessnews.net/articles/26985-study-gen-z-wants-more-transparency-in-food-sustainability-quality

Kabaja, B., Wojnarowska, M., Ćwiklicki, M., Buffagni, S. C., & Varese, E. (2023). Does environmental labelling still matter? Generation Z's purchasing decisions. *Sustainability*, 15(18), 13751. https://doi.org/10.3390/su151813751

Katanasho, G. (2024). Grill'd faces Australia's 'first-ever' fast food strike over low-pay, 'unfair' conditions claims. SBS. https://www.sbs.com.au/news/article/grilld-hit-with-australias-first-ever-fast-food-strike/zw3tlraqe

Khamitov, M., Rajavi, K., Huang, D-W., & Hong, Y. (2024). Consumer trust: Meta-analysis of 50 years of empirical research. *Journal of Consumer Research*, 51(1), 7–18. https://doi.org/10.1093/jcr/ucad065

Khan, R. & Ziah, J. (2021). Mapping Corporate Palm Oil: Gen Z for the Trees Uncovers over 300 Unlisted Mills. Rainforest Partnership. https://www.rainforestpartnership.org/blog/mapping-corporate-palm-oil-gen-z-for-the-trees

Lammers, J. (2025). How Gen Z Engages in Activism Online. American Student Assistance. https://www.asa.org/research/how-gen-z-engages-in-activism-online

Kymäläinen, T., Seisto, A., & Malila, R. (2021). Generation Z food waste, diet and consumption habits: A Finnish social design study with future consumers. *Sustainability*, 13(4), 2124. https://doi.org/10.3390/su13042124

Mahendra, S. (2024). Blockchain and AI for Food Traceability, AI Plus Info. https://www.aiplusinfo.com/blog/blockchain-and-ai-for-food-traceability/

Marinova., D. & Bogueva, D. (2022). *Food in a Planetary Emergency.* Springer, Singapore.

McKeever, M., O'Rourke, V., & Diffley, S. (2021). Generation Z: An exploration of their unique values driving brand affinity. Irish Academy of Management Conference. https://www.researchgate.net/publication/357187322_Generation_Z_an_exploration_of_their_unique_values_driving_brand_affinity

Mitchell, J. (2024). Gen Z Hungry for Transparency in Food Sustainability, Quality Amid Climate Concerns, According to New Survey. Quality Assurance and Food Safety. https://www.qualityassurancemag.com/news/gen-z-hungry-for-transparency-in-food-sustainability-quality-amid-climate-concerns-according-to-new-survey/

Mitchell, B. & Topic, M. (2023). Generation Z and consumer trends in environmental packaging. The Retail Institute, Leeds Beckett University. https://www.leedsbeckett.ac.uk/-/media/files/business-services/the-retail-institute/final_report-genz-compressed.pdf

Owens, J. (2023). Wendy's: the boycotted fast food chain arriving near you. Ethical consumer. https://www.ethicalconsumer.org/food-drink/wendys-boycott-fast-food

Packaging Technology Today (2024). Survey finds Gen Z is hungry for transparency in food sustainability and quality amid climate concerns. https://www.packagingtechtoday.com/markets/food/survey-finds-gen-z-is-hungry-for-transparency-in-food-sustainability-and-quality-amid-climate-concerns/

Rainforest Action Network (2013). Conflicting Palm Oil. https://www.ran.org/wp-content/uploads/rainforestactionnetwork/pages/2367/attachments/original/1405470759/conflict_palm_oil_lowres_%281%29.compressed.pdf

Rainforest Partnership. (2025). Gen Z for the Trees. https://www.rainforestpartnership.org/gen-z-for-the-trees

Rahn, W. M. & Transue, J. E. (1998). Social trust and value change: The decline of social capital in American youth, 1976–1995. *Political Psychology*, 19(3), 545–565. https://doi.org/10.1111/0162-895X.00117

Rand, S. (2025). Gen Z food trends: What today's young US consumers want. Attest. https://www.askattest.com/blog/research/gen-z-food-trends

Robichaud, Z. & Yu, H. (2021). Do young consumers care about ethical consumption? Modelling Gen Z's purchase intention towards fair trade coffee. *British Food Journal*, 124(9), 2740–2760. https://ijcrt.org/papers/IJCRT2403577.pdf

Rodriguez, I. (2024). Gen Z Activism: Digital Warriors Leading the Change. Woke Waves. https://www.wokewaves.com/posts/gen-z-digital-activism-leaders

Roodenburg, A. J. C., Hanssen, N., & van Santen, G. (2023). Marketing or transparency? A study into misleading labelling: With food experts. *Consumers and the Food Sector. Proceedings*, 91(1), 84. https://doi.org/10.3390/proceedings2023091084.

Roundtable on Sustainable Palm Oil (RSPO). (2023). Mobilising Gen Z to save the forests. Meet Rainforest Partnership's Roshan Khan. https://rspo.org/mobilising-gen-z-to-save-the-forests-meet-rainforest-partnerships-roshan-khan/

Salam, K. N., Singkeruang, A. W. T. F., Husni, M. F., Baharuddin, B., & Pratiwi, D. A. R. (2024). Gen-Z marketing strategies: Understanding consumer preferences and building sustainable relationships. *Golden Ratio of Mapping Idea and Literature Format*, 4(1). https://doi.org/10.52970/grmilf.v4i1.351

Storozhuk, A. (2024). How brands can adapt their content strategies to resonate with Gen-Z. Forbes. https://www.forbes.com/councils/forbesbusinesscouncil/2024/04/16/how-brands-can-adapt-their-content-strategies-to-resonate-with-gen-z/

The Citizen. (2024). Grill'd workers walk off the job, accusing fast food chain of underpayment, exploitation. https://www.thecitizen.org.au/articles/grilld-workers-walk-off-the-job-accusing-fast-food-chain-of-underpayment-exploitation

Theocharis, D. & Tsekouropoulos, G. (2025). Sustainable consumption and branding for Gen Z: How brand dimensions influence consumer behaviour and adoption of newly launched technological products. *Sustainability*, 17(9), 4124. https://doi.org/10.3390/su17094124

Thompson, P. B. (2023). The Ethics of Food Labels, From Silo to Spoon: Local and Global Food Ethics, New York. https://doi.org/10.1093/oso/9780197744727.003.0006

US Food and Drug Administration (US FDA) (2022). Food Code. https://www.fda.gov/food/retail-food-protection/fda-food-code

Van Niekerk, M. (2024). From source to stomach: How blockchain tracks food across the supply chain and saves lives. World Economic Forum. https://www.weforum.org/stories/2024/08/blockchain-food-supply-chain/

Wilson, S. (2021). Where brands are reaching Gen Z? Harvard Business Review. https://hbr.org/2021/03/where-brands-are-reaching-gen-z

Wood, J. (2022). Gen Z cares about sustainability more than anyone else — and is starting to make others feel the same. World Economic Forum. https://www.weforum.org/stories/2022/03/generation-z-sustainability-lifestyle-buying-decisions/

Worldwide Fund for Nature (WWF). (2013). RAN Launches Palm Oil Campaign Targeting US Snack Companies. https://wwf.panda.org/wwf_news/?208293/RAN-Launches-Palm-Oil-Campaign-Targeting-US-Snack-Companies

Chapter 12

The Cutting Edge: AI, AlphaFold and New Food Technology

"AI will be an integral part of solving the world's biggest problems, but it must be developed in a way that reflects human values."

— Satya Nadella, CEO of Microsoft

"The future of AI is not about replacing humans, it's about augmenting human capabilities."

— Sundar Pichai, CEO of Google

As researchers investigating Gen Z, we set out to explore these young people's evolving relationship with the fast-moving frontlines of technological change, where artificial intelligence, protein-folding breakthroughs like AlphaFold, and next-generation food systems intersect. What we uncovered is a generation standing at the cusp of transformation. We found a generation arriving at a pivotal moment, a technological revolution that is rapidly redefining and reshaping how food is grown, created, experienced and consumed.

Gen Z is not simply riding this wave of food innovation, it is helping steer it. Born into a world marked by climate disruption, social movements and digital saturation, it is as ethically grounded as it is digitally fluent. This is a generation driven by both rapid change and values, pairing technical curiosity with moral clarity. For this young cohort, the future of food is not just about efficiency or novelty, it is about justice, transparency and sustainability. They demand that innovation serve a purpose beyond profit, shaping a world where technology enhances, rather than erases, what it means to be human.

From recipe apps powered by artificial intelligence (AI) to sustainable food innovations like precision fermentation and lab-grown meats, Gen Z is interested and embracing food tech. However, they are doing it with eyes wide open. They are not just asking what is on their plate; they are asking how it got there, who is impacted, why it matters and what it says about the future they want to build. This blend of curiosity and caution aligns with the words of two global technology leaders. Microsoft's CEO Satya Nadella reminds us that AI will be part of solving the world's biggest challenges, but it will have to reflect human values. His call for human values in AI mirrors Gen Z's demand for transparency and fairness in food systems.

Google's Sundar Pichai also echoes this by emphasising that the role of AI is not to replace humans, but increase their capabilities. This idea about complementing human ability and not displacing it, is in particular relevant and resonates strongly with Gen Z. Gen Z does not want AI to take over their food choices, they want it to enhance them — to help them cook smarter, eat better and live more sustainably, without losing the joy, creativity and cultural roots of food. They are also inquisitive about what AI can offer to improve dietary choices (see Figure 12.1).

From AI that creates individual meal plans to robots inspecting food safety in real time, food-related AI is reshaping the industry with speed, precision and personalisation. We are drawn to an extraordinary intersection, where cutting-edge AI, molecular science and cultural values meet at the table. This technological leap matters deeply to Gen Z because it embodies the kind of innovation they support — science-driven, purpose-led and ethically aligned. AlphaFold — an AI database and system developed to predict protein structure from amino acid sequences that is freely available (AlphaFold, 2023), is not just a lab tool. It is a lens through which Gen Z can see their food values materialise. It helps create alternatives that are not only functional and scalable but respectful of planetary and bodily health.

Figure 12.1. Gen Z curious and unsure about AI in food.

Source: Infographic by authors.

This chapter uncovers how Gen Z is forging a food future that is digitally informed and deeply human. It is not just about what is next, but what is right. As one of the Gen Z's stated:

> *"We want tech that doesn't just make food faster — but smarter, cleaner and true to who we are."* (Female, born 2004)

This sentiment summarises the generational mindset that clearly shows AI should serve as a tool for enriching the cultural, ethical and ecological dimensions of food, not diminishing them. The promise of food technologies lies not only in their scientific courage but in their ability to align with and amplify the aspirations of a generation redefining the meaning of nourishment in the digital age. It is difficult to cover all aspects in which AI will affect food systems and dietary intake but we explore in more detail perspectives related to AI-driven innovation, contribution to personalised nutrition and well-being, AI-facilitated sustainability and ethical values, organisation of food delivery and safety as well as dealing with data. Ultimately, Gen Z's expectations regarding an AI-reinforced future of food are discussed.

Food and AI

Pairing food and AI is crucial for global transformation, because food systems sit at the intersection of some of the biggest global challenges and opportunities of our time: climate change, food security, health and technological transformation. This convergence is reshaping how we grow, distribute, consume and understand food. With the global population projected to exceed 9 billion by 2050, AI is helping optimise agricultural productivity, reduce waste and improve supply chain efficiency.

Precision agriculture and yield prediction models driven by AI are already enhancing food security in regions vulnerable to climate shocks (Sarku *et al.*, 2023; Ahmad *et al.*, 2024). Smarter resource use in agriculture, disease management and crop yield prediction are supported by AI (Khan *et al.*, 2025). Emissions reduction is also facilitated by streamlining logistics and minimising food waste (Marwala, 2024). Meal planning and product development based on individual health data, allergies and lifestyle goals can be tailored with AI. Tools, such as goFOOD™ and NutrifyAI, use image recognition and biometric data to deliver real-time nutritional analysis (Joshi *et al.*, 2024; Han *et al.*, 2024). Some claim that AI is revolutionising sustainability around the globe by changing our food systems (Khan *et al.*, 2025)

Food safety is another important aspect which is no longer a behind-the-scenes laboratory issue. It is a real-world problem that affects what Gen Z eats every day — from smoothies and snacks to takeout/takeaway and plant-based meals. Worries about adulterations, hidden additives, incorrect labelling, spoilage and contamination in everyday foods are increasing rapidly, making it harder to trust what is on the plate. Traditional testing methods are struggling to keep pace. They are often too slow, costly and not suited for the fast-moving, technology-driven food environment Gen Z navigates. Machine learning combined with AI is stepping in to change the game, bringing real-time monitoring, smart detection and more transparency to the food system (Balakrishnan *et al.*, 2025). For a generation that values health, technology and trust, AI could be the key to making food safer, faster and more reliable. Food safety can be enhanced by AI through detecting contaminants, predicting spoilage and improving traceability with blockchain and machine learning (Liu *et al.*, 2023).

Furthermore, AI accelerates research and development (R&D) for alternative proteins, including precision fermentation and cultivated meat. Start-ups, such as NotCo, use AI to analyse molecular structures of animal products to mimic and replicate them with plant-based ingredients (Callaway, 2023; Wang *et al.*, 2025). Perfect Day applies precision fermentation and AI to produce dairy proteins without animals (GEA, 2024).

In fact, AI is transforming the entire food systems from farm to fork across production, distribution and consumption. Around production, we see precision agriculture, autonomous farming, yield prediction and climate adaptation tools. Machine learning, deep learning and time series analysis can optimise agricultural processes, including crop selection, yield estimates, soil compatibility and water management requirements (Ali *et al.*, 2025). Precision agriculture, autonomous tractors and climate-adaptive crop models are improving yields and resilience (Bhagat *et al.*, 2022). For distribution, we see smart logistics, demand forecasting and waste reduction systems improving the responsiveness of supply chains (Yang *et al.*, 2025). In consumption, we see AI in meal planning apps, voice assistants, chatbot nutritionists and smart kitchen devices. Example case studies of start-ups companies are NotCo, Perfect Day and IBM Food Trust. Companies also leverage AI and blockchain to enhance traceability and safety (Dhal and Kar, 2025). Dining experience is similarly being transformed with the use of augmented reality which allows browsing through menus and interactions in real time (Styliaras, 2021).

Personalisation, Wellness and Food Identity

Gen Z's pursuit of personalised wellness is driving a new era of food technology, but it must be guided by ethical design, cultural sensitivity and scientific integrity. This generation's deep interest and investment in wellness, identity and ethical living is reshaping the food landscape, and AI is at the heart of this transformation.

As the majority of Gen Z consumers are prioritising health and sustainability in their food choices, the demand for personalised nutrition has surged (Halicka *et al.*, 2025), prompting innovation in AI-powered platforms like NutrifyAI and DNA-based diet apps that tailor recommendations to individual biomarkers, lifestyle goals and cultural preferences (Oewel *et al.*, 2024).

The development of customisable plant-based products and functional foods is also being revolutionised by AI. By analysing consumer feedback, sensory data and nutritional profiles, AI enables the creation of products that reflect Gen Z's values, whether it is gut-friendly kombucha, adaptogenic snacks or protein-rich meat alternatives. Research from the Australian Commonwealth Industrial and Research Organisation (CSIRO) shows that AI can improve the emulsification and heat stability of plant proteins, making them more versatile and appealing (Ral *et al.*, 2025).

However, personalisation is not just about health; it is also about identity. Gen Z views food as a form of self-expression and self-care. Examples of this are AI-enhanced apps that recommend meals based on mood, menstrual cycles or mental health data. They are gaining popularity, reinforcing the emotional connection between food and well-being. Conversely, this intimacy also raises ethical dilemmas. Studies warn of algorithmic bias, where AI systems may reinforce narrow beauty standards, misinterpret cultural diets or promote products based on commercial incentives rather than health outcomes (Bailey *et al.*, 2024).

There is also concern around greenwashing, where AI-generated sustainability claims lack verification. Without transparent algorithms and inclusive datasets, personalisation risks becoming a marketing tool rather than a wellness link. Ethical frameworks, including the European Union's AI Act and explainable AI models, e.g. Local Interpretable Model-Agnostic Explanations (LIME), are being explored to ensure fairness, autonomy and justice in nutritional technology. Parallels can be drawn with the application of AI in healthcare where it similarly holds great promise but its widespread adoption needs to be governed by ethical safeguards and regulatory oversight to balance potential benefits against risks (Mohammed and Malhotra, 2025).

Sustainability and Ethical Consumption

Gen Z is widely recognised as the most climate-conscious generation. Research across 31 countries shows over 70% of Gen Z being extremely worried about environmental harm caused by human activity with 38% feeling stressed or anxious most or all the time (Malmqvist, 2025). This anxiety translates into action. Gen Z is more likely than older cohorts to choose sustainable brands, demand ethical sourcing and scrutinise corporate green claims (Bardales Salguero *et al.*, 2024). Increasingly, AI technologies are being leveraged to meet these expectations, offering tools that enhance sustainability across the food system. A Gen Zer explains:

> *"It's not enough for a product to say it's 'eco-friendly'... you know. I want to know the receipts. If tech can help show the full journey, from soil where it was grown to its harvesting and transportation before placing it on the shelf, then I'm more likely to trust it. But it needs to tell the whole story, not just the polished, exciting parts."* (Male, born 2000)

Gen Z's demand for sustainability is supported by AI. A prime example are AI-powered food waste tracking systems. These tools use predictive analytics and real-time monitoring to reduce spoilage, optimise inventory and improve shelf-life management, especially in supermarkets and hospitality settings. Studies show that AI can significantly reduce food waste when integrated with smart sensors and inventory platforms (Khan *et al.*, 2025; Clark *et al.*, 2025; Onyeaka *et al.*, 2023). It also supports life cycle assessment by modelling environmental impacts across supply chains, enabling companies to make data-driven decisions that lower their carbon footprint (Onyeaka *et al.*, 2023).

Machine learning algorithms can verify sustainability claims, while blockchain ensures traceability from farm to fork. These technologies help combat greenwashing and build consumer trust

(Bhagat *et al.*, 2022; Balzarova, 2021; Wu *et al.*, 2022). However, concerns remain about algorithmic bias, especially when AI systems rely on incomplete or Western-centric datasets that may misrepresent global sustainability practices or exclude Indigenous food systems.

Ultimately, Gen Z's climate anxiety is not just emotional, it is a driver of systemic change. They expect AI to be a tool for ethical consumption, not just convenience. To meet this demand, developers must ensure that sustainability algorithms are inclusive, transparent and accountable.

AI and Food Delivery

Food delivery logistics are also transformed by AI by optimising routes, predicting demand spikes and reducing fuel consumption. Services like Uber Eats, Instacart and major grocery chains rely on machine learning algorithms and real-time traffic data to streamline operations. These systems not only improve delivery speed and customer satisfaction but also cut emissions and operational costs, aligning with Gen Z's climate-conscious values.

Logistics platforms powered by AI use reinforcement learning and predictive analytics to dynamically adjust delivery schedules and vehicle routing based on weather, traffic and order volume (Lalla-Ruiz and Mes, 2025; du Plessis *et al.*, 2025). Studies show that AI can reduce last-mile delivery emissions by up to 20%, especially when integrated with electric vehicle fleets and smart warehousing systems (Sodiya *et al.*, 2024).

Gen Z's preference for on-demand, eco-friendly services makes AI-driven logistics a key enabler of ethical convenience. However, caution is needed as over-optimisation can lead to worker exploitation in gig-based delivery platforms and algorithmic bias may prioritise high-income areas for faster service. Responsible design must balance efficiency with fairness and labour ethics to line up with Gen Z's values.

AI as a Guardian of Food Safety and Integrity

Food safety is another critical area where AI is making a significant impact. Gen Z expect transparency and accountability in how their food is sourced, stored and served, and AI is stepping up to meet those expectations. Computer vision systems and AI-based sensors can detect contaminants, monitor hygiene conditions and flag spoilage in real time (Liu *et al.*, 2023). These tools are increasingly used in processing plants, cold chains and retail environments to ensure compliance with standards like Hazard Analysis Critical Control Points (HACCP) and the food safety management systems ISO 22000 (Palakurti, 2022).

A further use of AI is to support traceability by linking supply chain data to blockchain systems, allowing consumers to verify origin, handling and safety protocols. Gen Z's demand for clean label transparency and ethical sourcing makes these technologies not just useful, but essential. However, concerns remain around data privacy, especially when biometric or behavioural data are used to personalise safety alerts or nutritional recommendations.

Data Dilemmas: Privacy, Trust and Transparency

The power of AI comes with ethical tensions, especially in food systems where personal health, cultural identity and consumption habits intersect. Increasingly collected with AI systems are sensitive biometric and behavioural food data, such as dietary patterns, gut microbiome profiles and emotional responses to food. According to the experts, by 2029 generative AI will not take over most human decision-making, but instead it will enhance it (Momentum, 2024). Around 80% of the choices people make will still involve human judgment, but AI will play a key supporting role by providing insights, suggestions or creative input to help people make better, faster and more informed decisions (Werran, 2024). Smart kitchen devices and wearables raise questions about consent, data ownership, surveillance and data use. For example, voice-activated nutrition assistants may record conversations or infer health conditions without explicit user awareness (Alexander *et al.*, 2024). Building trust requires multi-stakeholder engagement, including farmers, consumers and developers. Studies show that ethical AI adoption in agriculture and food systems hinges on clear data governance and participatory design (Alexander *et al.*, 2024).

There is a growing demand for transparent and interpretable algorithms in food recommendations and labelling (e.g. explainable AI). While protecting user privacy and working to reduce algorithmic bias, AI has the potential to encourage healthier eating habits by offering personalised nutrition guidance and smart food choices. When designed responsibly, AI can support well-being without compromising ethical standards or personal data. Techniques like LIME (Local Interpretable Model-Agnostic Explanations) are being explored to make AI decisions more understandable to consumers (Naithani *et al.*, 2025).

The EU AI Act is a landmark policy aiming to regulate high-risk AI applications, including those in health and food technology (European Union, 2024). It mandates transparency, human oversight and risk mitigation for AI systems that impact fundamental rights. Its applicability and compliance are across all businesses operating within the Act's jurisdiction, setting a global precedent for AI governance with non-compliances leading to significant penalties. The EU AI Act introduces a risk-based framework that categorises AI systems into four levels, including unacceptable, high, limited and minimal risk, with stricter rules for higher-risk applications. It mandates transparency, requiring businesses to clearly inform users when they are interacting with AI and to label AI-generated content. Companies must also conduct risk assessments, maintain detailed documentation and ensure compliance with data protection standards (Werran, 2024). The AI Act represents a significant step toward fostering trustworthy AI, balancing innovation with the protection of fundamental rights and public interest. Young people will be particularly affected as AI develops further and it is important for their rights to be protected.

Culturally Relevant and Inclusive AI

Gen Z demands food systems that reflect global diversity and cultural authenticity, and AI has both the potential to fulfill this vision and the risk of undermining it by reinforcing dominant food cultures. Its role in recipe generation, cultural preservation and the inclusion of minority cuisines is increasingly important. With the help of AI, culturally nuanced dietary needs can be supported by generating personalised recipes based on individual profiles, including gut microbiota, allergies, food intolerances and mental health data (Tsolakidis *et al.*, 2024). Gen Z is increasingly vocal about the need for food systems that reflect global diversity, authenticity and cultural integrity. As digital natives with a strong sense of social responsibility, many in Gen Z see AI as a tool that can either bridge cultural gaps or widen them. On one hand, AI holds the promise of supporting diverse dietary needs through personalised meal planning; on the other hand, when trained on biased or limited datasets, AI risks reinforcing dominant food norms while marginalising minority cuisines.

Religious dietary compliance is a further area where AI's growing role in cultural preservation is particularly evident. Halal and kosher food systems are already benefiting from AI and blockchain innovations, which are being used to verify ingredients, monitor processing and automatically generate compliant recipes. For example, AI-driven halal platforms now automate ingredient checks and certification workflows (Sunmola *et al.*, 2025; Ahmed, 2024), while Natural Language Processing (NLP) tools are helping to curate kosher recipes with cultural and spiritual sensitivity (Jha, 2024). Similarly, tools like NutriGenie and culturally adaptive chatbots offer tailored dietary support for halal, kosher and Indigenous foodways (Li *et al.*, 2024).

Despite this progress, structural limitations persist. Current AI-powered nutrition apps often misidentify culturally complex or blended meals, such as biryani, pozole or injera with stews, due to their reliance on databases built around Western food structures (Li *et al.*, 2024). These inaccuracies lead to distorted nutritional analysis and can perpetuate misrepresentations of non-Western diets (Li *et al.*, 2024). More critically, they signal a deeper issue: algorithmic homogenisation, the subtle erasure of culinary identities through systems that fail to reflect global diversity and are designed around dominant food norms. This is especially concerning to Gen Z, whose food choices are often rooted in heritage, ethics and identity. As one young Gen Z person reflected:

> *"I want my food apps to recognise the meals I grew up with, not just count calories, but actually understand what's on my plate. Technology shouldn't flatten our food cultures into something generic. It should help people celebrate where they come from."* (Male, born 1999)

These sentiments reveal a generation attuned to the risks of digital simplification and cultural oversight. To address these challenges, AI systems must be trained using inclusive, multicultural datasets and developed through participatory design processes that actively involve cultural experts and local communities. Ethical oversight mechanisms must also be put in place to guide algorithmic decision-making, ensuring respect for food heritage and spiritual values (Lugo-Morin, 2024). Ultimately, for AI to serve as a catalyst for food equity rather than erasure of cultural variety, it must be designed not just to process data, but to reflect and respect the diversity of the people it serves.

Gen Z's Expectations and the Future of AI in Food

Gen Z is rapidly emerging as a transformative force in the evolution of global food systems. This cohort, shaped by climate anxiety, digital fluency and social justice consciousness, is pushing beyond mere product innovation to demand ethical, transparent and inclusive applications of AI. For Gen Z, technology must not only serve individual convenience; it must also align with broader values such as sustainability, equity and cultural authenticity.

Rather than accepting technology as a fixed, top-down force, Gen Z envisions AI as a collaborative partner in reimagining regenerative and equitable food futures. Their engagement is increasingly participatory: they seek to co-create solutions, understand how digital systems function and hold developers and brands accountable. In this landscape, AI is not simply a tool, it becomes part of a social

Table 12.1. Gen Z values and AI applications.

Gen Z value	AI application
Personalisation	Nutritional recommender systems
Sustainability	Waste tracking, alternative protein R&D
Cultural diversity	Inclusive recipe generation, multicultural datasets
Transparency	Explainable AI, blockchain-based traceability
Participation	Co-design platforms, digital literacy tools

contract that reflects Gen Z's desire for long-term impact, not just short-term utility.

This generation expects AI to reflect its values in tangible ways. Table 12.1 shows some of the applications connected directly to Gen Z's values. Nevertheless, the actual development and use of AI have all the caveats of data management, governance and participatory design.

Transparency, in particular, stands as a central pillar. Gen Z is increasingly sceptical of "black box" algorithms that make food recommendations without clarity on how decisions are made. They expect AI to be explainable and interpretable, especially when it influences critical decisions such as dietary advice, sourcing claims or sustainability scores. Blockchain technologies, for example, are gaining traction for their ability to verify provenance and validate environmental claims, aligning with Gen Z's emphasis on accountability and truthfulness in food systems (Kuhl, 2025). As one Gen Z put it:

> *"If I don't know how an app decides what's healthy for me, I won't trust it, especially if it ignores my culture or values. I'm not interested in tech that just tells me what to eat. I want to be part of the process, understand it, question it, even change it."* (Female, born 2000)

These reflections illustrate that Gen Z's expectations extend far beyond functionality. They want to actively shape the systems they engage with, demanding a role in the design and governance of AI itself. This pushes brands, governments and developers to adopt new modes of interaction, ones that prioritise co-design, digital literacy and inclusive development practices.

To meet these demands, stakeholders must move beyond token inclusion and embrace mechanisms for participatory design, cultural representation and algorithmic transparency. Doing so not only builds trust but positions AI as a democratic partner in shaping resilient and inclusive food systems for generations to come. Failure to do so will alienate this young cohort and create unnecessary tensions as AI surely takes its place in the food of today and tomorrow.

Food Futures in the Age of AI

Artificial Intelligence is no longer a futuristic concept on the periphery of food systems. It is now embedded in Gen Z's daily food experiences, powering everything from smart kitchen appliances to personalised nutrition platforms. For this generation, AI is not just a tool for convenience or efficiency. It is becoming part of a broader cultural and ethical ecosystem. This generation expects technology to adapt to them, not the other way around. Food AI offers customisation at scale, but it also must align with Gen Z's values, including eco-friendliness, fairness and truthfulness.

The future of food, shaped by climate urgency and technological fluency, demands that AI evolve into a platform for empowerment, inclusion and co-creation. To build regenerative and ethical food futures, Gen Z calls for AI systems that are transparent, accountable and responsive. Their trust in AI hinges on explainability, clarity on how food recommendations are made, what data is collected and how biases are addressed. Tools like explainable AI and blockchain-backed traceability systems are increasingly seen as essential, not optional, in building systems that meet Gen Z's ethical expectations (Werran, 2024; Kuhl, 2025). These technologies offer verifiability and openness, helping to ensure that sustainability claims or ingredient sources are not just marketing slogans but traceable realities. As we understood from one Gen Zer:

> *"I want to trace my food's journey from soil to supermarket shelf, meet the farmer behind the harvest, understand its environmental impact, and know whether it respects the ethics I live by. Transparency isn't just a preference... it's part of my identity and activism. AI needs to be more than a clever tool... it should be a storyteller, a bridge between people and planet, and an advocate for truth in every bite I take. AI should help me uncover all of that, not hide it."* (Male, born 2001)

Moreover, Gen Z is moving from passive consumption to active co-creation. They are no longer content with apps that simply push recommendations. Instead, they seek platforms that encourage feedback, allow customisation and foster community-led innovation. Participatory design platforms — those that incorporate user input into app development and refinement, are gaining traction, especially when they reflect diverse perspectives and challenge traditional hierarchies of food knowledge (Elayat and Elalfy, 2025).

Ethical policies and regulations, such as the EU AI Act, provide a much-needed blueprint for regulating high-risk AI applications in all sectors and food systems in particular. These frameworks emphasise transparency, fairness and the prevention of discrimination, values that deeply align with Gen Z's commitment to equity, justice and social responsibility (European Union, 2024). For this generation, AI must not only be technically sound but also morally grounded. Governance structures are expected to act as both guard rails and enablers, protecting rights while fostering inclusive innovation.

However, for Gen Z ethical governance alone is not enough. The shift toward more inclusive food systems is also deeply cultural. Moving beyond technical efficiency, AI must reflect diverse food identities and culinary traditions, avoid algorithmic homogenisation and reinforcing dominant food norms (often Western, with significant environmental footprints and leading to high prevalence of non-communicable diseases). One of the most pressing concerns is the risk of misrepresenting or entirely omitting minority, Indigenous or mixed dishes due to inadequate or biased training data. When such gaps persist, AI systems may unintentionally erase cultural narratives embedded in food.

As a corrective, the development of inclusive datasets and culturally aware design practices becomes not just a technical requirement but a moral imperative. These tools are essential for preserving culinary heritage, respecting dietary varieties and fostering digital representation that honours the lived experiences of diverse communities (Li *et al.*, 2024). The lived reality of Gen Z increases this need as they do not view food merely as sustenance or trend but as identity, memory and belonging. As one Gen Z passionately explained to us:

> *"Tech should help preserve our food traditions, not erase them. I want to see my culture represented in what AI suggests, not just Western meals. My food tells a story... it's history, memory, identity. When I ask for a recipe, I'm not looking for the fastest, trendiest option... I want the flavours my grandmother grew up with, the ones I celebrate during holidays when I am visiting places, the ones that make me feel like myself. AI should learn our traditions, not overwrite them. It should uplift Indigenous cooking methods, recognise spices beyond salt and pepper, and honour preparation rituals that matter to our communities.... Representation in tech means seeing our dishes, our ingredients, our stories, not buried in a database, but brought to life with dignity... I think this is important not only to me, but also to many of my friends."* (Female, born 2004)

This perspective highlights a generational call for AI systems that do not merely interpret data, but honour cultural stories and social meanings. For Gen Z, the future of food and technology must be co-created, not imposed. Gen Z is not merely adapting to AI, they are advocating for its responsible evolution. Their climate anxiety, digital fluency and globally conscious mindset uniquely position them as active contributors to shaping ethical and sustainable food futures. To truly serve their needs and aspirations, AI must not just collect their data, it must engage their voices through inclusive design, ethical governance and meaningful dialogue.

In this context, AI must evolve into more than a system of logic and learning. It must become a space of inclusive interaction, cultural sensitivity and participatory design. Gen Z expects their voices to shape AI, from the governance structures that regulate it to the recipes it recommends. The future of food will not only be coded; it will be co-created with values, with care and with communities at the centre.

AlphaFold: The Cutting Edge

The convergence of artificial intelligence and molecular biology is transforming food technology at its deepest layers, unlocking possibilities once confined to speculative science fiction. At the forefront of this revolution is AlphaFold, a ground-breaking AI model and database developed by Google's DeepMind in 2020 which is capable of predicting protein structures with atomic-level precision. Originally designed to solve one of biology's grand challenges, the protein folding problem, AlphaFold is now influencing a wide array of sectors and food technology is emerging as one of the most promising frontiers.

One of AlphaFold's most transformative contributions to food systems is its ability to radically accelerate the discovery of novel alternative proteins. By accurately predicting the 3D structures of proteins derived from plants, fungi, algae and microbes, AlphaFold allows food scientists to identify candidates with ideal functional and nutritional characteristics, including desirable textures, flavours and nutritional properties. AlphaFold improves precision fermentation by helping design more efficient and resilient microbes, boosting the production of cultured meat, dairy proteins and functional ingredients. This supports cost-effective scaling, making sustainable food innovations more accessible to wider populations (Marcu *et al.*, 2022). While AlphaFold is still being validated and assessed, it continues to be adopted as a structure prediction tool (Kovalevskiy *et al.*, 2024). It facilitates the creation of next-generation meat and dairy alternatives that could potentially align with Gen Z's sustainability and animal welfare values. Protein engineering precision guided by AI also reduces costly trial-and-error phases and shortens the timeline from concept to commercialisation, which is particularly valuable as demand for sustainable protein sources intensifies (Varadi *et al.*, 2022).

Generative AI models are capable of designing entirely novel proteins with targeted functions (Barnett *et al.*, 2025). This opens the door to designing functional foods with embedded health benefits, e.g. proteins that bind toxins, enhance immunity or regulate metabolism. For Gen Z, who often prioritise foods that integrate wellness, identity and ethics, such precision aligns strongly with their future-forward values. AlphaFold also supports the design of customised enzymes for use in food processing. These enzymes can enhance taste, texture and nutritional value, while also contributing to shelf-life extension and waste reduction (Dobson *et al.*, 2025). For instance, enzymes engineered for improved lipid breakdown or protein cross-linking can enhance flavour release and reduce the need for additives. Applications like this align with Gen Z's demand for "clean label" products that are both high-performing and minimally processed.

Such innovation advances are of interest to Gen Z, especially if they are addressed with caution as technological efficiency alone is not enough. Their expectations extend beyond utility to values-based innovation. As voiced by a young Gen Z with whom we spoke, there is clear enthusiasm for protein technologies, but they should be respecting ecological boundaries and ethical labour practices. This enthusiasm however is tempered by a deep concern over the social, environmental and cultural implications of AI-driven food innovation. The appeal of speed and novelty must not come at the cost of transparency, inclusivity or scientific integrity:

> *"If we need meat alternatives at all, we need alternatives that actually are real not just feel real, flavourful, textured and emotionally satisfying, without compromising the health of our bodies or the planet. You know, convenience we all look to have shouldn't come at the cost of biodiversity, land degradation or exploitative labour practices. If AI can help us discover plant-based or other proteins that mimic the taste and experience of meat, that people would want to consume,...proteins all produced while using fewer resources and cutting down carbon emissions, then yes, it's a powerful ally. But we shouldn't rush into blindly adopting lab-grown or other solutions or algorithm-generated substitutes without thinking critically. Innovation must be transparent, inclusive and backed by*

> *science, not just slick branding or promises of sustainability. AI should speed up progress, not shortcut responsibility. We need to ask who's designing these technologies, what values are embedded in them, and how do they serve everyone, not just the wealthy or the West. If done right, AI can help create ethical, scalable food systems, but if misused, it can deepen inequalities, reinforce cultural bias and turn food into a product that's lost its soul. It's scary even to think about this."* (Male, born 2003)

What emerges from this young person's opinion is a call for AI not to be a mere disruptor, but a responsible catalyst, a means of enabling food systems that are both technologically advanced and socially just. Rather than replacing one form of unsustainable consumption with another masked in innovation, Gen Z advocates for a deliberate and reflective approach, one that critically examines who designs these technologies, whose interests they serve and whether they truly democratise food access or reinforce global inequities. The interview excerpt reflects a broader generational insistence that food should keep its cultural depth, ethical grounding and human meaning, even, or perhaps especially, when guided by algorithms.

Ultimately, AlphaFold and similar cutting-edge tools have enormous potential, but they must be embedded in governance structures and value systems that resist the commodification of food as a soulless product. For Gen Z, the future of food is not just high-tech, it is highly principled.

Implications for the Future of Food Technology

There is a paradigm shift in food innovation truly signalled by AI and its applications. Rather than merely substituting existing ingredients, AI can now design proteins from scratch, tailored to environmental constraints, consumer needs and cultural preferences. However, with great power comes responsibility. The deployment of these technologies must be guided by ethical frameworks, inclusive design and transparent governance, especially as they reshape not just what we eat, but how we understand food itself.

This paradigm shift in food innovation facilitated by AI is one that resonates deeply with Gen Z's values of sustainability, personalisation and social justice. The transformation is powered by tools like AlphaFold, generative AI, blockchain and other applications, which among others predict protein structures with atomic precision, generate novel proteins for functional foods and ethical meat alternatives and allow traceability of food (Kuhl, 2025).

For Gen Z, this is not just exciting, it is essential. The young people are not satisfied with step-by-step improvements, they want systemic transformation. They want smarter food systems by optimising supply chains, reducing waste and accelerating the development of climate-resilient crops and proteins, all of which AI can help to deliver (Khan *et al.*, 2025), but only if it listens to Gen Z as much as it calculates and optimises.

When it comes to responsibility, Gen Z is acutely aware of the ethical risks tied to AI, such as algorithmic bias, cultural erasure and opaque decision-making. They demand transparent governance, inclusive design and participatory innovation. Ethical frameworks like the EU AI Act and explainable AI models (e.g. LIME) are crucial to ensure that food technology respects privacy, equity and cultural diversity (Naithani *et al.*, 2025). Gen Z wants AI to assist in better food choices rather than impose them (see Figure 12.2).

Figure 12.2. Gen Z's expected role for AI in food.

Source: Infographic by authors.

Tomorrow's food landscape is not just for Gen Z, but it will be shaped by them. With tools like AI-enabled sensors or augmented reality-enhanced dining, we are entering an era of fully customised food experiences. Gen Z does not just want better food, they want smarter, safer and fairer food systems and nutrition. When used responsibly, AI and the new food technology offer an unprecedented opportunity to meet that vision at the molecular level and beyond. It is not just about what we eat, it is about how we design, govern and share the future of nourishment. Gen Z will embrace food AI... if it tastes good, does good and feels real.

References

Ahmad, A., Liew, A. X. W., Venturini, F., Kalogeras, A., Candiani, A., Di Benedetto, G. ... Martos V. (2024). AI can empower agriculture for global food security: Challenges and prospects in developing nations. *Frontiers in Artificial Intelligence*, 7, 1328530. https://10.3389/frai.2024.1328530

Ahmed, H. M. (2024). How Generative AI will transform the halal industry in 2025. Halal Times. https://www.halaltimes.com/how-generative-ai-will-transform-the-halal-industry-in-2025/

Alexander, C. S., Yarborough, M., & Smith, A. (2024). Who is responsible for 'responsible AI'?: Navigating challenges to build trust in AI agriculture and food system technology. *Precision Agriculture*, 25, 146–185. https://doi.org/10.1007/s11119-023-10063-3

Ali, Z., Muhammad, A., Lee, N., Waqar, M., & Lee, S.W. (2025). Artificial intelligence for sustainable agriculture: A comprehensive review of AI-driven technologies in crop production. *Sustainability*, 17, 2281. https://doi.org/10.3390/su17052281

AlphaFold. (2023). AlphaFold Protein Structure Database: Developed by Google DeepMind and EMBL-EBI. https://alphafold.ebi.ac.uk

Balakrishnan, P., Anny Leema, A. Jothiaruna, N., Assudani, P. J., Sankar, K., Kulkarni, M. B., & Bhaiyya, M. (2025). Artificial intelligence for food safety: From predictive models to real-world safeguards. *Trends in Food Science & Technology*, 163, 105153. https://doi.org/10.1016/j.tifs.2025.105153

Balzarova, M. A. (2021). Blockchain technology — a new era of ecolabelling schemes? *Corporate Governance*, 21(1), 159–174. https://doi.org/10.1108/CG-08-2020-0328

Bardales Salguero, R., Bogueva, D., & Marinova, D. (2024). Australia's university Generation Z and its concerns about climate change. *Sustainable Earth Reviews*, 7, 8. https://doi.org/10.1186/s42055-024-00075-w

Bailey, R. L., MacFarlane, A. J., Field, M. S., Tagkopoulos, I., Baranzini, S. E., Edwards, K. M. ... Stover, P. J. (2024). Artificial intelligence in food and nutrition evidence: The challenges and opportunities. *PNAS Nexus*, 3(12), pgae461. https://doi.org/10.1093/pnasnexus/pgae461

Barnett, A. J., Rajendra, K. C., Pandey, P., Somasiri, P., Fairfax, K. A., Hung, S., & Hewitt, A. W. (2025). Systematic comparison of Generative AI-Protein Models reveals fundamental differences between structural and sequence-based approaches. bioRxiv, the preprint server for biology. https://doi.org/10.1101/2025.03.23.644844

Bhagat, P. R., Naz, F., & Magda, R. (2022). Artificial intelligence solutions enabling sustainable agriculture: A bibliometric analysis. *PLoS One*, 17(6), e0268989. https://doi.org/10.1371/journal.pone.0268989

Callaway, E. (2023). AI tools are designing entirely new proteins that could transform medicine. https://www.nature.com/articles/d41586-023-02227-y

Clark, Q. M., Kanavikar, D. B., Clark, J., & Donnelly, P. J. (2025). Exploring the potential of AI-driven food waste management strategies used in the hospitality industry for application in household settings. *Frontiers in Artificial Intelligence*, 7, 1429477. https://doi.org/10.3389/frai.2024.1429477

Dhal, S. B. & Kar, D. (2025). Leveraging artificial intelligence and advanced food processing techniques for enhanced food safety, quality, and security: A comprehensive review. *Discover Applied Sciences*, 7, 75 https://doi.org/10.1007/s42452-025-06472-w

Dobson, L., Tusnády, G. E., & Tompa, P. (2025). Regularly updated benchmark sets for statistically correct evaluations of AlphaFold applications. *Briefings in Bioinformatics*, 26(2), bbaf104. https://doi.org/10.1093/bib/bbaf104

du Plessis, M. J., Gerber, R., Goedhals-Gerber, L. L., & van Eeden, J. (2025). Shaping the future of freight logistics: Use cases of artificial intelligence. *Sustainability*, 17(4), 1355. https://doi.org/10.3390/su17041355

Elayat, A. M. A. and Elalfy, R. M. (2025). Using SOR theory to examine the impact of AI Chatbot quality on Gen Z's satisfaction and advocacy within the fast-food sector. *Young Consumers*, 26(2), 352–383. https://doi.org/10.1108/YC-08-2024-2199

European Union. (2024). The EU Artificial Intelligence Act. Up-to-date developments and analyses of the EU AI Act. https://artificialintelligenceact.eu/

GEA. (2024). Precision fermentation advances sustainable dairy. https://www.gea.com/en/stories/how-precision-fermentation-can-advance-sustainable-dairy/

Halicka, E., Kaczorowska, J., Rejman, K., & Plichta, M. (2025). Investigating the consumer choices of Gen Z: A sustainable food system perspective. *Nutrients*, 17(3), 591. https://doi.org/10.3390/nu17030591

Han, M., Chen, J., & Zhou, Z. (2024). NutrifyAI: An AI-powered system for real-time food detection, nutritional analysis, and personalised meal recommendations. *Computer Vision and Pattern Recognition*. https://doi.org/10.48550/arXiv.2408.10532

Jha, A. (2024). Prompt optimization for AI-powered recipe generation: Challenges and insights. *International Journal for Research in Applied Science and Engineering Technology*. https://doi.org/10.22214/ijraset.2024.65150

Joshi, S., Bisht, B., Kumar, V., Jameel Pasha, S. B., Singh, N., & Kumar, B. (2024). Artificial intelligence assisted food science and nutrition perspective for smart nutrition research and healthcare. *Systems Microbiology and Biomanufacturing*, 4, 86–101. shttps://doi.org/10.1007/s43393-023-00200-4

Khan, G., Irshad, G., Ijaz, R., Javaid, S., Tahir, N., & Mehmood, S. (2025). Emboldening food security for global sustainability yoking artificial intelligence. *Discover Food*, 5, 2. https://doi.org/10.1007/s44187-025-00273-1

Kovalevskiy O., Mateos-Garcia, J., & Tunyasuvunakool, K. (2024). AlphaFold two years on: Validation and impact. *PNAS*, 121(34), e2315002121. https://doi.org/10.1073/pnas.231500212

Kuhl, E. (2025). AI for food: Accelerating and democratising discovery and innovation. *npj Science of Food*, 9, 82. https://doi.org/10.1038/s41538-025-00441-8

Lalla-Ruiz, E., & Mes, M. R. K. (2025). Optimisation and artificial intelligence in logistics management. *Annals of Operation Research*. https://doi.org/10.1007/s10479-025-06700-x

Li, X., Yin, A., Choi, H. Y., Chan, V., Allman-Farinelli, M., & Chen, J. (2024). Evaluating the quality and comparative validity of manual food logging and artificial intelligence-enabled food image recognition in apps for nutrition care. *Nutrients*, 16(15), 2573. https://doi.org/10.3390/nu16152573

Liu, Z., Wang, S., Zhang, Y., Feng, Y., Liu, J., & Zhu, H. (2023). Artificial intelligence in food safety: A decade review and bibliometric analysis. *Foods*, 12(6), 1242. https://doi.org/10.3390/foods12061242

Lugo-Morin, D. R. (2024). Artificial Intelligence on food vulnerability: Future implications within a framework of opportunities and challenges. *Societies*, 14(7), 106. https://doi.org/10.3390/soc14070106

Malmqvist, T. (2025). Why climate anxiety is hitting Gen Z the most and how brands can respond. Trellis. https://trellis.net/article/why-climate-anxiety-is-hitting-gen-z-the-most-and-how-brands-can-respond/

Marcu, Ş-B., Tăbîrcă S., & Tangney M. (2022). An overview of Alphafold's breakthrough. *Frontiers in Artificial Intelligence*, 5, 875587. https://doi.org/10.3389/frai.2022.875587

Marwala, T. (2024). Artificial Intelligence can transform global food security and climate action. United Nation University (UNU). https://unu.edu/article/artificial-intelligence-can-transform-global-food-security-and-climate-action

Mohammed, S., & Malhotra, N. (2025). Ethical and regulatory challenges in machine learning-based healthcare systems: A review of implementation barriers and future directions. *BenchCouncil Transactions on Benchmarks, Standards and Evaluations*, 5(1), 100215. https://doi.org/10.1016/j.tbench.2025.100215

Momentum. (2024). Experts predict generative AI's business impact over the next 5 years. The Business School Magazine, University of Queensland. https://business.uq.edu.au/momentum/predictions-generative-ai

Naithani, K., Raiwani, Y. P., Tiwari, S., & Chauhan, A. S. (2025). Ethical aspects of AI nutrition: Discussion of ethical considerations related to AI-based nutritional guidance in terms of privacy, data security, and potential biases. In: Kumar, R., Gupta, M., & Nayyar, A. (eds) *AI-based Nutritional Intervention in Polycystic Ovary Syndrome (PCOS).* Springer, Singapore. https://doi.org/10.1007/978-981-96-2120-0_11

Oewel, B., Guluzade, L., Zhu, J., & Huang, Y. (2024). The potential of Generative AI in personalised nutrition. Designing (with) AI for Wellbeing Workshop at CHI 2024. Hawaii, USA. https://www.researchgate.net/publication/379332664_The_Potential_of_Generative_AI_in_Personalized_Nutrition

Onyeaka, H., Tamasiga, P., Nwauzoma, U. M., Miri, T., Juliet, U. C., Nwaiwu, O., & Akinsemolu, A. A. (2023). Using artificial intelligence to tackle food waste and enhance the circular economy: Maximising resource efficiency and minimising environmental impact: A review. *Sustainability*, 15(13), 10482. https://doi.org/10.3390/su151310482

Palakurti, N. R. (2022). AI applications in food safety and quality control. *ESP Journal of Engineering & Technology Advancements*, 2(3), 48–61. https://doi.org/10.56472/25832646/JETA-V2I3P111

Ral, J-P., Wang C., & Jiang, J. (2025). Artificial Intelligence (AI) is set to transform how we develop future protein foods, by identifying less resource-intensive plant crop varieties, and the most productive ways to farm them and turn them into new products. CSIRO. https://www.csiro.au/en/news/All/Articles/2025/April/AI-for-plant-proteins

Sarku, R., Clemen, U. A., & Clemen, T. (2023). The application of artificial intelligence models for food security: A review. *Agriculture*, 13(10), 2037. https://doi.org/10.3390/agriculture13102037

Sodiya, E. O., Umoga, U. J., Amoo, O. O., & Atadoga, A. (2024). AI-driven warehouse automation: A comprehensive review of systems. *GSC Advanced Research and Reviews*, 18(02), 272–282. https://doi.org/10.30574/gscarr.2024.18.2.0063

Styliaras, G. D. (2021). Augmented reality in food promotion and analysis: Review and potentials. *Digital*, 1(4), 216–240. https://doi.org/10.3390/digital1040016

Sunmola, F., Baryannis, G., Tan, A., Co, K., & Papadakis, E. (2025). Holistic framework for blockchain-based halal compliance in supply chains enabled by artificial intelligence. *Systems*, 13(1), 21. https://doi.org/10.3390/systems13010021

Varadi, M., Anyango, S., Deshpande, M., Nair, S., Natassia, C., Yordanova, G. ... Velankar, S. (2022). AlphaFold protein structure database: Massively expanding the structural coverage of protein-sequence space with high-accuracy models. *Nucleic Acids Research*, 50(D1), D439–D444. https://doi.org/10.1093/nar/gkab1061

Tsolakidis, D., Gymnopoulos, L., & Dimitropoulos, K. (2024). Artificial intelligence and machine learning technologies for personalised nutrition: A review. *Informatics*, 11(3), 62. https://doi.org/10.3390/informatics11030062

Wang, D., Tan, Z., Gao, J., Zhang, S., Shen, J., & Lu, Y. (2025). AI4Protein: Transforming the future of protein design. *Science China Life Sciences*. https://doi.org/10.1007/s11427-024-2906-3

Werran, R. (2024). Navigating ethical challenges in an AI-enabled food industry. *Food Science and Technology*, 38(3), 40–43. https://doi.org/10.1002/fsat.3803_10.x

Wu, S. R., Shirkey, G., Celik, I., Shao, C., & Chen, J. (2022). A review on the adoption of AI, BC, and IoT in sustainability research. *Sustainability*, 14(13), 7851. https://doi.org/10.3390/su14137851

Yang, H., Jiao, W., Zouyi, L., Diao, H., & Xia, S. (2025). Artificial intelligence in the food industry: Innovations and applications. *Discover Artificial Intelligence*, 5, 60. https://doi.org/10.1007/s44163-025-00296-82025

Chapter 13

Packaging and the Planet: Sustainability beyond the Plate

Packaging of food these days is almost as important as the stuff that we put in our mouth. It may even be the case that the environmental footprint of packaging is higher than that of the foodstuff. Also, food is by essence biodegradable, can be composted or rot transformed by the planet's microorganisms while packaging, particularly if based on plastics is not. Although a lot of emphasis has been put on the human health aspects of packaging, its environmental impacts have been forgotten for long and only recently have come to the fore with the enormous scale of plastic pollution. Gen Z commonly expresses irritation with the use of plastics in supermarkets: "I went to the supermarket this morning to get some bits for our breakfast and everything was just in a plastic container and I was just feeling wretched" (Topic and Mitchel, 2019, p. 18). Hence, food packaging can trigger strong emotional responses (Avery Dennison, 2018). We saw this firsthand during our conversation with a Gen Zer:

> *"It's frustrating after so much talk about keeping the nature clean of plastic... You know, I went to grab brekkie [breakfast] stuff this morning and everythin' was wrapped in plastic. Felt pretty crap about it.... like we're just drowning in rubbish but no one's doing much. Food's natural, packaging's not, so why's it gotta be so wasteful?"* (Female, born 2005)

This frustration reflects a growing sentiment among Gen Z, who increasingly see excessive packaging as a disconnect between sustainability ideals and everyday reality. Their demand for less wasteful, simpler packaging is part of a broader push for brands to align with environmental values, not just in messaging, but in practice (Singh, 2025).

By definition, packaging describes "the materials used to wrap or protect goods" (Oxford Learner's Dictionaries, 2025, para. 1). The packaging encloses and protects the goods. How to properly package has become a science, technology, business, art and way of information communication and marketing. Packaged goods can be transported, stored, handled, sold and used, during which times their content is protected and preserved. Most commonly, it is expected that the packaging will be removed and disposed at a certain stage of the product's life.

Food packaging is particularly complex as it needs to deliver the content to the consumer in a safe way with maximum quality and attractiveness. Previously, food packaging had two major roles: first, a barrier that shields the packaged product from the external elements; and second, a source of information about the product, such as a list of ingredients, instructions related to preparation, expiry (or best before) date and informative pictures (Onyeaka and Nwabor, 2022). Wani *et al.* (2014) describe food packaging as serving the following functions: protection (from oxygen, moisture, microorganisms, dirt, chemical contaminants, toxins etc.); containment (to prevent mixing, bruising, act as transportation medium for non-solid foods, reduction of vibration and avoidance of mechanical shocks); convenience (easy opening, processing in oven or by microwave or retorting); and communication (labelling, nutritional information, manufacturer, price, product ingredients, product life, storage conditions etc.).

These basic requirements however have now changed and consumers, particularly Gen Z, have new expectations about the way their food is packaged. Many of Gen Z reached adulthood during the years of the COVID-19 pandemic and ordering meals online is as normal for them as it is to go out for shopping or to a restaurant. Although the specific food packaging requirements vary, there are some common expectations.

As Gen Z navigates a world of food delivery apps and online shopping, its expectations for packaging go beyond function. They want packaging to reflect their values. An opinion shared by a Gen Zer whom we interviewed captures the mindset of a generation shaped by digital convenience and heightened environmental awareness:

> *"Packaging these days is almost as important as what's inside. It's gotta look good, be easy to chuck in the bin or recycle, and not make me feel guilty about trashing the planet. I reckon if it ain't sustainable, it's not worth paying extra for."* (Male, born 2001)

These words reflect a broader trend. Gen Z sees packaging as part of the product experience, not just a container. According to a 2025 global survey by McKinsey, Gen Z is not only vocal about sustainability; they are willing to pay more for packaging that aligns with their environmental values (Green, 2025). Brands that fail to meet these expectations risk losing relevance with this socially conscious generation.

We have also witnessed the development of new packaging materials and technologies for food preservation which avoid the use of preservatives and other chemically-based solutions. Smart packaging using biobased and biodegradable polymers is competing with traditional materials, e.g. paper and glass, propelled by new demands, novel technologies and added functions (Salgado *et al.*, 2021). Packaging has now become integral to all aspects of food production and consumption (Gupta and Dudeja, 2016) and in many ways determines the relationship between young people and food. We explore some of these connections by discussing environmental concerns first, then functional packaging and the aesthetic expectations that young people have about the way their food is packaged.

Food Packaging and the Environment

Food is a major contributor to waste found on our planet, both on land and in water environments. Particularly worrying is the waste found in rivers and oceans as it is more difficult to clean and affecting species which often mistakenly consume it as food. Figure 13.1 shows that more than 50% of the most common items present in the world's

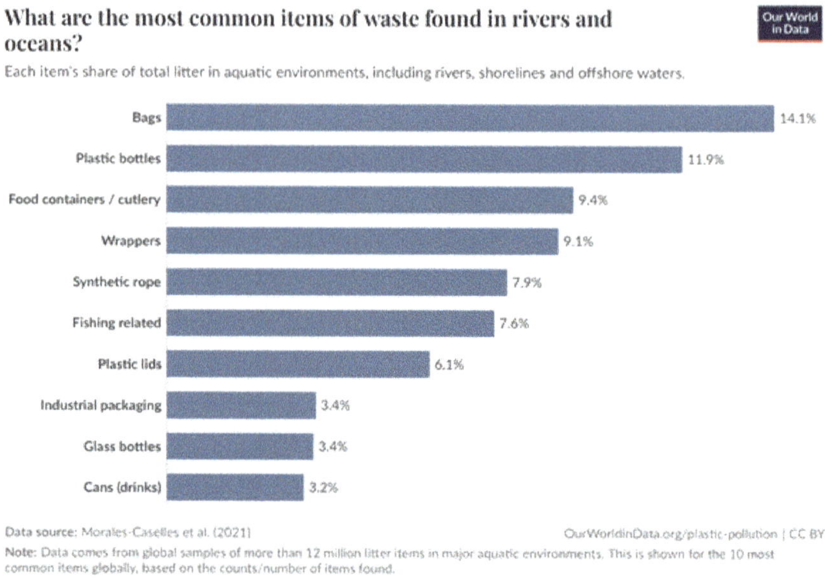

Figure 13.1. Most common items of waste found in the world's rivers and oceans.

rivers and oceans are related to food, including plastic and glass bottles, food containers and cutlery, cans and bags potentially used for shopping. Many of these items are plastic and they not only accumulate in the aquatic environments but also break down into microplastics and continue to float on the water surface endangering species and polluting their habitat. Gen Z is disproportionately contributing to this situation because of its poor recycling practices and food waste habits.

Another serious issue with food packaging is the use of per- or polyfluoroalkyl (PFAS) manufactured substances also named "forever chemicals" as they persist within the natural environment and are technically indestructible. Although the food industry uses these materials to make packaging resistant to heat, water and stains and hence preserve the quality of foodstuffs, these chemicals have contaminated the entire planet, including being found in all precipitations, and are associated with toxicological effects and serious health risks, such as cancers, liver problems, thyroid issues, lipid and insulin dysregulation, birth defects, kidney disease and overall decreased immunity (Fenton *et al.*, 2021). Additionally, most recycling schemes do not properly collect such soft packaging leaving the wrappers and other containers to end up in landfill or being incinerated further exacerbating the problems with PFAS. Although many countries around the world have outlawed the use of PFAS, their phasing out has been a slow process. It took Australia and New Zealand more than six years (between 2018 and 2025) to ban three of the most prominent PFAS: perfluorooctanesulfonic acid — PFOS, perfluorooctanoic acid — PFOA, and perfluorohexanesulfonic acid — PFHxS (Australian Government, 2022). Ironically, some packaging containing PFAS, such as moulded fibre bowls, are marketed as green and compostable (Perkins, 2023) making the environmental tasks of Gen Z even harder. This contradiction fuels Gen Z's growing distrust in the so-called "eco" solutions, especially when their good intentions backfire:

> *"It's messed up. You think you're doing the right thing by choosing the 'eco' bowl, but turns out it's full of forever chemicals! It's like trying to be green is a trap... can't win. I am so scared that many things are like this — covered up under various eco-friendly claims"* (Male, born 2000)

The communicated view acts as a raw, emotional snapshot of the broader issue. This sense of betrayal reinforces Gen Z's frustration with greenwashing and fuels their demand for greater transparency in packaging materials. It personalises Gen Z's disillusionment with sustainability marketing, showing how even well-intentioned choices can feel sabotaged by hidden risks. The mention of "forever chemicals" like PFAS in compostable bowls directly connects to the expressed sense of betrayal, where trying to do the right thing ends up feeling like a trap. This tension between eco-conscious intent and misleading execution is exactly what previous studies have brought to light (Chung, 2023).

Moreover, any assumptions that food containers that can be recycled would be environmentally acceptable to Gen Z need to be treated with caution because Gen Zers are not instinctive recyclers. Although Gen Z is perceived as the most environmentally- or sustainability-oriented generation, it seems to lack in recycling skills, knowledge and willingness to do so. A survey of university students in the UK showed that among all generations, Gen Z has the least confidence in recycling and is least likely to know how to find such advice (DS Smith, 2022). Almost all, namely 92%, of UK Gen Z would throw something in the rubbish bin, including food waste and plastic bags instead of recycling or composting it (DS Smith, 2025). Despite their best intention to follow the rules, Gen Z are very often confused:

> *"Honestly, I try to recycle but half the time I don't even know what goes where. It's a bit unclear to me... One council says one thing, the next where my grandparents live say another... it's a bloody maze. Just chucking it feels easier... But most of the time I feel I may not exactly be doing the right thing..."* (Female, born 2006)

Such uncertainty reflects a core obstacle to Gen Z's sustainable intentions. The overcomplexity of recycling systems discourages follow-through even when values align. By comparison, the Baby Boomers are much more likely to recycle, particularly paper and cardboard (Dennis, 2024).

The Gen Z young people are looking for simple clear instructions and dislike the complexity of the current systems. They look for convenience and recycling bins available in public spaces. A boy from UK explains: "People on the whole are quite lazy. I am... happy just to throw it in the most convenient place rather than think about what they are doing", while a girl from the same survey sees things more optimistically: "I think, in 5 to 10 years' time, it will be a lot simpler. I think there will be less things to recycle anyway because we will be using more things that we can re-use, such as glass jars and filling those up in shops, and loose vegetables instead of vegetables in plastics..." (DS Smith, 2022). Overall, Gen Z understands that the current attitudes to recycling need to change but they look for convenience, clarity and accessibility to close the gap between their habits and the better practices of older generations (Keenan Recycling, 2024). Food packaging is not an exception and contributes significantly to plastic pollution and habitat contamination, issues quite high on the environmental radar of young people.

In Australia and other western countries, the situation is very similar to that in the UK. For example, the Western Australian government introduced a "Containers for Change" program that rewards financially people for returning their drink containers (10 cents per container). Research data shows that Gen Z's participation in the "Containers for Change" is half the rate of older generations (Containers for Change, 2022). Again, young people are looking for convenience, including convenient locations, and are unlikely to go out of their way for recycling as they are time-poor and often do not drive. They also think that the responsibility for recycling lays with the government and local councils and not with them. In the USA, although Gen Z understands the benefits from recycling, it is not inspired to do so because of the low rate of recyclables actually being recycled and the different practices across states (Yubo Team, n.d.).

To make things even worse from an environmental point of view, Gen Z appears to be the generation that wastes the most food. Even in the Czech Republic, where the levels of food waste by the households are very low, Gen Z is the biggest culprit (Mendel University, 2023). Moreover, the level of food waste is double when Gen Zers live outside of the family home where they also fail to properly segregate their rubbish (Mendel University, 2024). In Canada, Gen Z spends the most on food but also throws away the most (Greater Victoria News Staff, 2023). An Australian study similarly shows a negative correlation between age and amount of food waste with possible explanations being food management skills related to shopping, product storing, preserving and using leftovers (Karunasena *et al.*, 2021). Gen Z are estimated to waste AUS$ 1,446 (more than US $900) of food per year in Australia (Duffy, 2019) and even higher CAN$ 2000 (around US $ 1400) in Canada (Greater Victoria News Staff, 2023). As most wasted food is packaged, this contributes to the environmental problems resulting from dietary behaviours. A contributing factor to food waste is the fear-of-missing-out (FOMO) behaviour and the impulsive buying of Gen Z who want new and exciting meals, are reluctant to use any leftovers or to plan in advance (Askew, 2020). It is something that is not new to Gen Z.

As one individual from this cohort pointed out:

> *"I'll admit it... I buy way more than I need, especially if it's something trending on TikTok. Then it just sits in the fridge 'til it dies. I know it's wasteful, but it's hard to stop... and most of the time I just don't have the time to consume it being busy with stuff I do on a daily basis. At the same time, I am afraid not to have enough food in case I am stuck at home... like we all were during COVID. It's a very complex thing."* (Male, born 1998)

The emotional pull of novelty, often driven by social media, contributes to avoidable food waste, making packaging waste an even greater issue in Gen Z's consumption patterns.

A logical response to Gen Z's behaviours is to change the nature of packaging to reduce its environmental footprint. Gen Z is not only open to choosing a plant-based meal when it is the default option on the college menu, these young people also expect the packaging to be ethical and environmentally friendly (Owings, 2023). It is more realistic to improve the materials used for packaging than to anticipate an entire demographic cohort to change, particularly as the trends exhibited in Gen Z have been foreshadowed by the Millennials. It seems that Gen Z consumers are reluctant to change their lifestyle but instead demand packaging solutions that are more sustainable.

The alternative to fossil fuel-based packaging (derived from petroleum or natural gas) are products that are produced sustainably and are biodegradable. Such promising packaging options are attractive to Gen Z as they respond to their social and environmental values. The field of food packaging is constantly changing as novel technologies offer new opportunities. High-tech solutions come onto the market co-existing with nostalgia about simple natural ways of consuming food. We explore some of these trends and their relationships with Gen Z consumers.

Biodegradable Packaging

Significant efforts are made to recycle food packaging produced from synthetic plastics to avoid them becoming waste as they cannot naturally undergo a biological or chemical degradation. Many of these conventional plastics are subject to physical degradation and break down into nanoplastics resulting in soil, water and ocean pollution that is even more difficult to control or clean up. There is also the class of bio-plastics which are produced from renewable biomaterials, such as corn, starch or cellulose, but cannot be naturally decomposed. An example is bio-PET — polyethylene terephthalate produced from sugar cane, which is not biodegradable because its chemical structure is the same as PET and microbes cannot break down the strong polymer bonds. Such bio-plastics are not seen as a proper environmental solution and Gen Z can react strongly against greenwashing by claims from product brands (Segel and Hatami, 2023). This frustration with misleading sustainability claims is endorsed by many Gen Zers, who are quick to call out marketing spin. As one of them stressed out:

> *"Just slapping a 'bio' label on plastic doesn't cut it. If it's not breaking down naturally, then it's still clogging up the planet... don't greenwash us, we're not stupid,... we can figure it out ourselves."* (Female, born 2004)

This opinion illustrates the sharp environmental literacy among Gen Z particularly in Australia, who increasingly demand authenticity and proof, not just branding, when it comes to sustainability.

By comparison, any packaging that naturally decomposes, disintegrates or rots under the action of bacteria, algae and fungi is considered biodegradable. Paper falls into this category and has been and continues to be used to wrap food, but often it cannot serve all the other functions expected from packaging. The term biodegradable packaging is now used to describe new materials, particularly biopolymers (or bio-based polymers), that can replace petroleum-based plastics and are used not just to wrap food but also to preserve and contain it, such as bags, trays, coatings, films, gels and foam products. Such biodegradable polymers can also be used in 3D printing. Shaikh *et al.* (2021) distinguish between different categories of polymers used as biodegradable packaging based on the starting material, namely: extracted from biomass (such as polysaccharides, including starch, cellulose and chitin, and proteins), synthesised from monomers (e.g. polylactic acid or polylactide) and produced from microorganisms (including natural and genetically modified). In decomposing, biodegradable packaging generates biomass, water, carbon dioxide and other natural inorganic compounds, which do not represent waste. Many of these biopolymers can be composted.

Beyond the composting ability, young Gen Z consumers are also evaluating the full life cycle of packaging materials, from production to disposal. We discussed this aspect with one of them:

> *"Look, everyone talks about how it's compostable and all, but that's just the tail end of the story. It's not just about chucking it in the compost.... what's the energy footprint to make it in the first place?.... What about the effort it takes to make it in the first place? If it's chewing through heaps of energy or resources just to be called 'eco', then we're not really winning, are we? Sustainability's gotta be the whole package,... it has to go all the way, not just stop at the bin, start to finish,... not just a feel-good label when you chuck it in the bin."* (Male, born 2000)

Gen Z holds a more nuanced view of sustainability and is beginning to question the entire supply chain rather than just the end-of-life stage of packaging. The various biodegradable polymers have different properties, such as tensile strength, water vapor transmission rate, oxygen transmission rate, elongation at break, thermal stability and melting temperature, which influence their application in food packaging (Shaikh *et al.*, 2021).

From a sustainability point of view, in addition to decomposition properties, qualities, such as toxicity, energy (and energy type) used in production, social acceptability and land use, also need to be considered. For the new materials to properly integrate in the planet's biogeochemical cycles (Lucas *et al.*, 2008), they need to undergo the following conversions under the action of microorganisms: biodeterioration — break down into tiny fractions; depolymerisation — break down to oligomers, dimers and monomers with the release of enzymes and other catalytic agents by the microorganisms; assimilation — microbial metabolism is integrated in the cells' cytoplasm to produce energy, new biomass, storage vesicles, primary and secondary metabolites; and mineralisation — some metabolites are secreted by microbial cells in the extracellular surrounding releasing carbon dioxide, methane, nitrogen, water and different salts in the environment (Lucas *et al.*, 2008; Shaikh *et al.*, 2021).

The biodegradable materials used for food packaging have different decomposition times and they cannot just be thrown into the natural environment. They may take very long in open environments and a proper treatment requires special conditions, such as in industrial facilities (AllThings.Bio, 2018). It is also important to be aware of the specific requirements of the processes and facilities used to collect, process, recycle and compost waste in the local area where people live as there are vast differences in standards and technologies. This complexity can often be confusing. Some Gen Z consumers express frustration with the lack of consistency in composting systems and want clearer guidance that matches real-world practice:

> *"We're told it's compostable, but no one tells you it won't break down unless it's in an industrial composter. I've seen 'biodegradable' stuff still hanging around in bushland or beaches... here around the campsite... it's not magic. If councils and brands aren't aligned, we're just being greenwashed."* (Female, born 2003)

This comment reflects growing eco-literacy and impatience among Gen Z, who is demanding honesty, infrastructure alignment and practical education around sustainable materials. Consumer understanding and reaction to biodegradable packaging is essential and young people's acceptance can encourage further innovation. A survey of Gen Z in Indonesia showed 94% wanted eco-friendly packaging and 93% were willing to pay more for it (Foopak, 2023). Environmental packaging of drinks was also found to positively influence green purchasing intention by Gen Z in India (Meet *et al.*, 2024). There is now an increasing number of new packaging materials and options that offer biodegradable and compostable ways to package food with the aim to avoid the use of fossil fuel-based plastics. If produced with renewable energy, such materials contribute to lowering the overall environmental footprint of food. Examples include packaging manufactured from the bio-based polylactic acid (PLA) or polyhydroxylalkanoates (PHA) produced naturally by microorganisms, including through bacterial fermentation. Such biodegradable polymers are commonly used as films to line up paper containers resulting in the entire product being compostable.

Compostable Packaging

Biodegradable packaging is generally designed to naturally degrade in modern landfills which is a better option than fossil fuel-based plastics. However, landfills still require land, produce methane (often captured for energy production) and leachate contaminating groundwaters, emit odour and create health risks. The "tsunami" of waste that goes to landfills can be overwhelming and solutions that reduce the volume of disposed materials are most needed. Compostable packaging is such a solution.

Compostable packaging is a special class of biodegradable packaging products that are compostable, which means they decompose into organic material (humus), water and carbon dioxide within a certain, relatively short period of time under specific conditions during clearly defined processes (e.g. aerobic digestion in industrial settings with controlled temperature, usually above 50°C, and moist conditions). Compostable packaging is designed for industrial installations and is not suitable for home composting. For the packaging to be compostable, it needs to satisfy conditions, such as: 90% of the packaging to have rotted within 90 days; not generate any toxic substances; and to not negatively impact the rest of the composting process (Harder, 2024). Biodegradation during composting is similar to the natural degradation of materials, such as leaves, grass and food scraps (Shaikh *et al.*, 2021). Being able to dispose such compostable packaging in a convenient appropriate place is essential for Gen Z who generally, and similarly to recycling, are less keen to do composting at home (Corey, 2021).

Edible Packaging

Edible packaging is another category of biodegradable packaging made from compounds that are safe for people to eat (Nair *et al.*, 2023). It covers films and coatings applied directly to the food to improve or preserve its qualities as well as containers that can be consumed (as is the case of ice cream cones). Edible films and coatings are obtained from the same formulation, e.g. proteins, polysaccharides, lipids or composites; films are applied as solid sheets while coatings are applied in liquid form through coating, spraying or immersing (Nair *et al.*, 2023). Both, edible films and coatings become integrated with the food and consumers have the choice to remove or consume them. Examples include chitosan (a polysaccharide made from the outer skeleton of shellfish) and gelatine (a protein made from animal collagen, present in cartilage, bone and skin).

Edible films and coatings have long been used in the food industry to preserve products, prevent moisture loss, oxidative rancidity, increase storage life and improve physical appearance as well as for pharmaceuticals and nutraceuticals. They can also replace the use of conventional plastics as a barrier between the food product and the external environment.

Given the unfamiliarity of edible materials in the context of packaging, establishing consumer trust and acceptance remains a key hurdle.

> *"I can say most of us genuinely care about the environment, yeah? But eating something that used to be the wrapping... that's a weird one. I mean, we've spent our whole lives chucking packaging in the bin, not on our plates. If we're gonna start chewing the stuff that holds our food, then it needs to look decent, taste alright, and not make us feel like we're crossing some hygiene line. It's all about how it's introduced... good design, clear messaging, and making sure it feels more like part of the meal, not part of the trash."* (Male, born 2004)

This insight underscores the importance of not only developing edible packaging technologies but also ensuring consumer education and psychological readiness, particularly among Gen Z, who seek transparency and safety in food innovations.

Nanotechnology also offers new promising ways to provide bioactives, antimicrobials, vitamins, antioxidants, nutrients and other components to improve the functionality of edible packaging (Trajkovska Petkovska *et al.*, 2021). A study of young consumers in

Slovakia showed that all of them would purchase a food product with edible packaging rather than plastic-packaged product (Horská *et al.*, 2021).

Newer edible packaging options go beyond the food product itself and extend to becoming its dispenser, wrapper or eating utensil. Examples include hard cookie coffee cups, fruit-like casings, candy wrappers made of potato fibres or rice paper, consumable straws made of pasta or sweet sugar, cutlery from rice flour, wheat and sorghum, and edible pouch-like bottles made of sodium alginate and calcium chloride (Chen, 2023). Despite the growing array of edible packaging innovations, concerns about functionality and sensory appeal remain central to consumer uptake:

> *"Edible straws and cookie cups? That's wild... I mean, cool idea and all, but it's gotta actually work, yeah? If it turns soggy or tastes off, no one's gonna stick with it. I'm keen as, but it better feel like a proper part of the meal, not just some gimmick."* (Male, born 2005)

This argument indeed captures the Gen Z's open-mindedness, but it also points to something deeper. They are not just looking for clever ideas, they want stuff that works. Practicality matters. Gen Z needs clear proof of performance before they make it part of their daily use. Cool concepts are great, but functionality, taste and trust are the deal-breakers. There is preference for neutral, subtle or tasteless flavours to avoid compromising the main food product. Edible packaging needs to have good barrier qualities — to protect from the external environment, carrier properties — to transport and manage the release of active substances, mechanical capabilities — to serve the expected functions, and edibility — to be accepted by consumers (Abdulla *et al.*, 2024).

Young consumers need to trust and feel safe with eating edible packaging rather than the custom of throwing it away. As the food product will be directly consumed, there is still the need to provide external, preferably biodegradable packaging, particularly as many Gen Z are shopping online even for foodstuff. Edible packaging can be more costly and more sensitive to temperature and heat requiring specific transportation and storage conditions. Notwithstanding this, its market size has been consistently growing (Chen, 2023) to reach US$2 billion in 2024 (Marchese, 2022) with many new innovative ideas and improved technologies.

Functional Packaging

According to its purpose, packaging traditionally serves well-defined functions of protecting the goods. These conventional tasks have now been broadened by adding new functions beneficial for the consumers, food manufacturers, distributors and sellers. Functional packaging provides value-added benefits (Packaging Strategies, 2018) by transforming the foodstuff in its appearance and quality, often giving it a longer expiration time (Allpack, 2025). With new technological solutions and materials, there are many ways to achieve this. For example, smart or intelligent packaging may use sensors to monitor and indicate the freshness of the food or the state of the environment surrounding the product (e.g. temperature and moisture). The growing food delivery industry has new expectations from packaging, such as impact-resistant containers that can maintain temperature during transit, have tamper-evident seals and are stackable (Vicky, 2024). However, Gen Z are looking for packaging ways that are also biodegradable and environmentally friendly. For example, flexible packaging — wraps, pouches, stickpacks and bags, can be more efficient for the food product but Gen Z expects them to also be compostable.

Modified Atmosphere Packaging

Modified atmosphere packaging is “the packaging of a perishable product in an atmosphere which has been modified so that its composition is other than that of air” (Shaikh *et al.*, 2021, p. 512). The modified atmosphere helps preserve the freshness of the food without the need to freeze it or add preservatives to the list of ingredients. Carbon dioxide, oxygen and nitrogen are the three main gases added in different mixtures to modify the atmosphere in the packaging (Gym Meals, n.d.). Another way to alter the atmosphere of the food packaging is through vacuum packaging, that is by removing as much air as possible (Crompton, 2012). Creating a high-humidity environment also helps in preventing food to shrink and preserve its freshness (Shaikh *et al.*, 2021).

There are numerous applications of modified atmosphere packaging ranging from shredded lettuce and freshly cut vegetables to berries and tomatoes. When biodegradable polymers and renewable sources of energy are used, the packaging becomes more sustainable. It is also possible to combine modified atmosphere packaging with edible packaging to achieve better outcomes, particularly for meat-based products (Muller *et al.*, 2017). The overall process is quite complex in terms of preparation and can be more expensive. Mathematical modelling and simulation are used to determine the exact parameters of the modified atmosphere packaging to achieve optimal results depending on the specific food product and packaging material used as well as to estimate its shelf life (Rashvand *et al.*, 2023). Various food products — meat, seafood, bakery, vegetables and fruits, dairy, coffee, ready-made meals and snacks, have different requirements (Czerwiński *et al.*, 2021). The ultimate aim is to avoid food wastage and deliver safe and nutritious products.

Active Packaging

The European definition for active packaging is "materials and articles . . . designed to deliberately incorporate components that would release or absorb substances into or from the packaged food or the environment surrounding the food" (EC, 2004, p. 3). Such type of packaging covers technologies which release certain substances that act as antimicrobial, antioxidant or flavour-enhancing agents, as well as systems that absorb oxygen, carbon dioxide, moisture, odour or ethylene (Yildirim *et al.*, 2018). Examples include antimicrobial packaging of baking products with oregano-based essential oils and the use of emitting sachets of eugenol in cellulose pillow packages for iceberg lettuce (Shaikh *et al.*, 2021). A study of university participants in Romania showed the desirability of using biopolymer film for active packaging of food (Dabija *et al.*, 2024).

Reusable Packaging

Another aspect of the way food is packaged, transported and consumed is single-use versus reusable packaging. This equally applies for traditional products, such as dairy milk, juices and soft drinks, as well as the new plant-based alternatives. Reusable packaging is designed for lasting use in a system that also ensures its retrieval and return for continuous use (Reusable Packaging Association, 2025). It is part of a circular economy in which there is a repeated loop between making, using and reusing the packaging.

With food products being highly heterogenous, any assessment of the suitability of packaging needs to take into consideration factors, such as production and processing, logistics, shelf life, nature of the foodstuff, consumer safety, health and well-being considerations (Nilsson *et al.*, 2024). Different contexts and conditions need to be factored in, increasing further the complexity of providing healthy and nutritious food while reducing waste. Table 13.1 is an example of the multitude of factors that influence the sustainability of reusable packaging (Bradley and Corsini, 2023).

It seems that there is not enough research evidence and comparisons between single-use and reusable packaging to clearly identify environmental and safety impacts (Nilsson *et al.*, 2024). While waste from single-use packaging is a significant issue, cleaning and returning reusable packaging have their own challenges. A life-cycle analysis of the environmental impact of plastic food containers shows that a container needs to be reused 6 times to have a lower global warming potential than a new single-use container (Yadav *et al.*, 2024). This is a big ask, particularly for Gen Z who are reluctant to go the extra mile for recycling. Convenience is a defining factor in their food choices (Chen *et al.*, 2024). For reusable packaging to be a success, it would also require appropriate infrastructure, legislation and alignment between different jurisdictions, including federal, state

Table 13.1. Factors influencing the sustainability of reusable packaging.

Sustainability aspect	Factor	Considerations
Environmental	1. Material selection and production	Material type, recycled content, packaging and product weight
	2. Use	Useful volume
	3. Sorting and cleaning	Energy and water use
	4. Transportation	Packaging's volume ratio and distance
	5. Shrinkage	Rates of return, deterioration and loss
	6. End of life	Disposal scenarios
Social	7. Convenience	Ease of use, cleaning and maintenance
	8. Usability	Of the packaging service and its functions
	9. Awareness	Labelling and marketing
	10. Consumer perception	Related to functional design, aesthetics, service experience, value for money, product quality, hygiene and sustainability
	11. Consumer behaviour	Includes consumer knowledge and education, familiarity with the food, availability of low waste alternatives, peer pressure, market trends and type of reuse
	12. Consumer characteristics	Gender, age, culture, educational level, disposable income, lifestyle, attitudes and values
Economic	13. Policies and legislation	Economic incentives, strategies to increase producer responsibility, encourage consumer behaviour change and technical viability
	14. Material, infrastructure and operational costs	Costs of storage, transportation, operation, infrastructure, material and manufacturing, consumables
	15. Labour	Handling, sorting, cleaning and maintenance, system management
	16. Customer retention	Brand loyalty, financial incentives, deposit return schemes, loyalty schemes and subscription services

(*Continued*)

Table 13.1. (*Continued*)

Sustainability aspect	Factor	Considerations
Technical	17. Logistics	Supply chain management and complexity, system cycle time, standardisation, third party providers
	18. Management	Information flow, stakeholder collaboration
	19. Maintenance and infrastructure	Inspection and refurbishing, retail space
	20. Technology	E.g. tracking technology
	21. Product features	Standardisation of packaging, product protection
	22. Standards and protocols	Hygiene and safety standards, contamination, quality standards

Source: Bradley and Corsini (2023).

and local government in places, such as USA, Australia and Canada (Sperber, 2024).

There do not appear to be any strict rules with the exception of the fact that packaging needs to be minimised as many food products are currently overpackaged. Also, the use of appropriate safe biodegradable materials would overall reduce the environmental impact of packaging.

The appearance of the reusable packaging systems, e.g. in a supermarket, and the sensory interactions when using them, can influence the attractiveness of this food-purchasing option (Miao *et al.*, 2023). New reusable food wraps and other containers are now appearing on the market made from materials, such as beeswax and mushrooms, that are more attractive to the environmentally-oriented young consumers. Food packaging made from recycled plastics and other responsively sourced materials, is similarly of interest although there may be a gap between intentions and actual behaviour (Polyportis *et al.*, 2022). Gen Z has already whole-heartedly embraced reusable water bottles to reduce the use of plastics and also as a way of self-expression and as a social status reflected in the choice of design, colours, brand and other features (Asgard, 2024; Phally, 2023; Tuang, 2024).

Aesthetics of Packaging

In addition to the environmental credentials and ethical sourcing of the food packaging, its aesthetics need to be appealing to the Gen Z consumer. Its design has to be bold and eye-catching to attract attention amongst many competing products online and on the supermarket shelf. Visually attractive bold colours are preferred but also nostalgic retro-looking designs with artistic elements and natural colours, such as beige, green and brown. Gen Z sees a link between colours and the healthiness of a food product as explained in the comments of a Germain youth: "you could present it much healthier, for example with the colour green... Blue is not appealing. I would not associate it with healthy, the colour purple is better, because it is it reminds of berries and has a healthier effect" (Meyerding and Ahrens, 2024, p. 15) as well as an Irish youth: "I'd associate white with clean and the brown being more earthy, green being healthy, obviously and red being bad because that's just because it's associated" (Meyerding and Ahrens, 2024, p. 16). Australian Gen Z consumers echo similar views, showing a strong connection between visual design and the product's perceived health and integrity.

> *"You know, if it's bright red and shiny, it screams fake to me. But something earthy, with greens or pastels, feels like it's actually good for you — and made with care. The colours just set the vibe straight away, yeah? If it looks like it belongs in nature, I'm way more likely to trust it."* (Female, born 1998)

This perspective emphasises the importance of colour psychology in packaging design, where natural tones signal wellness, authenticity and trustworthiness to Gen Zs.

Playful, positively intended printed elements should combine with QR (quick response) codes, NFC (near-field communication) tags, augmented reality features, e.g. 3D images of a dish (ZealAR, 2025), web addresses and other digital elements from where the consumer can find more information about the product. This includes list of ingredients, health credentials, personalised nutritional value, provenance, ethical sourcing and sustainability certification as well as recipes, information about the company supplying the food and the creators of the selected artistic images and design. Allowing for consumer engagement and interaction is similarly important for the inquisitive and self-expressing Gen Z. Digital touchpoints on packaging are not just expected; they are welcomed by young consumers who enjoy interactive and informative design elements:

> *"I love when there's a QR code on the box that takes you somewhere, like a food producing story, a recipe, or even a sneak peek behind the scenes. It's not just food anymore, it's like you're part of the journey. Makes the whole thing feel personal, not just another plastic-wrapped snack chucked in the trolley."* (Male, born 2002)

In Gen Z's appetite for immersive experiences packaging becomes a medium for storytelling, education and connection, not just a container. Using digital links helps reduce the volume of packaging and minimise waste without the added burden of recycling.

Gen Z values authenticity rather than curated content and they prefer to see real customers sharing genuine experiences in any advertising (Keplinger, 2024). They are also influenced by social media, including friends and influencers whose views and values they share.

The actual design of the containers should be ergonomic, aesthetic and convenient to use. Creativity and innovation can help designing packaging that has added functions, for example, being the eating or cooking utensil, being easy to open, to be re-used with a reclosure option or for a different purpose, e.g. as a storage container or carry bag. With Gen Z being health-oriented, the appearance of the packaging is associated with its healthiness (e.g. Meyerding and Ahrens, 2024). The name of the food product can also be seen as indicative of its healthiness as is the case of granola or the use of berries (Meyerding and Ahrens, 2024). Quality, safety, efficacy and attractive appearance need to go hand in hand and this is what Gen Z expects.

A survey in India shows that Gen Z shoppers are 1.5 times more likely to research the product they buy compared to the Millennials

Figure 13.2. Example of dopaminergic design.

(Rooprai, 2024). It is critical for packaging to convey clear accurate and transparent information about the food product without any claims that could be perceived as misleading or greenwashing, given how easy it is for Gen Z to query and fact-check bold statements. In fact, in marketing already exists the notion of Gen Z-friendly packaging that offers all these desired features synergistically together with high-quality substance (Volarevic, 2025). The food's packaging needs to evoke a dopaminergic reaction of excitement followed by genuine satisfaction with the product (Volarevic, 2025). Insights from neuroaesthetics help designers create shapes, patterns and colours (see Figure 13.2) that leave lasting impressions in the human brain and create the desire to engage with the food product (Biggs, 2024).

Conclusion

Gen Z grew up with digital technology through the years of COVID-19 and is now facing the challenges of climate change combined with widespread plastics pollution. This is also the highest educated generation which has vast information on its fingertips.

The tech-savvy Gen Z is socially conscious, very diverse and often rallies behind causes of injustice. With its growing purchasing power, Gen Z is pushing the food and beverage industry to be more sustainable (Bar-Cohen, 2023). Even though Gen Z consumers are environmentally oriented, this not always translates in eco-friendly behaviours, particularly when it comes to recycling. This attitude-behaviour gap (Lisboa *et al.*, 2022) is essential in understanding how food packaging in particular should facilitate the outcomes desired by these young people. They are open to support better packaging by paying a higher price and also in favour of edible packaging to eliminate waste and plastic pollution.

Bold minimal packaging with the use of biodegradable materials is what Gen Z wants but these young people also expect to be able to digitally connect with the food. Convenience is a major factor in the way they assess their food options (Martinez, 2024) and packaging needs to cater for this. Rather than expecting Gen Z to recycle, food manufacturers and selling outlets need to conceptually change their practices by shifting to sustainable materials for functional and active packaging that preserves the freshness and nutritional values of the food. Some are already doing this successfully; others will need to catch up with the trend. With their choices and expectations, these young consumers are defining the future of food as well as food packaging.

References

Abdulla, S. F., Shams, R., & Dash, K. K. (2024). Edible packaging as sustainable alternative to synthetic plastic: A comprehensive review. *Environmental Science and Pollution Research International.* htpps://doi.org/10.1007/s11356-024-32806-z

AllThings.Bio. (2018). Facts or Myth "Biodegradability". https://www.allthings.bio/fact-or-myth/facts-or-myth-biodegradability/

Allpack. (2025). What Is Functional Packaging? https://www.allpackchina.com/what-is-functional-packaging/

Asgard, Z. (2024). The cult of the water bottle. The Spectator. https://www.spectator.co.uk/article/the-cult-of-the-water-bottle/

Askew, K. (2020). FOMO drives Gen Z food waste: Why under 25s don't love their leftovers. https://www.foodnavigator.com/Article/2020/11/02/FOMO-drives-Gen-Z-food-waste-Why-under-25s-don-t-love-their-leftovers/

Australian Government. (2022). PFAS: Per- and polyfluoroalkyl substances: FAQs. https://www.pfas.gov.au/about-pfas/faq

Avery Dennison. (2018). Consumer perceptions of functional packaging: An Avery Dennison White Paper. https://label.averydennison.com/content/dam/averydennison/lpm-responsive/asia-pacific/en-sa/documents/promotional/reclosure/wp-consumer-perceptions-of-functional-packaging-sapssa.pdf

Bar-Cohen, B. (2023). Why Gen Z Is Driving Food and Beverage Brands to Become More Sustainable. https://www.linkedin.com/pulse/why-gen-z-driving-food-beverage-brands-become-fpetc/

Biggs, H. (2024). Dopamine Design — why the latest trend feels like a flashback. https://paperandspark.com.au/dopamine-design-why-the-latest-trend-feels-like-a-flashback/

Bradley, C. G. & Corsini, L. (2023). A literature review and analytical framework of the sustainability of reusable packaging. *Sustainable Production and Consumption*, 37, 126–141. https://doi.org/10.1016/j.spc.2023.02.009

Chen, R. (2023). Edible Packaging — The Fast Approaching Sustainable Packaging Trend. https://pakfactory.com/blog/edible-packaging-the-fast-approaching-sustainable-packaging-trend

Chen, J., Xu, A., Tang, D., & Zheng, M. (2024). Divergence and convergence: A cross-generational study on local food consumption. *Scientific Reports*, 14(1), 13463. https://doi.org10.1038/s41598-024-64284-1

Chung, E. (2023). Compostable takeout bowls contain 'forever chemicals,' study finds. CBC. https://www.cbc.ca/news/science/pfas-compostable-food-packaging-1.6794550

Containers for Change. (2022). Tips to help Millennials and Gen Z prioritise recycling. https://www.containersforchange.com.au/qld/tips-help-millennials-and-gen-z-prioritise-recycling

Corey, S. (2021). Do Gen Z and millennials care about their environmental impact? https://savanta.com/knowledge-centre/view/do-gen-z-and-millennials-care-about-their-environmental-impact

Crompton, G. (2012). Packaging machinery and line operations. In Anne Emblem, A. & Emblem, H. (eds) *Packaging Technology: Fundamentals, Materials and Processes*, pp. 490–537. Woodhead Publishing, Sawston, UK.

Czerwiński, K., Rydzkowski, T., Wróblewska-Krepsztul, J., & Thakur, V. K. (2021). Towards impact of Modified Atmosphere Packaging

(MAP) on shelf-life of polymer-film-packed food products: Challenges and sustainable developments. *Coatings*, 11(12), 1504. https://doi.org/10.3390/coatings11121504

Dabija, D.-C., Pocol, C.-B., Mititean, P., & Semeniuc, C. A. (2024). Consumers' perception of biopolymer films for active packaging: From aesthetic appeal to nutritional value and experiential consumption. *Journal of Food Quality*. https://doi.org/10.1155/jfq/1255122

Dennis, P. (2024). Baby Boomers more likely to recycle than Gen Z, research shows. Circular. https://www.circularonline.co.uk/news/baby-boomers-more-likely-to-recycle-than-gen-z-research-shows/#

DS Smith. (2022). New research shows Gen Z lack recycling confidence. https://www.dssmith.com/media/our-stories/2022/8/new-research-from-ds-smith-shows-gen-z-lack-recycling-confidence-compared-to-their-older-counterparts

DS Smith. (2025). Are Gen Zs the worst at recycling? https://www.dssmith.com/media/our-stories/2024/9/are-gen-z-the-worst-at-recycling

Duffy, E. (2019). Millennials and Gen Z bin over $1,000 of food every year. https://www.savings.com.au/news/millennials-and-gen-z-bin-over-1000-of-food-every-year

European Commission (EC). (2004). Regulation (EC) No. 1935/2004 of the European Parliament and the Council. https://eur-lex.europa.eu/legal-content/EN/TXT/PDF/?uri=CELEX:02004R1935-20210327

Fenton, S. E., Ducatman, A, Boobis, A., DeWitt. J. C., Lau, C., Ng, C., Smith, J. S., & Roberts, S. M. (2021). Per- and polyfluoroalkyl substance toxicity and human health review: Current state of knowledge and strategies for informing future research. *Environmental Toxicology and Chemistry*, 40(3), 606–630. https://doi.org/10.1002/etc.4890

Foopak. (2023). Survey of Gen Z's interest level in eco-friendly packaging. https://www.foopak.com/survey-of-gen-zs-interest-level-in-eco-friendly-packaging

Greater Victoria News Staff. (2023). Gen Z in Canada spends the most on food — and wastes the most — says survey. https://www.pqbnews.com/news/gen-z-in-canada-spends-the-most-on-food-and-wastes-the-most-says-survey-1289339

Green, E. (2025). Gen Z willing to pay more for sustainable packaging, McKinsey reveals. Sigma Earth. https://sigmaearth.com/gen-z-willing-to-pay-more-for-sustainable-packaging-mckinsey-reveals/

Gupta, R. K. & Dudeja, P. (2016). Food packaging. In Gupta, R. K., Dudeja, & Minhas, S. (eds) *Food Safety in the 21st Century: Public Health Perspective*, pp. 547–553. Academic Press, Cambridge, MA.

Gym Meals. (n.d.). Fresher meals — what is modified atmospheric packaging (MAP)? https://www.gmdcanberra.com.au/blogs/news/fresher-meals-what-is-modified-atmospheric-packaging-map

Harder, L. (2024). Compostable packaging — all the advantages and disadvantages. https://www.packaging-warehouse.com/en/magazine/compostable-packaging-all-the-advantages-and-disadvantages-119.php

Hilton Segel, L. & Hatami, H. (2023). Mind the Gap. McKinsey & Company. https://www.mckinsey.com/ /media/mckinsey/email/genz/2022/12/06/2022-12-06b.html

Horská, E., Šedík, P., Mušinská, C., Savitskaya, T., Grinshpan, D., & Kačániová, M. (2021). Acceptability of edible food packaging in Slovakia: A case study on young generation. *Frontiers in Sustainable Food Systems*, 5. https://doi.org/10.3389/fsufs.2021.720700

Karunasena, G. G., Ananda, J., & Pearson, D. (2021). Generational differences in food management skills and their impact on food waste in households. *Resources, Conservation and Recycling*, 175, 105890. https://doi.org/10.1016/j.resconrec.2021.105890

Keenan Recycling. (2024). Gen Z: The least engaged generation in recycling. https://keenanrecycling.co.uk/gen-z-the-least-engaged-generation-in-recycling/

Keplinger, M. (2024). Boomer, Meet Zoomer: Packaging Design for Generation Z. https://www.smashbrand.com/articles/boomer-meet-zoomer-packaging-design-for-generation-z/

Lisboa, A., Vitorino, L., & Antunes, R. (2022). Gen Zers' intention to purchase products with sustainable packaging: an alternative perspective to the attitude-behaviour gap. *Journal of Marketing Management*, 38(9–10), 967–992. https://doi.org/10.1080/0267257X.2022.2083214

Lucas, N., Bienaime, C., Belloy, C., Queneudec, M., Silvestre, F., & Nava-Saucedo, J.-E. (2008). Polymer biodegradation: Mechanisms and estimation techniques — A review. *Chemosphere*, 73(4), 429–442. https://doi.org/10.1016/j.chemosphere.2008.06.064.

Marchese, K. (2022). Edible Packaging: The four fab start-ups tackling plastic pollution. https://designwanted.com/edible-packaging-four-startups/

Martinez, T. (2024). Tracking the Latest Gen Z Food Trends in 2025. https://pos.toasttab.com/blog/on-the-line/gen-z-food-trends-and-statistics

Meet, R. K., Kundu, N., & Ahluwalia, I. S. (2024). Does socio demographic, green washing, and marketing mix factors influence Gen Z purchase intention towards environmentally friendly packaged drinks? Evidence from emerging economy. *Journal of Cleaner Production*, 434, 140357. https://doi.org/10.1016/j.jclepro.2023.140357

Mendel University. (2023). Generation Z wastes the most food, experts want to find a way to change that. https://mendelu.cz/en/generation-z-wastes-the-most-food-experts-want-to-find-a-way-to-change-that/#

Mendel University. (2024). How is Generation Z Doing with Food Waste? https://mendelu.cz/en/how-is-generation-z-doing-with-food-waste/

Meyerding, S. G. H. & Ahrens, S. (2024). Generation Z's perception of food healthiness: The case of Kellogg's cereals — A qualitative study in Germany and Ireland. *Cleaner and Responsible Consumption*, 14, 100218. https://doi.org/10.1016/j.clrc.2024.100218

Miao, X., Magnier, L., & Mugge, R. (2023). Switching to reuse? An exploration of consumers' perceptions and behaviour towards reusable packaging systems. *Resources, Conservation and Recycling*, 193, 106972. https://doi.org/10.1016/j.resconrec.2023.106972.

Muller, J., González-Martínez, C., & Chiralt, A. (2017). Combination of poly (lactic) acid and starch for biodegradable food packaging. *Materials*, 10(8), 952. https://doi.org/10.3390/ma10080952

Nair, S. S., Trafiałek, J., & Kolanowski, W. (2023). Edible packaging: A technological update for the sustainable future of the food industry. *Applied Sciences*, 13(14), 8234. https://doi.org/10.3390/app13148234

Nilsson, F., Silva, N., & Schelin, J. (2024). Single-use versus reusable packaging for perishable liquid foods — Exploring evidence from research on climate impact and food safety. *Resources, Conservation and Recycling*, 207, 107655. https://doi.org/10.1016/j.resconrec.2024.107655

Onyeaka, H. N. & Nwabor, O. F. (2022). *Food Preservation and Safety of Natural Products*. Academic Press, Cambridge, MA.

Owings, T. (2023). Study: Plant-Based Attains Popularity with Gen Z. The Food Institute. https://foodinstitute.com/focus/study-plant-based-attains-popularity-with-gen-z/

Oxford Learner's Dictionaries. (2025). packaging noun. Oxford University Press. https://www.oxfordlearnersdictionaries.com/definition/english/packaging

Packaging Strategies. (2018). What Consumers Seek in Functional Packaging. https://www.packagingstrategies.com/articles/90655-what-consumers-seek-in-functional-packaging

Perkins, T. (2023). Toxic PFAS chemicals used in packaging can end up in food, study finds. The Guardian. https://www.theguardian.com/environment/2023/apr/17/pfas-forever-chemicals-food-containers-study

Phally, E. (2023). Reusable bottles aren't just accessories. https://dailytrojan.com/2023/09/26/reusable-bottles-arent-just-accessories/

Polyportis, A., Mugge, R., & Magnier, L. (2022). Consumer acceptance of products made from recycled materials: A scoping review. *Resources, Conservation and Recycling*, 186, 106533. https://doi.org/10.1016/j.resconrec.2022.106533

Rashvand, M., Matera, A., Altieri, G., Genovese, F., Fadiji, T., Opara, U. L., Mohamadifar, M. A., Feyissa, A. H., & Di Renzo, G. C. (2023). Recent advances in the potential of modeling and simulation to assess the performance of modified atmosphere packaging (MAP) systems for the fresh agricultural product: Challenges and development. *Trends in Food Science & Technology*, 136, 48–63. https://doi.org/10.1016/j.tifs.2023.04.012

Reusable Packaging Association. (2025). What is Reusable Packaging? https://reusables.org/what-is-reusable-packaging/

Rooprai, A. (2024). Unpacking the future: How Gen Z is shaping packaging trends. https://www.whatpackaging.co.in/features/unpacking-the-future-how-gen-z-is-shaping-packaging-trends-58430

Salgado, P. R., Di Giorgio, L., Musso, Y. S., & Mauri, A. N. (2021). Recent developments in smart food packaging focused on biobased and biodegradable polymers. *Frontiers in Sustainable Food Systems*, 5. https://doi.org/10.3389/fsufs.2021.630393

Shaikh, S., Yaqoob, M., & Aggarwal, P. (2021). An overview of biodegradable packaging in food industry. *Current Research in Food Science*, 4, 503–520. https://doi.org/10.1016/j.crfs.2021.07.005

Singh, K. (2025). Gen Z speaks: Wasteful packaging. What Packaging. https://www.whatpackaging.co.in/news/gen-z-speaks-wasteful-packaging-58852

Sperber, B. (2024). Gen Z Wants a Circular Economy, But at What Price? https://www.packagingdigest.com/sustainability/gen-z-wants-a-circular-economy-but-at-what-price-

Topic, M. & Mitchel, B. (2019). Generation Z & consumer trends in environmental packaging. Project Report. The Retail Institute, Leeds, UK. https://eprints.leedsbeckett.ac.uk/id/eprint/6066/1/GenerationZAndConsumerTrendsInEnvironmentalPackagingPV-TOPIC.pdf

Trajkovska Petkoska, A., Daniloski, D., D'Cunha, N. M., Naumovski, N., Broach, A. T. (2021). Edible packaging: Sustainable solutions and novel trends in food packaging. *Food Research International*, 140, 109981. https://doi.org/10.1016/j.foodres.2020.109981

Tuang, A. (2024). Opinion: The social status of water bottles: How Gen Z trends are shaped by cultural capital. https://highschool.latimes.com/opinion/opinion-the-social-status-of-water-bottles

Vicky, A. (2024). Must-have food packaging solutions your bakery can't live without. https://plasticcontainercity.com/blog/post/must-have-food-packaging-solutions-your-bakery-cant-live-without

Volarevic, L. (2025). The power of packaging in winning over Gen Z. https://www.linkedin.com/pulse/power-packaging-winning-over-gen-z-lana-volarevic-deduf/

Wani, A. A., Singh, P., & Langowski, H.-C. (2014). Food technologies: Packaging. In Motarjemi, Y. (ed.) *Encyclopedia of Food Safety*, Vol. 3, pp. 211–218. Academic Press, Cambridge, MA.

Yadav, P., Silvenius, F., Katajajuuri, J.-M., Leinonen, I. (2024). Life cycle assessment of reusable plastic food packaging. *Journal of Cleaner Production*, 448, 141529. https://doi.org/10.1016/j.jclepro.2024.141529

Yildirim, S., Röcker, B., Pettersen, M. K., Nilsen-Nygaard, J., Ayhan, Z., Rutkaite, R., Radusin, T., Suminska, P., Marcos, B., & Coma, V. (2018). Active packaging applications for food. *Comprehensive Reviews of Food Science and Food Safety*, 17(1), 165–199. https://doi.org/10.1111/1541-4337.12322

Yubo Team. (n.d.). The Importance of Recycling to Gen Z. https://www.yubo.live/blog/the-importance-of-recycling-to-gen-z

ZealAR. (2025). The Growing Trend of Augmented Reality in the Food Industry. https://zealar.com.au/augmented-reality-in-food-industry/

Chapter 14

Gen Z's Sober Curiosity beyond Nutrition

"I drink to make other people more interesting."

— Ernest Hemingway

The words of Ernest Hemingway, the American novelist, journalist and Nobel Prize winner for literature: "I drink to make other people more interesting", reflect a mix of wit and introspection. On the surface, it humorously suggests that alcohol enhances social interactions and alters our perception of others, acting as a social lubricant. However, it also hints at profound dissatisfaction and shows a deeper truth about how he may have used alcohol to handle social situations or personal feelings. The statement captures irony, self-awareness and even cynicism, offering insight into Hemingway's perspective on human connections.

Unlike Hemingway's perspective, which suggests that alcohol can make social interactions more engaging, Gen Z is actively redefining its role in socialising. Rather than relying on alcohol to enhance conversations or make others seem more interesting, these young people prioritise genuine connections built on mindfulness and authenticity. For this generation, meaningful interactions do not require tipsiness or intoxication but rather a conscious and intentional approach to socialising. This shift reflects a broader cultural transformation, where emotional well-being and self-awareness take precedence over other generations' traditional drinking norms. The chapter explores the evolving relationship with alcohol and its impact on social dynamics starting with a historic perspective and moving onto Gen Z's attitudes.

Drinks as Social and Cultural Catalysts: Lessons from History

Drinks, like food, extend beyond their primary purpose of offering sustenance or pleasure, holding a deeper cultural, social and emotional significance. They act as catalysts for connection, functioning as social linkages that foster bonds between individuals and groups. Sharing drinks, whether it is a cup of tea in a quiet conversation, a celebratory toast at a gathering, or coffee in the company of friends or colleagues, creates moments of togetherness and a sense of belonging, reinforcing social structures (Douglas, 2003).

Among drinks, alcohol has played a particularly pivotal social role in human civilisation. It has been an integral part of many cultures for millennia, serving diverse functions such as aiding in religious practices and rituals, providing nutritional and medicinal benefits, satisfying thirst, promoting relaxation, enhancing social connections, elevating dining experiences and enriching the quality of life (Hanson, 2013; Wang and Liu, 2022).

During the Neolithic period (10,000–4,500 BC), the establishment of permanent settlements in Mesopotamia facilitated grain cultivation and the brewing of beer from barley, while in Asia and Eurasia, people began fermenting grapes and other fruits to produce wine (McGovern *et al.*, 1997). Beer in ancient Mesopotamia was far more than a dietary staple as it symbolised abundance and fertility, reinforced cultural identity and community ties, functioned as compensation for labour, and played a central role in social rituals such as feasts and festivals (Paulette, 2021). It also shaped social interactions by altering moods and stimulating conversations, emotions and sensory experiences (Paulette, 2021). Highlighting its important societal role, the significance of beer was even codified in the world's earliest laws, such as the Code of Hammurabi, which regulated drinking establishments (Yale Law School, n.d.).

Similarly, in ancient Greece and Rome, alcohol, particularly wine, was at the heart of cultural practices and provided men in the higher echelons of society with a formal social context for drinking (Cook *et al.*, 2007; Thompson, 2010). Greek symposia and Roman banquets were key events where wine was consumed in a regulated fashion, often diluted with water, to help its role in fostering community and intellectual exchange. Alcoholic beverages gained popularity for their nutritional value and germ-resistant properties compared to water and became of cultural significance. From ancient China and Egypt to 19th-century Europe and highland Peru, societies assigned religious, political and social meanings to alcohol, solidifying its place in human history (McGovern *et al.*, 1997).

Throughout history, alcohol has influenced human behaviour and social norms, though it has also posed challenges (Castro *et al.*, 2014). In response, societies have established cultural guidelines to regulate its use (Schmidt and Room, 1999). After a prohibition period between 1920 and 1933 in the United States, alcohol consumption steadily increased (Dufour, 1995). In the late 1940s, a United States' survey found that men and urban residents consumed alcohol more frequently than women and rural populations. According to Kraus *et al.* (2020), alcohol consumption fluctuated with a downward trend in the period between the two World Wars, after which it surged post-World War II, peaking in the mid-1970s. In the 1980s, Western alcohol consumption reached saturation levels (Global Alcohol Policy Alliance, n.d.) with increasing awareness of alcohol-related chronic medical conditions, including alcoholic liver disease (i.e. alcoholic cirrhosis). These changes were influenced by global events like wars, economic downturns and government policies shaped by external pressures (Kraus *et al.*, 2020). Consumption however increased with higher education levels and socioeconomic status, with shifts in gender roles post-World War II leading to a notable rise in alcohol use among women (Castro *et al.*, 2014).

Religious affiliation impacted abstinence rates, with Protestants abstaining most often, followed by Catholics and Jews (Castro *et al.*, 2014). By the 1970s, studies in Mexican and Peruvian villages highlighted alcohol's role in fostering community bonds, particularly among men. In these contexts, declining an invitation to drink was seen as rejecting social ties. In some Mexican villages, alcohol use also reflected cultural ideals like machismo, associating drinking with displays of competitive masculinity (Castro *et al.*, 2014).

Between 1990 and 2013, alcohol consumption became increasingly understood through a cultural lens, with societal norms shaping expectations about its effects and varying definitions of serious drinking and drunkenness. Cultural norms heavily influenced how alcohol consumption and its effects were perceived, with differing views on serious drinking and dependence symptoms (Schmidt and Room, 1999; Room *et al.*, 2009). Schmidt and Room (1999) identified four cultural factors: severity thresholds, perceptions of problem behaviours, causal explanations and unique cultural interpretations of alcohol-related symptoms. This highlights the deep connection between culture and alcohol use.

Adolescent alcohol use has been a particular research focus, revealing notable cross-national and cultural differences. Schmid *et al.* (2003) found that greater overall alcohol consumption and distilled spirit intake were linked to higher drunkenness rates, with southern European countries showing fewer instances compared to Russia, Scandinavia and the Baltic nations, which illustrated distinct "drinking cultures". Furthermore, Link (2008) identified stronger influences of peer alcohol use and perceived risk among American adolescents compared to German counterparts, highlighting the need to study alcohol use within specific cultural contexts.

In resent years alcohol consumption among younger generations has significantly declined in the high-income countries, marked by a shift toward moderation in Millennials and abstinence in Gen Z (Kraus *et al.*, 2020; Ludmir, 2023). While Baby Boomers favoured heavy drinking patterns, Millennials and now the Gen Z cohort are adopting more selective habits (Burgess *et al.*, 2022). This decline in alcohol consumption occurs alongside challenges related to worsening mental health among young adults and reflects broader generational changes, including increased individual choice, societal pressures and prolonged adolescence with reduced autonomy. This shift has led to Gen Z being called a "sober generation". Nevertheless, the reasons behind this overall positive decline remain uncertain, leaving researchers, policymakers and the public to look for explanation (Raninen and Livingston, 2018) which this chapter is trying to give.

Gen Z and Rethinking the Alcohol's Legacy

Despite its deep-rooted historical and cultural roles, alcohol consumption patterns are undergoing profound changes today and we are witnessing youth-specific changes (Kraus *et al.*, 2020). The new generation is driving a shift in how beverages are perceived. Gen Z's sober curiosity reflects a break from traditions where alcohol symbolised social connection and cultural values, signalling a broader rethinking of its role and heritage in contemporary society. In the words of a female Gen Zer:

> *"You know before... back at the time, drinking used to be a given, something you just did to fit in. And there was a big peer pressure to comply with this... not to feel isolated. But for us, it's different. We're questioning why alcohol has to be at the centre of every social event. There is no need... It's not about rebelling against tradition, but about making choices that actually align with how we want to live."* (Female, born 2001)

Gen Z is transforming old, traditional drinking habits, consuming significantly less alcohol compared to previous generations (Månsson *et al.*, 2022). A study in Sweden shows that alcohol abstinence among Gen Z has increased significantly, with nearly half of the participants choosing not to drink (Englund, 2019). Similar results were revealed in research by Gallup which shows that only 62% of adults under the age of 35 report drinking alcohol, a notable decline from 72% two decades ago (Fares and Panetta, 2024). Additionally, this demographic drinks less frequently and is less inclined to do so excessively (Saad, 2023; Witherington, 2025). Their consumption of beer and wine has dropped by approximately one-third, as many Gen Zers are turning to alcohol-free alternatives, which are gaining popularity among young consumers (Buchholz, 2023). Similarly, a report from the United Kingdom highlights that young consumers aged 20 to 24 are significantly less inclined to prioritise spending on alcoholic beverages for home consumption — nearly half the likelihood compared to the Baby Boomers aged 75 and older (Witherington, 2025). This indicates a generational shift in drinking habits and purchasing priorities. It is not surprising that one of the Gen Zers shared:

> *"Our grandparents drank, our parents drank... sometimes like it was a competition between them. I saw so many of my friends' parents drinking every day with no occasion at all... like this, just from the love to alcohol or boredom.... who knows... And now, looking back, all they have are stories of embarrassing moments and bad decisions. Maybe it's time to break that cycle and prove that you don't need to be drunk to have a good time."* (Male, born 2004)

Framing habitual and excessive drinking of past generations as both unnecessary and often regretful, is an important critical perspective of Gen Z. By acknowledging the normalisation of alcohol in older generations and its potential consequences, Gen Z advocates for a cultural shift, one that prioritises mindful consumption and challenges the idea that alcohol is essential for social enjoyment. On the other hand, Gen Z youth value socialising and acknowledge its importance in creating shared experiences:

> *"I've never really been drawn to drinking, and I don't feel the need to drink to fit in. But sometimes, I do wish I could just have a drink... doesn't even have to be alcoholic... just something to share with others and feel part of the social moment. It's hard to explain... it's not about the drink itself, but more about connecting in that shared experience."* (Female, born 2000)

This generation's actions speak for themselves. Research by Mintel (Witherington, 2025) states, about one-third of individuals aged 18–24 abstain from alcohol entirely, while those who do drink often see it as a treat, a way to relax or a means of celebrating special occasions. Unlike older generations, who frequently use alcohol as a regular and necessary refreshment or to enhance meals, Gen Z tends to approach alcohol more sparingly, treating it as a sporadic indulgence rather than a habitual part of their lifestyle (Witherington, 2025). This difference highlights a generational shift in how alcohol is perceived and consumed. One of the Gen Z individuals we spoke to, explained it well:

> *"I don't really see alcohol as something I need to enjoy a night out or unwind after a long day, like my parents or grandparents. I can't see the purpose for it.... For me, it's something more of a rare indulgence... something for special occasions rather than a routine. A lot of my friends feel the same way. We'd rather spend our money on experiences, buying nice food, clothes or even better-quality drinks, even if they're alcohol-free. You know, it's not about avoiding alcohol completely... don't get me wrong, but about kind of making more mindful choices."* (Male, born 1999)

Alcohol use is typically linked to social expectations and behaviours associated with emerging adulthood (Månsson *et al.*, 2022). However, this is not the case for Gen Z. Their adoption of a non-alcoholic lifestyle represents a significant cultural transformation. Gen Z is redefining their relationship with alcohol, moving away from traditional norms of drinking for leisure or refreshment (Richards, 2023). Instead, they are making more intentional choices, with many opting for soberness or treating alcohol as an occasional indulgence rather than a regular habit passed down by previous generations. This shift underscores their focus on values and priorities that matter more to them:

> *"For us, drinking less isn't just about avoiding hangovers. It's about living better. We care about our health, our mindset, and even the world around us. It's not just a trend... It's a choice that fits the way we want to live... It's good for me and I like it. I feel much better this way."* (Female, born 2005)

Gen Z's broader values of health, mindfulness and sustainability are reshaping not only drinking trends but also the social and cultural expectations surrounding alcohol. Their "sober curiosity" (Fares and Panetta, 2024) is driven not only by the physical benefits of avoiding alcohol, but by a mix of their own changing social patterns, cultural, health and wellness consciousness, mental health and lifestyle factors.

The shift in drinking habits among young people is not just about personal preference. As Månsson *et al.* (2022) point out, it also reflects the meaning attached to alcohol, drinking and getting drunk as well as Gen Z's place in society and the broader cultural changes shaping this generation.

Some researchers believe that Gen Z's decline in drinking is part of a broader trend toward moderation, with lower smoking rates and healthier lifestyle choices (Burgess *et al.*, 2022). Others link this shift to cautious parenting and a culture that prioritises safety. These patterns are evident in countries with traditionally high levels of alcohol consumption, such as UK (Gayle, 2016; Health Survey for England, 2016) and United States, highlighting a generational move toward more mindful living (Burgess *et al.*, 2022).

While traditional risk behaviours like drinking and substance use are declining, mental health challenges such as depression and self-harm are rising among young people. Studies in the UK and United States show that Gen Z experiences significantly higher rates of depressive symptoms than the Millennials (Burgess *et al.*, 2022). Additionally, Gen Z reports higher levels of loneliness, especially among young renters with weak social ties (ONS, 2018) and as a whole, younger adults are likely to feel lonelier (Duffy, 2021, p. 87).

A global survey of 10,000 young people revealed a growing widespread sense of uncertainty about their future, with 75% expressing fear about limited opportunities and 40% hesitation to start families compared to the Baby Boomer generation, who grew up in a time of economic stability and optimism (Marks *et al.*, 2021). For the Boomers, this environment supported larger families, often including three or more children. These concerns align with Gen Z's increased focus on mental well-being and mindfulness. Unlike Boomers, who often embraced alcohol as a social norm, many Gen Z individuals see reducing alcohol consumption as essential for emotional clarity and self-care. According to the research by Mintel, nearly two-thirds of 18- to 24-year-olds expressed concern about alcohol's emotional impact and were eager to adopt mindful drinking habits, reflecting the broader generational move toward prioritising mental health and balanced lifestyles (Witherington, 2025).

Alcohol, Mental Well-being and Mindfulness

With mental health being a priority, Gen Z sees reducing alcohol consumption as a way to support their emotional well-being and mindfulness. Many Gen Z individuals are focused on mental clarity, emotional stability and self-care. They may perceive alcohol as a hindrance to their well-being rather than just a dietary choice or a way to have a good time. Nearly two-thirds of 18- to 24-year-olds are concerned about alcohol's emotional impact, with many eager to learn mindful drinking habits (Witherington, 2025). For Gen Z, the mental impact of drinking is a significant concern, with 86% considering mental health as important as physical health when it comes to alcohol consumption (Richards, 2023). Anxiety about alcohol consumption seems a key factor contributing to the decline in youth drinking:

> *"For me, drinking just isn't worth it if it messes with my mood or leaves me feeling anxious the next day. I don't want to feel like sh**. I'd rather wake up feeling good and in control than deal with that. A lot of my friends think the same. We're not totally against alcohol, but we're way more mindful about when and how we drink. It's really about making choices that support our mental well-being, not just going along with what past generations did. We can see our parents and other people's drinking culture... it's their thing, not ours."* (Female, born 2005).

Gen Z is increasingly opting for low- and non-alcoholic drinks (Fuller, n.d.; Kraus *et al.*, 2020), with about a quarter citing lower calorie content and added health benefits, such as prebiotics and vitamins (Witherington, 2025). Compared to other age groups, Gen Z is far more drawn to these options with nutrition being a key motivator for their "sober curiosity" (Fares and Panetta, 2024). A 2022 Consumer Report by the now closed online alcohol-ordering company Drizly revealed that 38% of Gen Z and 25% of Millennials chose more non-alcoholic drinks, with the main reason behind this choice a desire for a healthier lifestyle and reducing alcohol consumption (Alverson, 2024).

While the non-alcoholic drink options are popular, they are not the sole choice for non-drinkers. Research by Gallup suggests that less than half of Gen Zers actually consume non-alcoholic drinks, indicating a preference for other beverages, including carbonated soft drinks and fruit juices (Witherington, 2025). Nearly 86% of Gen Z believe that their mental health is just as important as their physical health when it comes to making decisions about drinking alcohol (Fuller, n.d.; Hines, 2024).

Cost-Conscious Choices

Cost is another key factor driving alcohol moderation in Gen Z. These young people are highly budget-conscious, likely due to growing up during recessions and crises. Gen Z is particularly careful about spending on alcohol. They pay attention to how often they buy it, how much they drink, which brands they choose and where they shop, reflecting financial awareness (Dhawan, 2023). Their focus on saving money for emergencies and the future has helped them navigate employment insecurity and inflation (Alverson, 2024). Many young people see sobriety as a budget-friendly and financially savvy choice. When shopping, they are budget-conscious, prefer using their smartphones to find the best options, and enjoy discovering new products (Alverson, 2024). They often research brands and retailers online, relying on social media and delivery services for convenience (Alverson, 2024). Nearly a third of Gen Zers are opting for low or non-alcoholic types of beverages because of their affordability (Witherington, 2025). As one Gen Zer explained:

> *"I'd much rather spend my money on experiences than waste it on overpriced drinks. When we do drink, we usually get together at each others' homes first to have pre-s (meaning a drink or two before going out). We have a couple of drinks — vodka with something or wine, before heading out. Once we're at clubs or events, we don't feel the need to keep drinking. Honestly, sticking to cheaper alcohol-free options means I can still have something in my hand, enjoy the night, have fun, the next day wake up feeling good, and save money all at once. It's a win-win."* (Female, born 2001)

Gen Z's approach to drinking reflects a broader shift in priorities, where financial mindfulness, well-being and social experiences take precedence over traditional alcohol consumption. Gen Z, on average, spends around $46 per month on alcohol, with most keeping their expenses low, between $1 and $20 (Dhawan, 2023). Spending also varies by gender, with males spending nearly $69 per month, almost double the $35 spent by females (Dhawan, 2023). This gap suggests that while alcohol remains part of social life for many Gen Z consumers, it is not a top priority, especially for women, who may be more mindful of health, budget or lifestyle choices.

Rather than viewing alcohol as central to a night out, Gen Zers strategically manage their intake, opting for casual drinks in familiar settings before heading to social venues where drinking becomes secondary. This strategy not only aligns with their cost-conscious mindset but also highlights a redefinition of fun, where presence and shared moments matter more than continuous drinking, after which you may not have clear memories of what had happened. By embracing alcohol-free alternatives, they maintain social participation without compromising their health, energy or budgets, reinforcing a more intentional and balanced lifestyle. These evolving preferences have prompted businesses to start to adapt and expand their alcohol-free offerings, while social media influencers, trends and movements like #SoberLife continue to reshape and normalise a culture of mindful drinking (Fares and Panetta, 2024). A diverse, affordable and high-quality soft drink selection is key to attracting Gen Z's mindful and budget-conscious consumers along with fun spaces with events like karaoke, themed nights and live music. Live sports are also a major draw. To stay relevant, pubs and restaurants must offer diverse drink choices, a vibrant atmosphere and engaging experiences (CGA, 2024).

Gen Z is influenced by price also when trying new alcoholic drinks, with 47% of them saying it is the most important factor (Alverson, 2024). In comparison, the older generations — Millennials, Generation X and Boomers, are more likely to rely on recommendations from friends and family when choosing a brand. It seems that for Gen Z, the price matters more than what others suggest (Alverson, 2024).

Cultural Shifts, Social Media and Anxiety

Alcohol is commonly enjoyed during social gatherings and celebrations, and many people also use it as a way to relax or relieve stress. The decline in alcohol consumption among younger generations is driven by multiple cultural and societal shifts, including the rise of social media, evolving gender roles, changing expressions of masculinity and femininity, and an increasing focus on health and fitness trends (Törrönen *et al.*, 2019). Unlike previous generations, for whom alcohol often symbolised social belonging or rebellion, Gen Z prioritises intentional and inclusive social experiences that do not necessarily revolve around drinking (Fares and Panetta, 2024).

A key factor behind this shift is Gen Z's cautious and self-aware approach to alcohol. As the silent film icon Charlie Chaplin once remarked: "A man's true character comes out when he's drunk" (Goodreads, 2025). For Gen Z, this is precisely the concern (van de Ven and Lee, 2025). Raised in an era where every action can be documented, shared and scrutinised online, Gen Zers are hyper-conscious of their digital footprint. Unlike earlier generations, who may have viewed drunken misadventures as humorous or even rites of passage, Gen Z recognises the long-term consequences of losing control.

There are also differences in how social media are consumed. Older generations tend to use social media to share updates or showcase what they have done, like an announcement platform. In contrast, Gen Z uses social media more interactively (Alverson, 2024). They react, learn and engage in conversations all at once, doing so in a much more fluid and dynamic way (Alverson, 2024). A social media image is always on their mind when they socialise and alcohol is perceived as making them vulnerable (Carnegie, 2022).

Many young adults today see alcohol not as a source of relaxation but as a potential source of stress, largely due to the omnipresence of social media (Richards, 2023). The fear of having their drunken behaviour recorded and shared online contributes to growing anxiety around drinking. Viral videos and social media callouts have amplified their fear of public embarrassment, reinforcing the idea that excessive drinking is more of a risk than a reward. A study from the Netherlands (Hendriks *et al.*, 2017) found that young people tend to share moderate representations of alcohol consumption on social media rather than extreme depictions.

Beyond simply reflecting drinking culture, social media also dictates how alcohol is consumed and displayed. Many young people post alcohol-related content primarily for entertainment purposes (Hendriks *et al.*, 2017), signalling that drinking has, in some cases, become performative rather than purely social. However, this heightened awareness of online image also fuels a shift toward moderation. A study by the marketing company Red Brick Road found that 49% of Gen Z constantly worry about their online presence when they go out drinking, making it difficult for them to fully relax (Richards, 2023). For many, adopting a mindful drinking approach or avoiding alcohol altogether offers a sense of control over how they are perceived and prevents the anxiety of having embarrassing moments permanently documented.

Peer influence also plays a crucial role in reshaping drinking norms. Unlike previous generations, where peer pressure often encouraged alcohol consumption, many within Gen Z now support each other in making sober choices. Törrönen *et al.* (2019) suggest that drinking is no longer seen as an essential milestone of adulthood, with reduced peer pressure to drink and greater opportunities for alternative activities. Social circles that emphasise wellness, productivity and mental clarity are growing, making non-drinking or mindful drinking socially acceptable, if not preferable. The rise of movements like #SoberCurious and #MindfulDrinking reflects this collective shift toward moderation, with influencers and celebrities normalising alcohol-free lifestyles (Fares and Panetta, 2024).

Recognising this generational shift, businesses and familiar brands have begun adapting (see Figure 14.1). Alcohol companies are expanding their portfolios to include non-alcoholic versions of popular drinks, while hospitality venues are catering to sober lifestyles

Figure 14.1. Non-alcoholic drinks in a supermarket in Perth, Western Australia.

with alcohol-free bars, wellness-focused events and sophisticated mocktail menus. The market for premium alcohol-free beverages is booming, with Gen Z demanding alternatives that fit their social needs without compromising their health-conscious values (Ludmir, 2023). This response from the industry highlights the growing commercial impact of Gen Z's changing drinking habits, as companies start to move to align with their preferences rather than resist them.

In a time when privacy and personal image are closely connected, Gen Zers are mindful of both their online identity and their real-life actions. Drinking is no longer just a casual choice. It is a carefully considered decision shaped by concerns about reputation, self-image and risk. This shift reflects a broader focus on control and intentional living, with social media playing a key role in both encouraging and discouraging alcohol use.

Ethical and Environmental Concerns

Some Gen Zers associate alcohol production with sustainability issues, labour ethics or corporate responsibility, influencing their decision to consume less or not at all. Drizly's 2023 Consumer Trend Report found that 55% of Gen Z prioritise unique and interesting packaging when choosing alcohol brands (Alverson, 2024). Additionally, 31% of Gen Z said they would avoid products with unclear ingredient transparency (Alverson, 2024). Gen Z values a balance between affordable products and those that align with their ethical beliefs. They prioritise items that support communities and are environmentally responsible, even if it means spending a little more. However, the abundance of options can overwhelm them, especially in the beverage sector. Gen Z tends to snack more than cook, often treating beverages as snacks (Alverson, 2024).

Gen Z's focus on wellness and self-care, combined with their preference for brands that reflect their values, requires companies to prioritise community influence, targeted messaging and thoughtful positioning to build trust. By aligning with these priorities, brands can effectively connect with this generation and foster consumer loyalty.

Personal Identity, Health and Control

Young people are embracing moderation strategies like damp drinking — meaning reducing alcohol without cutting it out altogether, and zebra striping — meaning alternating between alcoholic and non-alcoholic drinks, which encourage gradual change and self-compassion (van de Ven and Lee, 2025). A Gen Zer explains:

> *"I don't see drinking as a big part of my life like it was for older generations. I still go out, but I don't want to waste the next day feeling awful. A lot of my friends are the same... To be honest, we'll have a drink or two, but we mix it up with non-alcoholic options. It's more about being social than getting drunk. Plus, there are so many good alcohol-free drinks now that it doesn't feel like you're missing out."* (Female, born 2001)

While health concerns play a role, shifting social and economic pressures also influence Gen Z's drinking habits. Many Gen Z individuals prioritise career and personal growth, seeing excessive drinking and its side effects, such as hangovers and anxiety, as obstacles to their ambitions (van de Ven and Lee, 2025). The rise of better-tasting zero-alcohol options, such as alcohol-free beer, wine and mocktails, has further supported this shift, making it easier to socialise without the stigma of abstaining. This trend reflects a broader cultural move toward mindful consumption and balance. It is captured by the Australian national statistics showing that only 29% of young Australians drink alcohol monthly (AIHW, 2024). When Gen Z drinks, they do so less frequently and in smaller quantities, compared to previous generations at the same age (van de Ven and Lee, 2025).

Gen Z values community and belonging, often seeking advice from their networks before making decisions (Alverson, 2024). At the end of the day, the relationship with alcohol is about your own desire, identity and control. Avoiding alcohol can be a way to assert independence, align with personal values and reject peer pressure or traditional social norms around drinking.

Gen Z's Drink Preferences

Not all Gen Z members avoid alcohol, but when they do drink, clear spirits like vodka, tequila and pre-mixed cocktails are their top choices (Hines, 2024; Witherington, 2025). This preference reflects Gen Z's inclination toward lighter and more versatile options that align with their lifestyle, which emphasises convenience and modernity. Vodka favoured by 88.5% of Gen Z drinkers, is the top choice, with tequila following at 71.2% (Dhawan, 2023). These statistics show a strong demand for spirits that can be adapted into cocktails and ready-to-drink formats.

Additionally, spirit-based seltzers dominate among younger consumers' preferences, with 25% of the market share coming from those aged 21 to 34, the highest among any age group (Statista, 2025). This trend highlights the popularity of beverages combining refreshing flavours with health-conscious branding, which appeals to Gen Z's growing focus on wellness. Gen Z's drinking habits diverge significantly from Baby Boomers, who prefer malt whiskies, cognacs and traditional beers like Guinness (Devane-Waters, 2024). In contrast, Gen Z's preference for lighter, fruitier beverages gravitates toward beer alternatives and craft beers such as Beavertown Neck Oil and Blue Moon, demonstrating a marked departure from more traditional drink options favoured by older age groups (Devane-Waters, 2024). This shift away from traditional drinks points up a generational transformation in how alcohol fits into social and lifestyle contexts. Businesses catering to this trend with innovative, light and flavourful drinks are likely to resonate strongly with Gen Z consumers.

According to Drizly's report on Gen Z's purchasing trends, this generation has tried wine and gin during the pandemic, but they prefer drinks like tequila, Champagne and pre-mixed cocktails (Alverson, 2024). Beer alternatives are also popular, with Gen Z gravitating toward lighter, fruity flavours and options with lower sugar and calorie content (Alverson, 2024). Ready-to-drink cocktails are growing segments within the market, competing with hard seltzers, and they are an increasingly popular choice among Gen Z consumers (Alverson, 2024). One Gen Z expressed a similar opinion:

> *"Honestly, I think people are too stuck on old drinks like wine or gin. Gen Z is all about convenience, and that's why pre-mixed cocktails and ready-to-drink drinks are such a hit. We like having options that are easy, fun and don't take too much effort to enjoy. Tequila, vodka and champagne are just more exciting, and I think we gravitate toward those because they're fresh and different. Plus, the focus on lighter, lower-sugar options really fits with how we're all thinking more about health now. I'm definitely all in on these trendy drinks. They make socialising more fun without all the hassle."* (Male, born 2003)

This young person's opinion highlights several key trends in Gen Z's evolving drinking habits, in favour of convenient, health-conscious drinks, including ready-to-drink cocktails, over traditional options, reflecting their modern, fast-paced and mindful lifestyles.

In Australia, Gen Z shows a preference for more traditional drinks, such as white clear spirits like vodka and ready-to-drink options, reflecting a balance between traditional and newer drink options (Berry, 2023). There is a notable trend of declining alcohol consumption among Gen Z, with Australia ranking among the top 15 countries where alcohol abstention rates are high for this age group. In Japan, 63% of Gen Z reported not consuming alcohol in the past six months, followed by 54% in the US and 44% in Canada (Berry, 2023). These abstention rates exceed those of the total adult population. Many Gen Z individuals are moving away from traditional drinking habits, showing increased rates of abstention and moderation compared to older generations. Among those who drink, preferences are shifting. Gen Z is underrepresented in high-volume categories like beer and wine but overrepresented in ready-to-drink options, white spirit-based cocktails, liqueurs and aperitifs (Berry, 2023). Notably, this generation's preference for lighter and more modern alcohol options contrasts with Millennials, who are more inclined toward whisky-based cocktails. This generational shift underscores growing demand for innovative and tailored beverage categories.

Despite drinking fewer fizzy soft drinks at bars compared to Millennials, Gen Z consumes more of these beverages at home than older generations (Witherington, 2025), with their top 10 most loved drink brands being all soft drink options (Devane-Waters, 2024). This suggests a shift toward refined and sophisticated soft drinks. Additionally, Gen Z often chooses energy drinks during nights out, as they provide an energy boost similar to the stimulating effects of alcohol but without the crash (Witherington, 2025).

> *"I often stay sober. I just don't feel the need to drink. It's convenient to be able to drive yourself or some of my friends home if needed. I think a lot of us are looking for drinks that feel special but don't rely on alcohol to make them fun. I personally enjoy having something that stands out, like a juice with a unique flavour... or something more exciting than the regular stuff. I also like having energy drinks on nights out because they keep me going, but without the alcohol crash. If brands could make those energy-boosting, low-alcohol drinks feel a bit fancier and still taste*

> *great, like adding a tiny bit of Aperol in your drink, I'd be all in. It's all about having the experience without the hangover, and finding the right balance between flavour and energy!"* (Male, born 2000)

To cater for this growing sober-curious market, brands should focus on premium low- and non-alcoholic options, such as fruit juices with sophisticated flavours, rather than traditional soft drinks (Witherington, 2025). This approach appeals to Gen Z's desire for luxury and enjoyment while maintaining a focus on health-conscious choices. Incorporating energy-lifting ingredients into these drinks could further attract consumers seeking a boost without the side effects of alcohol. Interestingly, energy drinks like Red Bull are particularly popular among Gen Z but less so among Baby Boomers, who did not grow up with these products (Devane-Walters, 2024). This trend is largely driven by marketing targeted at younger consumers.

The rise of health and wellness trends has driven the popularity of mindful drinking and the "NoLo" (no and low-alcohol) movement, with Gen Z leading the demand for these alternatives (Myles *et al.*, 2023). This trend is expected to continue, with the market for non-alcoholic beverages projected to grow by 25% between 2022 and 2026 (Ludmir, 2023). Major beverage companies, including Anheuser-Busch InBev, are focusing on expanding their non- and low-alcohol offerings to meet this growing demand (Ludmir, 2023). With the introduction of non-alcoholic spirits, wine and beer, companies that adapt to these shifting preferences will be well-positioned for success in the evolving market led by Gen Z's preferences.

Although the reduction in alcohol consumption can be seen as a positive change for Gen Z's health, there are other worrying trends. For example, there is increased use of marijuana among American young adults (Saad, 2023), potentially substituting it for alcohol, and this can be associated with depression and social anxiety (National Academies of Sciences, Engineering and Medicine, 2017). Every young generation will need to pave its future among many obstacles on the way and Gen Z will not be an exception, although its challenges are bigger and more complex than ever before.

Conclusion

Gen Z is driving significant potential changes in the beverage alcohol sector, compelling brands to adjust their approach to both marketing and product development. This generation is more health-conscious than ever, with many choosing to completely abstain from alcohol, while others prefer drinking less frequently and in smaller quantities. With a strong focus on personal well-being, both mental and physical, Gen Z is reshaping the drinking landscape, encouraging brands to innovate with options that cater to their desire for healthier, balanced lifestyle with more mindful choices.

References

Alverson, C. (2024). Generation Z shakes things up in beverage. Younger generation brings unique habits to beverage consumption. Beverage Industry. https://www.bevindustry.com/articles/96378-generation-z-shakes-things-up-in-beverage

Australian Institute of Health and Welfare (AIHW). (2024). National Drug Strategy Household Survey 2022–2023: Young people's consumption of alcohol. https://www.aihw.gov.au/reports/alcohol/young-people-alcohol

Berry, K. (2023). Gen Z's "very different relationship" with alcohol. Food and Drink Business. https://www.foodanddrinkbusiness.com.au/news/gen-z-s-very-different-relationship-with-alcohol

Buchholz, K. (2023). Gen Z: The More Sober Generation, Alcohol consumption, Statista. https://www.statista.com/chart/30783/alcohol-consumption-by-generation/

Burgess, A., Yeomans, H., & Fenton, L. (2022). 'More options . . . less time' in the 'hustle culture' of 'generation sensible': Individualization and drinking decline among twenty-first century young adults. *The British Journal of Society*, 73, 4, 903–918. https://doi.org/10.1111/1468-4446.12964

Carnegie, M. (2022). Why Gen Zers are growing up sober curious. BBC. https://www.bbc.com/worklife/article/20220920-why-gen-zers-are-growing-up-sober-curious#

Castro, F. G., Barrera, M. Jr, Mena, L. A., & Aguirre, K. M. (2014). Culture and alcohol use: Historical and sociocultural themes from 75 years of alcohol research. *Journal of Studies on Alcohol and Drugs Supplements*, 75(17), 36–49. https://doi.org/10.15288/jsads.2014.s17.36.

Cook, C. C., Tarbet, H., & Ball D. (2007). Classically intoxicated: Correlations between quantity of alcohol consumed and alcohol related problems in a classical Greek text. *BMJ*, 335(7633), 1302–1304. https://doi.org/10.1136/bmj.39420.333565

Curren Goodden Associates (CGA). (2024). How Gen Z is redefining high tempo, low alcohol pub culture. https://cgastrategy.com/gen-z-high-tempo-pub-culture/

Devane-Waters, L. (2024). What's in your glass? A look at the top drinks across generations. Savanta. https://savanta.com/knowledge-centre/view/generational-love-for-drinks

Dhawan, A. (2023). Gen Z and its attitude to alcohol. Food and Drink Business. https://www.foodanddrinkbusiness.com.au/news/gen-z-and-its-attitude-to-alcohol

Douglas, M. (2003). *Constructive Drinking: Perspectives on Drink From Anthropology.* Routledge.

Duffy, B. (2021). *Generations: Does When You're Born Shape Who You Are?* Atlantic Books.

Dufour, M. C. (1995). Twenty-five years of alcohol epidemiology: Trends, techniques, and transitions. *Alcohol Health and Research World*, 19(1), 77–84.

Englund, A. (2019). Skolelevers drogvanor 2019 [Alcohol and drug use among students 2019]. CAN report no 187. The Swedish Council for Information on Alcohol and Other Drugs, Stockholm.

Fares, O. H., & Panetta, M. (2024). The rise of 'sober curiosity': Why Gen Zers are reducing their alcohol consumption. The Conversation. https://theconversation.com/the-rise-of-sober-curiosity-why-gen-zers-are-reducing-their-alcohol-consumption-243775

Fuller, K. (n.d.). Alcohol consumption in Gen Z, Millennials, Gen X, and Baby Boomers. Alcohol Help. https://www.alcoholhelp.com/blog/alcohol-consumption-generations

Gayle, D. (2016). Smoking and drinking among young people at lowest level on record. The Guardian. https://www.theguardian.com/society/2016/dec/15/smoking-drinking-young-people-lowest-level-on-record-england

Global Alcohol Policy Alliance. (n.d.). Global Alcohol Policy Alliance early history. https://globalgapa.org/about-us/history-and-the-globe/global-alcohol-policy-alliance-early-history/

Goodreads. (2025). Charlie Chaplin > Quotes > Quotable Quote. https://www.goodreads.com/quotes/175759-a-man-s-true-character-comes-out-when-he-s-drunk

Hanson, D. J. (2013). Historical evolution of alcohol consumption in society. In: Boyle, P., Boffetta, P., Lowenfels, A. B., Burns, H., Brawley, O.,

Zatonski, W., & Rehm, J. (eds), *Alcohol: Science, Policy and Public Health*, pp. 3–12. Oxford Academic. https://doi.org/10.1093/acprof:oso/9780199655786.003.0001

Health Survey for England. (2016). NHS digital. https://digital.nhs.uk/data-and-information/publications/statistical/health-survey-for-england/health-survey-for-england-2015

Hendriks, H., Gebhardt, W., & Putte B. (2017). Alcohol-related posts from young people on social networking sites: Contents and motivation. *Cyberpsychology, Behaviour, and Social Networking*, 20(7), pp. 428–435. https://doi.org/10.1089/cyber.2016.0640

Hines, C. (2024). How Gen Z's drinking habits are shaping the future of bev-alc. https://www.provi.com/blog/gen-z-drinking-habits-beverage-industry

Kraus, L., Room, R., Livingston, M., Pennay, A., Holmes, J., & Torronen, J. (2020). Long waves of consumption or a unique social generation? Exploring recent declines in youth drinking. *Addiction Research and Theory*, 28(3), 183–193. https://doi.org/10.1080/16066359.2019.16294

Link, T. C. (2008). Youthful intoxication: A cross-cultural study of drinking among German and American adolescents. *Journal of Studies on Alcohol and Drugs*, 69, 362–370. https://doi.org/10.15288/jsad.2008.69.362.

Ludmir, C. (2023). Why Gen Z is drinking less and what this means for the alcohol industry. Forbes. https://www.forbes.com/sites/claraludmir/2023/06/27/why-genz-is-drinking-less-and-what-this-means-for-the-alcohol-industry/

Marks, E., Hickman, C., Pihkala, P., Clayton, S., Lewandowski, E. R., Mayall, E. E., Wray, B., Mellor, C., & van Susteren, L. (2021). Young people's voices on climate anxiety, government betrayal and moral injury: A global phenomenon. *The Lancet Planetary Health*, 5, 12, e863–e873. https://doi.org/10.1016/S2542-5196(21)00278-3

McGovern, P. E., Hartung, U., Badler, V. R., Glusker, D. L., & Exner, L. J. (1997). The beginnings of winemaking and viniculture in the ancient Near East and Egypt. *Expedition Magazine*, 39(1). https://www.penn.museum/sites/expedition/the-beginnings-of-winemaking-and-viniculture-in-the-ancient-near-east-and-egypt/

Månsson, J., Samuelsson, E., & Törrönen, J. (2022). Doing adulthood — Doing alcohol: What happens when the "sober generation" grows up? *Journal of Youth Studies*, 25(1), 84–99. https://www.tandfonline.com/doi/pdf/10.1080/13676261.2020.1844173

Myles, C. C., Vander Weil, B., Watson, B., & Wiley, D. S. (2023). "Sober curious" or "semi-sober"? An exploration of the moderation movement in the United States as "trendy teetotalism" or "neo-temperance". In Patterson, M. W. & Hoalst-Pullen, N. (eds), *The Geography of Beer*, pp. 279–291. Springer. https://doi.org/10.1007/978-3-031-39008-1_22

National Academies of Sciences, Engineering and Medicine. (2017). The health effects of cannabis and cannabinoids: The current state of evidence and recommendations for research. The National Academies Press. https://nap.nationalacademies.org/catalog/24625/the-health-effects-of-cannabis-and-cannabinoids-the-current-state

Office for National Statistics (ONS). (2018). Loneliness — what characteristics and circumstances are associated with feeling lonely? Analysis of characteristics and circumstances associated with loneliness in England using the Community Life Survey, 2016 to 2017. Office for National Statistics. https://backup.ons.gov.uk/wp-content/uploads/sites/3/2018/04/Loneliness-What-characteristics-and-circumstances-are-associated-with-feeling-lonely-.pdf

Paulette, T. (2021). Inebriation and the early state: Beer and the politics of affect in Mesopotamia. *Journal of Anthropological Archaeology*, 63, 101330. https://doi.org/10.1016/j.jaa.2021.101330

Raninen, J., & Livingston, M. (2018). Exploring the changing landscape of youth drinking: We are still drawing the map. *Drug and Alcohol Review*, 37, S6–S8. https://doi.org/10.1111/dar.12808

Richards, R. (2023). Is the alcohol industry drying up? World Finance. https://www.worldfinance.com/special-reports/is-the-alcohol-industry-drying-up

Room, R., Osterberg, E., Ramstedt, M., & Rehm, J. (2009). Explaining change and stasis in alcohol consumption. *Addiction Research and Theory*, 17(6), 562–576. https://doi.org/10.3109/16066350802626966

Saad, L. (2023). Young adults in U.S. drinking less than in prior decades. Gallup. https://news.gallup.com/poll/509690/young-adults-drinking-less-prior-decades.aspx

Schmid, H., Ter Bogt, T., Godeau, E., Hublet, A., Dias, S. F., & Fotiou, A. (2003). Drunkenness among young people: A cross-national comparison. *Journal of Studies on Alcohol*, 64, 650–661. https://doi.org/10.15288/jsa.2003.64.650

Schmidt, L., & Room, R. (1999). Cross-cultural applicability in international classifications and research on alcohol dependence. *Journal of Studies on Alcohol*, 60(4), 448–462. https://doi.org/10.15288/jsa.1999.60.448

Statista. (2025). Share of tequila consumers in the United States in 2024, by age. https://www.statista.com/statistics/1349936/tequila-consumers-by-age-us/

Thompson, S. W. (2010). "Daimon drink": Ancient Greek and Roman explanations for drunkenness. *Christian Spirituality and Science*, 8(1), 7–24. https://core.ac.uk/download/pdf/234108619.pdf

Törrönen, J., Roumeliotis, F., Samuelsson, E., Kraus, L., & Room, R. (2019). Why are young people drink less than earlier? Identifying and specifying social mechanisms with a pragmatist approach. *International Journal of Drug Policy*, 64, 13–20. https://doi.org/10.1016/j.drugpo.2018.12.001.

van de Ven, K., & Lee, N. (2025). With 'damp drinking' and 'zebra striping', Gen Z are embracing moderation — not abstinence — from alcohol. The Conversation. https://theconversation.com/with-damp-drinking-and-zebra-striping-gen-z-are-embracing-moderation-not-abstinence-from-alcohol-246250

Wang, J., & Liu, L. (2022). Introduction: Alcohol, rituals, and politics in the ancient world. *Journal of Anthropological Archaeology*, 65, 101397. https://doi.org/10.1016/j.jaa.2022.101397

Witherington, A. (2025). Gen Z: The Sober Curious Generation, Mintel. https://www.mintel.com/insights/food-and-drink/gen-z-sober-curious-generation

Yale Law School (n.d). The Code of Hammurabi. The Avalon Project — Documents in law, history and diplomacy. https://avalon.law.yale.edu/ancient/hamframe.asp

Chapter 15

Redefining the Future: Gen Z and the Food Landscape of 2050

By the year 2050, all of Gen Z will have entered midlife and will be adults between 39 and 55 years of age. Many of them will be parents as well as professionals and technology experts, policymakers, entrepreneurs and digital pioneers, steering everyday life with a wealth of knowledge and experience. Many will hold influential roles in shaping food systems, health policies, technological innovation and responses to climate and other environmental challenges. They will no longer be the disruptors of tradition, breaking new ground in the food landscape. While Generations Alpha and Beta will be imposing their lifestyles and dietary preferences, pushing forward new cultural norms and dietary identities, Gen Z will be the bridge connecting the transformative ideals of the start of the 21st century with the unpredictable realities of a new planetary era. Generation Gamma will be just emerging and starting its journey, raised on the foundations Gen Z helped to lay.

The world in 2050 will likely be defined by intensified climate impacts, technological acceleration and probably space exploration (Gourani, 2023). For those of us who are alive to see it, 2050 may be a year of reckoning and reflecting — a time to ask: What did we get right and what did we get wrong in the 2020s? Did we act on the climate crisis and manage to protect the planet or simply learn to adapt to the changes? Did we mitigate irreversible loss, address severe climate impacts or did we simply shift the burden to future generations? Would we still be able to recognise the food items on the kitchen table, in the local shop or on the restaurant menu and the systems that brought them to us? Will we still be ordering our favourite foodstuff online and will we be growing food ourselves? Will our kitchen tables hold 3D-printed meals, lab-grown meats, AI-personalised nutrition packs or hydroponically grown heritage vegetables? Will our food come from rooftop farms, vertical greenhouses, ocean crops or even orbital stations cultivating lettuces in low gravity? Will "local" and "seasonal" still be meaningful terms in a digitised, global food web? Also, will we still be eating together or will mealtime become another fragmented, automated ritual in an attention-deficient world?

How will nutrition change? How might nutrition evolve? Will we witness the rise of new or the return of old diseases threatening both human health and food production systems? What will food look like across different parts and regions of the world, shaped by climate, culture and technology? Could lettuces, grains or lab-cultured proteins be cultivated on lunar or Martian outposts, or grown on other planets as humanity pushes into space?

It is difficult to guess what the food landscape will look like in 2050. The truth is that the food landscape of 2050 remains unpredictable, shaped by variables we cannot fully anticipate. In fact, will we still be eating in the way we understand it today, or will nourishment take entirely new forms, perhaps more functional, programmable or even disembodied?! One certain thing is that by 2050, Gen Z will have helped shape and provided the answers to all of these questions through their values, innovations and decisions as consumers, employees, voters, scientists and caregivers.

This chapter is a speculative journey into how Gen Z might redefine the future of food. Whenever we think about the future, we need to start from where we are at the moment. Our point of departure is the present — a moment marked by what many describe as both a polycrisis (multiple crises unfolding simultaneously) and a permacrisis (a prolonged state of instability). We then discuss how to forecast the future of food while allowing for uncertainties. These overlapping global challenges, from climate shocks and supply chain fragilities to food insecurity and digital disruption, set the stage for imagining alternative futures. We explore possible food scenarios, not as definitive forecasts, but as open-ended narratives that embrace uncertainty. Rather than predict, we probe; rather than prescribe, we reflect. Ultimately, we conclude expressing a positive outlook with cautious optimism that Gen Z, with its blend of digital fluency, social consciousness and adaptive spirit, is uniquely positioned to co-create a more just, resilient society and the ability to shape a better food future.

The Current Crisis

We both originally come from Bulgaria — the homeland of the blind clairvoyant Baba Vanga, who famously predicted that the year 2025 would be the start of humanity's downfall (Kukreti, 2025). While Baba Vanga did not say anything specifically about the year 2050, her prophecies carry an enduring warning that unless people fundamentally change their actions and choices, humanity's destiny and fate will remain grim. Hence, according to Baba Vanga, people have agency in shaping their own future and that of the planet. She also left room for hope. Baba Vanga explains: "The consciousness of people will change. Difficult times will bring unity" (Oakes, 2025, para. 8). For this to happen, through collective awareness and action, people need to shift and get away from the current practices and trajectories in living on this planet.

The future is not fixed. Nowhere is this need for transformation more urgent or more visible then in our food systems and dietary preferences. Food production and consumption are no longer merely questions of personal preference or cultural tradition, they are entangled with global challenges spanning across a complex interplay between climate, economic, technological, social, geopolitical and health factors. They reflect a complex and sophisticated relationship between humans, animals and food substrates, such as soils, water and air. The question that stands before us can be uttered as: Is Gen Z going to be the generation that shifts humanity's destiny?

What makes things even more challenging is that in 2025, when we are writing this book, the world is widely described as being in a polycrisis (Morin and Kern, 1999) — a convergence of multiple, interconnected crises that amplify one another in unpredictable ways. As the Cambridge Dictionary (2025b) defines it, this is "a time of great disagreement, confusion, or suffering that is caused by many different problems happening at the same time so that they together have a very big effect". Even more disheartening is that this polycrisis is also a permacrisis — "a long period of great difficulty, confusion, or suffering that seems to have no end" (Cambridge Dictionary, 2025a). The polycrisis situation stretches beyond the COVID-19 pandemic. It encompasses the persistent rise of non-communicable diseases, accelerating climate change, biodiversity loss, deforestation, soil depletion, wildfires, widespread plastics pollution, forever chemicals. antimicrobial resistance, socio-political disruptions, war, armed conflicts, population movements, forced migrations, shortening human life expectancies, species extinctions, a widening gap between the rich and the poor, the challenges of artificial intelligence-related disruptions and unprecedented concentration of wealth and power... These cascading challenges are not isolated; they feed into each other. The environmental, economic, political, social, technological, health and ethical dimensions of the polycrisis have diverse, extensive and enduring consequences, often reinforcing each other and creating domino effects that destabilise both human and ecological systems (Rahman *et al.*, 2024). They go to the nature of human beings and their place on Earth.

This is not just a crisis of systems, it is a crisis of meaning. As Cantor (2025, para. 18) reflects, "[w]hen the world feels like a nightmare, identifying the condition gives us something to hold on to — a kind of understanding amid the chaos". For us, this anchor is food. Not only because it is the basic necessity for human survival, but because it is a symbolic and material expression of our values, identities and responsibilities. Food is at the core of many of the synergistic and persisting problems we face influencing the polycrisis and the permacrisis. We are what we eat in many different ways. It is not just human health and state of mind that are influenced by the food choices we make; what is also affected is how we relate to one another and all other species alive, extinct or killed by 2050.

In the Anthropocene, all geophysical processes on the planet are shaped by human activities (Steffen *et al.*, 2011) — humans have become the dominant force of nature, moulding and determining the evolution of the Earth system (Hamilton, 2015). Eating is the most essential human activity. Among all escalating trends of the Anthropocene, food has been a major player behind the exponentially growing use of fertilisers, water consumption, destruction of native habitats, global deforestation, soil deterioration, methane and other greenhouse gas emissions, habitat loss... However, eating is not only a way to survive; it "is inherently an ethical act: in choosing what to eat, we engage in moral decision-making and our decisions contribute to societal and environmental outcomes. As individuals and societies, our decisions about food reflect our aesthetic, moral, cultural, and religious values" (Fanzo *et al.*, 2020, p. 2). We are what we eat, yes — but we also become who we are through how we eat. Our food choices influence not only our health and well-being, but our relationships with others, with animals, plants and the planet itself. Every bite carries the weight of invisible labour, emissions, suffering, innovation and tradition. It is a political, cultural and planetary stance.

By 2050, Gen Z will have had decades to shape the food landscape — through the businesses they build, the policies they influence, the technologies they develop, the causes they support and the values they embody. They would have demonstrated their moral obligations to the planet and its state. What is there to expect? Will the polycrisis continue to exist as a permacrisis or will the moral, financial, political and decision-making power of Gen Z transform the food landscape of our world as well as the world itself? Will they lead us out of the polycrisis or prolong it into an age of irreversible loss? Against this framework of uncertainty and complexity, we turn our sight forward, with curiosity and cautious hope toward the possible food futures that Gen Z may help create.

Forecasting Food's Future

Ironically, despite food being one of humanity's most fundamental needs and one of the most pressing global challenges, it often remains sidelined in most literature exploring future-defining megatrends. Food seems to be an unacknowledged issue. For example, the Pictet Group, a Swiss multinational private bank and financial services company, has been providing strategic insights into the future for more than two centuries. In its 2025 Megatrends report, food does not appear as a central future-shaping area despite identifying agriculture as the sector most exposed to a "double trouble", having the highest biodiversity impact and being the most dependent on nature (Pictet, 2025). Yet, even in this context, the downstream impacts on food systems are not fully acknowledged. The Megatrends Watch Institute (2025), an independent global think tank, similarly does not seem to emphasise the importance of food in shaping the future; neither does Forbes, the American business magazine, which also fails to give food its due weight in futurist analyses, focusing more on technology, markets and workforce transformation (Gourani, 2023). A rare and notable exception comes from Australia's Commonwealth Scientific and Industrial Research Organisation (CSIRO), whose 2022 report talks about alternative proteins expected to be a key area of future growth, particularly across Asia (CSIRO, 2022). However, even this insight tends to be framed more in terms of innovation and opportunity than as part of a broader, systemic food transformation. By contrast, reports produced by the food industry itself often tend to look only to the immediate future, prioritising more the short term and reflecting mainly current market trends over long-range thinking. What is missing across most of these narratives is a recognition of food's profound intersection with climate, culture, equity and technology, and how food can either accelerate collapse or foster resilience.

With Gen Z increasingly becoming the most important food consumers globally, their evolving values, choices and behaviours will impact any future food trends. The characteristics of Gen Z will be reflected in this. For example, Gen Z are guided mainly by social media (Rand, 2025), including influences by their peers. They also demonstrate that ethics should be part of the trends shaping the years to 2050 as represented by their core values of affordability, health, sustainability and convenience (Reinoso, 2024). For Gen Z, food is not merely fuel, it is a statement of identity, values and justice.

So, how do we forecast food's future when uncertainty dominates and conventional trend analyses fall short? Thinking about the food in 2050 needs to embrace uncertainty while identifying the full range of current and future possibilities (Saffo, 2007). According to Saffo (2007), the six rules of forecasting provide a useful guide for navigating an unpredictable future, especially when it comes to something as multifaceted as food:

1. Define the cone of uncertainty.
2. Identify the S-curve of uptake.
3. Embrace new things that do not fit in the original thinking.
4. Hold strong opinions of what can work.
5. Look back twice as far as you look forward.
6. Know when not to make a forecast.

Applying Saffo's approach to food allows us to move beyond trend-spotting and into scenario thinking, imagining multiple, diverging food futures shaped by values, technologies, planetary boundaries and generational agency. In this chapter, we invite the reader to imagine how Gen Z could redefine the global food landscape by 2050, not through certainty, but through a creative and critical engagement with possibility.

The Cone of Uncertainty

The aim of the cone of uncertainty (see Figure 15.1) is to identify the current food options, established and emerging, and the factors that will be influencing future developments. It is important to keep in mind that there will be many unknown new options or directions (represented with dotted ovals on Figure 15.1) as our food science and technology advance in the next 25 years. It is highly likely that the new options, such as plant-based alternative proteins or dairy products produced with precision fermentation, will be mixed with the existing traditional foods, including plants, livestock-based products, fungi, algae and insects. This is represented in Figure 15.1 by one combination oval, but in fact, the options there are numerous.

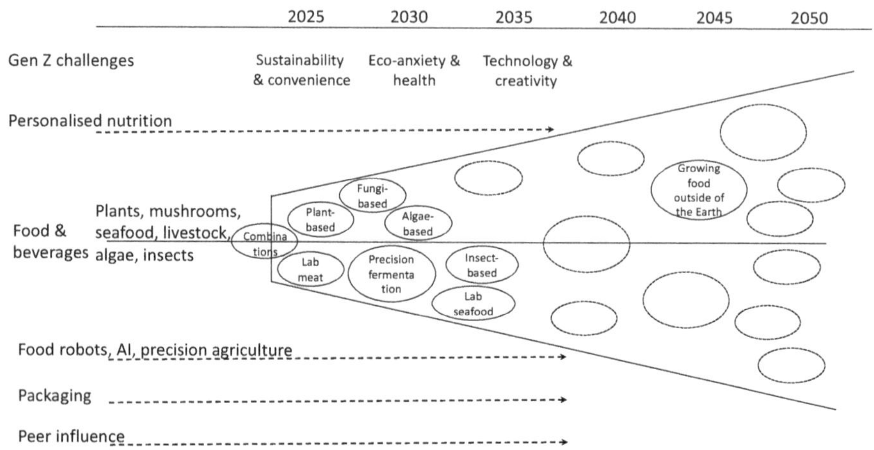

Figure 15.1. The cone of food uncertainty.

Some of the options in the cone of uncertainty may seem like wild cards, e.g. growing food on another planet. Almost all judgements about the future rely on informed intuition and the cone is a way to visualise and narrow down this complex environment.

In addition to the options represented in the cone, Gen Z will shape the food space in the short term with expectations about sustainability and convenience. Mental and other health concerns, such as eco-anxiety and food safety considerations associated with Gen Z are similarly likely to influence the food options. With many of Gen Z becoming professionals, they will be contributing to any technological progress and innovations in this area. The use of technology in food production, including artificial intelligence, robots, sensors and precision methods, will be another dimension of the food landscape, as will be food packaging. With Gen Z being technologically and globally connected, peer influences should similarly be taken into consideration in the cone of food uncertainty.

The S-Curve of Uptake

In innovation diffusion theory, uptake is even more important than the invention of a new product or service (Rogers, 2003). The cumulative diffusion generally follows an S-shape curve, starting slowly, then accelerating until it reaches a turning point and starts to slow down. Some S-curves may reach 100% saturation with all people adopting the new product; others may temper off or drop very quickly as consumers do not embrace a particular food option. The S-curves also may look different, depending on the geographical place or population group. For example, insects may be a good example with low intake in the West but greater popularity for populations in Africa or traditional Australian Aboriginal people. There is a role for the food industry to understand the nature of the S-curve for a particular product and consequently redirect its efforts. It is also helpful to recognise emerging S-curves which essentially will add new options in the cone of uncertainty.

Research on the consumer preferences of Gen Z should be able to indicate how the S-curve will unfold. This is a very different approach to the commonly used projection methods based only on extrapolating the current trends. Linear extrapolation in particular can give misleading expectations as it may have been the case with plant-based meat analogues. Another feature of the S-curves of uptake is that they tend to go slower at the start, then accelerate but drop off quickly once the inflection point of reduced interest is reached (Saffo, 2007).

Embracing New, Unexpected Food Ideas

Noticing an emerging S-curve requires an open mind as well as the ability to spot unexpected developments rather than dismiss them. Such new ideas for producing food may come from technological developments in other areas. For example, growing tomatoes in desert areas using desalinated water became possible because of advances in solar photovoltaic (PV) systems which made the energy-intense desalination feasible and commercially viable. Advances in AI, robotics, drones and machine learning are likely to create new, unexpected food production and consumption options.

Express Strong Opinions of What Can Work

Conflicting evidence about what consumers like or may gradually embrace is an added challenge in forecasting the future food landscape. As human beings, we are conditioned to go with the established norms and strong evidence about what consumer preferences are and get locked in particular trajectories, such as over-reliance on livestock-derived products. Weak evidence, such as for example, the initial uptake of plant-based milks, may in fact signal a new strong trend as has been the case with Gen Z's preferences for soya, almond, oats and other non-dairy milks. This helps break the status quo and create new fashion trends or a new trajectory that may define the future of the dairy industry. The Gen Z consumers have been loud and explicit in expressing their preferences and expectations for plant-based milk options which has changed the practices of many coffee shops, restaurants and supermarkets. Similarly, vegetarian and vegan options are now commonly available because of strong consumer opinions and demands.

Look Back Twice as Far as You Look Forward

The past has many lessons and insights for the future. For example, it was common for food to be grown near people's homes in the 20th century, particularly during times of war shortages. The interest in growing own food in urban environments resurrected in the 2000s with home (backyard and front yard), school and community gardens. During the COVID-19 pandemic, people started doing this even within the constraints of apartment living. Home-grown food is expected to be fresher and healthier with less transport-related emissions and reduced waste. Many are already arguing that cities are transforming into food hubs (O'Hara, 2015). Combining urban agriculture and green infrastructure can bring new life into cities where the majority of the world's population now lives (Evans *et al.*, 2022). Could it be that by 2050, people will be growing the majority of their food themselves or in their locality using traditional and new technologies? Some of the future options may actually stem from older practices.

Past events hold lessons for the future. Some may be success stories; others may be failures. It is important to acknowledge both and generate insights. As knowledge about nutrition expands, some parallels will be easier to make. Gen Z will be learning from history and previous generations, but will also be interpreting the lessons through its moral compass.

Know When Not to Make a Forecast

It is important to also acknowledge that through the millennia of human existence some food sources have not changed. A good example is soyabeans. Referred by the Chinese as a “miracle plant”, soya has been used as a food source, but the plant also contributes to nitrogen fixation in the soils. It may be best not to look for replacement of the soya products, but instead expand their use as a proven and highly beneficial food source. As Soffa (2007, para. 43) writes: “Even in periods of dramatic, rapid transformation, there are vastly more elements that do not change than new things that emerge.”

Careful exertion of observation, knowledge and intuition helps look into the future and what we see is probably very different from what you see. It is also important not to be stuck in the originally traced cone of uncertainty and expand or narrow it accordingly as new evidence emerges. How the future landscape is seen depends very much on the person who looks. Some may see the world through rose-tinted glasses, others may find it to be doom and gloom. Forecasts should not be taken too seriously unless they become the expression of the views and actions of a distinctive and influential cohort of people, such as Gen Z. Its distinctive features will influence how the 2050 food landscape will shape, including considerations about health, sustainability, convenience and technological advances. A globally connected Gen Z and its values, combined with emerging food technologies, will shape a more sustainable, ethical and innovative food system which we hope will heal the current crisis.

Gen Z’s Ability to Shape a Better Food Future

A World Bank report (2024) put forward three strategic pathways to get out of the global polycrisis, namely eradicating poverty by fostering investment in human, physical and financial capital, prioritising income growth that reduces vulnerability and creates synergies, and focussing on mitigation while managing the costs. The report has linked these pathways to the levels of national income, reinforcing an economic lens on systemic challenges. These strategies however do not address the essence of the polycrisis as a crash in shared value systems caused by disagreement, confusion and suffering. They remain anchored in traditional growth paradigms and do not reflect the global, responsible and inclusive ethics espoused by Gen Z. In contrast, Gen Z offers a new ethical framework that prioritises sustainability, collective responsibility and social justice — principles largely absent from mainstream policy discourse.

Despite facing unprecedented challenges, from climate anxieties about their future and that of the planet to economic instability, Gen Z remains optimistic, resilient and forward-looking. According to the Stanford Report they are "highly collaborative, self-reliant and pragmatic" (Stanford Report, 2022, para 1). As the Stanford University researcher Roberta Katz explains: "a typical Gen Zer is a self-driver who deeply cares about others, strives for a diverse community, is highly collaborative and social, values flexibility, relevance, authenticity and non-hierarchical leadership, and, while dismayed about inherited issues like climate change, has a pragmatic attitude about the work that has to be done to address those issues" (Stanford Report, 2022, para 9). All these qualities position Gen Z to be able to lead the transformation of the food system and shape a better food future.

Most importantly, Gen Z are trend breakers and system challengers. They are different in the way they work and redefine work culture — 68% of the managers report they have difficulties dealing with and adapting to Gen Z's expectations (Ipsos, 2024), the way they receive and consume the news — 60% primarily from social media (Digital News Report, 2024), bypassing traditional gatekeepers. Their preference for living at a slower pace and mental wellness is evident, with 67% of Gen Z being worried and expressing concern about their mental and physical health (Wealth & Finance, 2024). They are also deeply impact-driven, interested in making an impact on society (Deloitte, 2025). For them, purpose is prioritised over profit and they seek to make a tangible difference in society.

This generation is also very sceptical of corporate greenwashing and about companies making claims about sustainability. A striking 88% of American Gen Zers do not trust environment, social and governance (ESG) statements made by big brands (Hilton Segel and Hatami, 2023), demanding transparency and measurable impact. Politically, they are disengaged. Their political interests rarely align with the established major parties (Chowdhury, 2025) and they replace the work of "big" politics — governments and politicians, with "everyday politics" which reflect personal values and lived experiences and beliefs (Abidin and Zeng, 2021). As true global citizens, they have a declining interest in nationalism and low desire to participate in military service, identify with or fight for their nation (Hoffmann and Beamond, 2024), which further reflects a shift toward global citizenship and ethical individualism.

Their food choices mirror this ethos. Gen Z's attitude towards food also reflects the desire to do good and break with previous patterns. As a male Gen Zer explains: "There's some room for individual personal impact, even though it's ultimately a small piece of the puzzle. Everyone should do common sense things... and generally trying to eat less meat. Being an early adopter of plant-based products can also help. I think there's some advancement that I can be at the forefront of" (Deloitte, 2025, p. 48).

Long-term thinking is essential and the best way to move beyond the permacrisis, a state of perpetual instability. As futurist Tinworth argues, we must open our eyes to a broader horizon and embrace strategic foresight (Tinworth, 2023). Is Gen Z capable to stop focussing on short-term fixes and act strategically? Gen Z is already doing this. Faced with compounding environmental, social, economic pressures and health challenges, Gen Z is redefining the meaning of urgency not as panic, but as proactive, values-driven action. They chose to trust their own abilities, values, knowledge, networks and lived experience to guide change.

The future of food is not just about innovation. It is about ownership. Gen Z is claiming that space, reshaping consumption, production and policy through their preferences and principles. Food's future belongs to them and their preferences which only they should control. They have the tools, insight, knowledge and power and the conviction to lead us and the world out of the crisis and into a more just, sustainable and nourishing future.

References

Abidin, C. & Zeng, J. (2021). 'OK Boomer': how a TikTok meme traces the rise of Gen Z political consciousness. The Conversation. https://theconversation.com/ok-boomer-how-a-tiktok-meme-traces-the-rise-of-gen-z-political-consciousness-165811

Cambridge Dictionary. (2025a). Permacrisis. https://dictionary.cambridge.org/dictionary/english/permacrisis

Cambridge Dictionary. (2025b). Polycrisis. https://dictionary.cambridge.org/us/dictionary/english/polycrisis

Cantor, M. (2025). What is this era of calamity we're in? Some say 'polycrisis' captures it. The Guardian. https://www.theguardian.com/us-news/2025/mar/06/polycrisis-disasters-politics

Chowdhury, I. (2025). This election, young people held the most political power. Here's how they voted. https://theconversation.com/this-election-young-people-held-the-most-political-power-heres-how-they-voted-255769

Commonwealth Scientific and Industrial Research Organisation (CSIRO). (2022). Our Future World: Global megatrends impacting the way we live over coming decades. https://www.csiro.au/en/research/technology-space/data/Our-Future-World

Deloitte. (2025). 2025 Gen Z and Millennial Survey: Growth and the pursuit of money, meaning, and well-being. https://www.deloitte.com/content/dam/assets-shared/docs/campaigns/2025/2025-genz-millennial-survey.pdf

Digital News Report. (2024). Australia 2024: AI, social media, misinformation and distrust — what the data tells us about the news landscape in 2024. https://www.canberra.edu.au/about-uc/media/media-releases/2024/june/digital-news-report-australia-2024-ai,-social-media,-misinformation-and-distrust-what-the-data-tells-us-about-the-news-landscape-in-2024

Evans, D. L., Falagán, N., Hardman, C. A., Kourmpetli, S., Liu, L., Mead, B. R., & Davies, J. A. C. (2022). Ecosystem service delivery by urban agriculture and green infrastructure — a systematic review. *Ecosystem Services*, 54, 101405. https://doi.org/10.1016/j.ecoser.2022.101405

Fanzo, J., Hood, A., & Davis, C. (2020). Eating our way through the Anthropocene. *Physiology & Behavior*, 222, 112929. https://doi.org/10.1016/j.physbeh.2020.112929

Gourani, S. (2023). 2050 Unveiled: A Glimpse into the Future. Forbes. https://www.forbes.com/sites/soulaimagourani/2023/12/09/2050-unveiled-a-glimpse-into-the-future/

Hamilton, C. (2015). Getting the Anthropocene so wrong. *The Anthropocene Review*, 2(2), 102–107. https://doi.org/10.1177/2053019615584974

Hilton Segel, L. & Hatami, H. (2023). Mind the Gap: Curated Reads for Gen Z — and Their Z-Curious Colleagues. https://www.mckinsey.com/%7E/media/mckinsey/email/genz/2022/12/06/2022-12-06b.html

Hoffmann, R. & Beamond, M. T. (2024). Gen Z is turning away from military service in record numbers. We're trying to understand why. https://theconversation.com/gen-z-is-turning-away-from-military-service-in-record-numbers-were-trying-to-understand-why-230671

Ipsos. (2024). Are they really that liberal and progressive? How the complexity of Gen Z challenges market understanding. https://www.ipsos.com/en/flair-collection/are-they-really-liberal-and-progressive-how-complexity-gen-z-challenges-market-understanding

Megatrends Watch Institute. (2025). Megatrend intelligence. https://www.megatrends2050.com/megatrend-intelligence.html

Morin, E. & Kern, A. B. (1999). *Homeland Earth: A Manifesto for the New Millennium*. Hampton Press.

O'Hara, S. (2015). Food security: The urban food hubs solution. Solutions. http://docs.udc.edu/causes/reports/Urban%20Food%20Hubs%20Solution%20to%20Food%20Insecurity.pdf

Oakes, B. (2025). AI asked which of Baba Vanga's predictions is most likely to come true. https://uk.style.yahoo.com/ai-asked-baba-vanga-predictions-145522157.html

Pictet. (2025). Megatrending 2025: Opportunities ahead. https://www.pictet.com/us/en/insights/megatrending-2025-investment-opportunities

Rahman, M. S. A., Awang, M., & Jagun, Z. T. (2024). Polycrisis: Factors, impacts, and responses in the housing market. *Renewable and Sustainable Energy Reviews*, 202, 114713. https://doi.org/10.1016/j.rser.2024.114713

Rand, S. (2025). Gen Z food trends: What today's young US consumers want. https://www.askattest.com/blog/research/gen-z-food-trends

Reinoso, K. L. (2024). Gen Z in 2025: What drives the future of food. https://tastewise.io/blog/gen-z-in-2025-what-drives-the-future-of-food#The_trends_shaping_Gen_Zs_F_B_preferences

Rogers, E. M. (2003). *Diffusion of Innovations*. Free Press.

Saffo, P. (2007). Six Rules of Effective Forecasting. *Harvard Business Review Magazine.* https://hbr.org/2007/07/six-rules-for-effective-forecasting

Stanford Report. (2022). Gen Z are not 'coddled.' They are highly collaborative, self-reliant and pragmatic, according to new Stanford-affiliated research. https://news.stanford.edu/stories/2022/01/know-gen-z

Steffen, W., Grinevald, J., Crutzen, P., & McNeill, J. (2011). The Anthropocene: Conceptual and historical perspectives. *Philosophical Transactions of the Royal Society A*, 369, 842–867. http://doi.org/10.1098/rsta.2010.0327

Tinworth, A. (2023). Long-term thinking is our best weapon against the permacrisis. https://nextconf.eu/2023/04/long-term-thinking-is-our-best-weapon-against-the-permacrisis/

Wealth & Finance. (2024). GenZnomics: 10 GenZ Beliefs & Behaviors Driving the Global Economy. https://wealthandfinance.digital/genznomics-10-genz-beliefs-behaviors-driving-the-global-economy/

World Bank. (2024). Poverty, Prosperity and Planet Report 2024: Pathways out of the Polycrisis. https://openknowledge.worldbank.org/server/api/core/bitstreams/ed189cbd-46e7-4c9f-ab78-a68d50df5e67/content

Chapter 16

Passing the Plate: How Generation Alpha Could Adapt and Learn from Gen Z's Food Trends

> *"What one generation tolerates, the next generation will embrace."*
>
> — John Wesley

The above quote by the theologian John Wesley refers to the evolution of societal norms and cultural acceptance across generations. It suggests that behaviours, values or ideas that one generation merely tolerates, perhaps reluctantly or passively, often become fully embraced and accepted by the next. This has been the case with many of the attitudes and behaviours flowing from the Millennials to Gen Z, such as the use of technology, including generative artificial intelligence (GenAI), their high expectations from any employers and environmental concerns (Deloitte, 2025). Relating this to the new Generation Alpha (born between 2011 and 2024) and Gen Z, we can see the pattern of how technology use and innovation are evolving. Gen Z has championed sustainability and ethical food sourcing, advocating for transparency and environmentally conscious choices. In its turn, Generation Alpha is likely to fully embrace these values, integrating them with cutting-edge advancements, such as AI-driven nutrition, 3D-printed meals and cultured food, making personalisation and ethical sourcing the norm rather than the exception. Wesley's insight is timeless, showing how gradual acceptance in one period often leads to full adoption in the next. We see this unfolding as Generation Alpha shapes food, technology and digital culture, embodying ideas and requirements once introduced by Gen Z but still being navigated by them with caution.

As Gen Z matures into adulthood and Generation Alpha begins shaping the world around them, we are witnessing a generational relay that could transform the global food system. Both cohorts, though distinct in their social and technological environments, are united by their share of foundational values around wellness, technological innovation and authenticity as well as by the continuing ecological, climate change and geopolitical challenges. While Gen Z has already redrawn the contours of food culture through social media-driven trends, embracing plant-rich wholefood eating and a commitment to prioritising ethical consumption, Generation Alpha is poised to inherit these foundations and expand this legacy further, integrating technology into even more seamless and radical ways. It is the first generation exposed to screentime entirely during its formative years for which some referred to them as screenagers (McCrindle, 2025). They are faster in adapting and understanding technology (Howarth, 2025). Their relationship with food is likely to be even more intimately intertwined with technology, personalisation and sustainability.

Born entirely during the 21st century, most of Generation Alpha will live to see the 22nd century by when the planet's human population numbers are likely to stabilise making the food security problem somehow better defined. Although it is difficult to describe fully Generation Alpha as many of them are still very young, we already have some insights (Iberdrola, 2025). Their main characteristics, namely being hyperconnected, independent, visual, technological and diverse, in terms of ethnicity, gender, tastes, lifestyles and points of view (Iberdrola, 2025), will also define their relationship with food. On the other hand, some of the drawbacks already manifested in Gen Z are likely to deepen in Generation Alpha. According to Iberdrola (2025), this includes having a reduced attention span and concentration because of working on multiple screens, being time-poor as a result of spending too much time on their devices, different type of creativity and imagination due to playing less with physical toys and most likely, increased mental health problems (Haidt, 2024). As Generation Alpha spent a lot of time at home with their parents during the COVID-19 pandemic, some became interested in food, both in how to grow and how to prepare it. Generation Alpha is also very socially oriented with 61% of the American youngsters saying that helping people is their number one priority (Connell, n.d.). In the remaining of the chapter, we look at Generation Alpha mainly as consumers but many of them may in fact engage in food production. They will be part of the innovations to come and will be shaping the future food landscape in new ways, in some aspects unimaginable to previous generations.

A New Generation of Foodies on the Horizon

Gen Z has redefined how society interacts with food and played a critical role in transforming food into a medium or a vehicle for expression, identity, activism and storytelling. Now, attention is shifting toward Generation Alpha, the first cohort deeply shaped by immersive technologies, artificial intelligence and ubiquitous digital environment and engagement. Numbering over two billion globally, with the highest populations in India, China and Indonesia (Kumar, 2024), Generation Alpha represents more than a demographic shift. They signal a jump forward in cultural and technological orientation (Eastlake, 2025).

Raised primarily by Millennials (70%), Generation X (21%) and Gen Z adult (8%) parents (Kumar, 2024), Generation Alpha are digital-first individuals. Technology for them is a given, it is not a tool or distraction, but a deeply embedded part of their daily lives (Moors, 2024). This generation's children have never known a world without smart devices, streaming content, smart phones and speakers, e-games or voice-assisted search. Their reality is algorithmically curated and AI-mediated. They are immersed in a landscape where AI-enhanced devices, voice search and real-time connectivity are not innovations but norms. This digital immersion profoundly influences their sense of self, relationships and expectations, especially when it comes to food.

Environmental consciousness is another defining feature of this cohort. Growing up within escalating climate discourse, Generation Alpha has a heightened awareness of sustainability. Research shows that 46% of them are interested in the environment or climate change issues (Trifonova and Moran, n.d.), reflecting an early engagement with ecological topics that often took years to develop in older cohorts. Generation Alpha is likely to adopt more plant-based options as they understand the negative impacts of animal-based industry, often due to stories and information from the social media (Eastlake, 2025).

The families of Generation Alpha are also fostering broader conversations. Research suggests 46% discuss mental health at home, 45% address body image and over a third tackle race gender norms and climate change (Feger, 2023). This socially aware upbringing is likely to cultivate consumers who are not only ethically engaged but demand systemic accountability from food producers. As such, Generation Alpha are likely to value ethical consumption and demand climate-conscious practices across all industries, including food.

Eastlake (2025) projects that Generation Alpha may become the first generation to conduct their entire shopping and food ordering experience online, signalling a future where convenience, digital personalisation and transparency will dominate. In this context, food is no longer something passively consumed but interactively selected, digitally traced and often co-created, particularly when it comes to selecting ingredients, flavours and condiments. Their influence on the future of food is not just possible, it is inevitable.

Gen Z's Legacy and Generation Alpha's Launchpad

Gen Z's contribution to the food system is both cultural and behavioural. They have laid the cultural groundwork by championing wellness, sustainability and digital savviness as key drivers in food choices. They have advanced flexitarianism, normalised food activism and embraced functional foods like kombucha, kefir or seaweed snacks. Furthermore, they have reoriented food around social narratives and tech-savvy behaviours. Gen Z have used platforms like TikTok and Instagram not just to document meals and provide cooking hacks but to generate viral trends and support niche movements, from foraging to food justice. Their relationship with food is deeply visual, narrative-driven and socially networked.

Generation Alpha, however, is launching from a more advanced technological and cultural base and they enter the stage with even higher expectations. Where Gen Z adapted to the emerging digital world, Generation Alpha has been digitally programmed by design from the cradle. One Gen Z captured this distinction well with humour:

> *"We remember life before TikTok. My little sister doesn't even know how to use YouTube without voice search. She just assumes everything works by talking to it — screens, speakers, Alexa, Siri... even the microwave once! It's like she was born expecting the world to listen and respond instantly."* (Female, born 2002)

This observation captures the key generational shift in how Generation Alpha instinctively interacts with technology compared to Gen Z, who had to adapt as it evolved. Technology is no longer supplemental, an add-on; it is foundational and fundamental. While Gen Z learned to integrate technology into their lives, Generation Alpha expects seamless, intuitive interactions across all domains, including food.

As such, we can expect Generation Alpha to push for more advanced digital food solutions. Their ecosystem already includes AI-powered fridges, robotics, wearables, streaming media, gamified nutrition apps, augmented reality (AR)-enhanced dining, and even early exposure to food produced through 3D printing or cellular agriculture. These tools are not novelties but part of their daily experience and expectations, including what goes on their plate.

The evolution from Gen Z to Generation Alpha marks not just a generational handover, but a systems-level transformation in how food is perceived, produced and consumed. While Gen Z disrupted traditional models by prioritising ethics, wellness and digital visibility, Generation Alpha is poised to internalise and normalise these changes. Food companies, policymakers and educators must therefore rethink how they engage with youth-led food cultures. Generation Alpha's preference for immersive, tech-integrated experiences means that static labels, generic marketing and opaque supply chains will quickly become obsolete. Instead, brands will need to deliver dynamic, interactive experiences that align with real-time data, user-generated content and digital identity.

Additionally, there is a growing need to educate and support Generation Alpha in critically navigating a food system that will be increasingly shaped by algorithmic decisions and synthetic biology. Issues of digital sovereignty, data ethics and nutritional literacy will become central. Generation Alpha may be empowered consumers, but they will also face new risks, algorithmic manipulation, hyper-personalisation leading to dietary silos or ethical blind spots around food technologies.

Technology as a Driver of Food Innovation

The evolution of food culture is accelerating, driven by rapid technological advancements and shifting generational preferences. Generation Alpha, the first truly digitally-native generation, is poised to redefine the way we experience food, from its sourcing to its personalisation. When compared to Gen Z, their approach to food and dining reveals both stark contrasts and unexpected continuities. Especially the implications of Generation Alpha's relationship with technology are profound. Generation Alpha's tech fluency is not just behavioural; it is visionary. Five emerging scientific domains — alternative proteins, cellular agriculture, genomics, microbiome science and synthetic biology, are being accelerated by AI, machine learning, quantum computing and sensor technologies. These fields are converging to create what Hunter (2022) terms a "techxponential" food future. Generation Alpha is the first generation poised to grow up as consumers, contributors and even co-creators in this ecosystem.

Data already shows that 63% of Generation Alpha value having the latest technology, with many desiring innovations that do not yet exist, compared to only 31% of Gen Z (King, 2024; Razorfish, 2023).

This signals a deeper, generational expectation of hyper-innovation. For them, the boundaries between digital and physical, entertainment and nourishment, or science and cuisine are dissolving. They will likely favour brands and food systems that can offer tech-enabled personalisation, transparency and ethical alignment.

Self-expression is also vital. An overwhelming 92% of Generation Alpha believe authenticity is critical (Razorfish, 2023), making them more likely than their Gen Z predecessors to assert their preferences and expect the world to accommodate them. Unlike Gen Z, who used food and online platforms to escape rigid social structures, Generation Alpha may use them to curate reality, influence peers and craft identities in real time.

Food choices fully reflect this. Sixty-three percent of Generation Alpha already influence family meal decisions (Eastlake, 2025), and their adventurous palate is embraced by parents, 78% of whom are open to their children trying diverse cuisines early in life. These numbers suggest Generation Alpha will grow up expecting what Gen Z already started — diverse, inclusive and globally inspired menus, delivered with the speed and convenience of technology.

Moreover, Hunter (2022) predicts that Generation Alpha will not merely accept technology in food systems, they will demand it. While 77% of Gen Z are already comfortable with technology in food production (Moors, 2024), Generation Alpha will expect deeper personalisation, from microbiome-based nutrition to emotion-driven meal design. For this cohort, similarly to Gen Z, food must be satisfying on multiple levels: nutritional, ethical, sensory and emotional (Bogueva and Marinova, 2020).

Authenticity vs Innovation

Generation Alpha is inheriting Gen Z's call for authenticity, but adding layers of interactivity, play and scientific precision. As one teen from Sydney said:

> *"If my food can be smart and kind to the planet and still taste great. That's the future I want... And if it can be personalised just for me, even better!"* (Male, born 2011)

Both generations value authenticity but express it differently in relation to food. Gen Z, shaped by sustainability concerns, focuses on organic, ethical sourcing and transparency. Meanwhile, Generation Alpha will demand "true and meaningful personalisation" (Moors, 2024). Cultivated and 3D-printed foods may become common, but like Gen Z this could happen only if they maintain taste, texture, nutrition and emotional resonance, ensuring that innovation never compromises authenticity.

Generation Alpha, similarly to Gen Z, are not merely passive consumers of futuristic cuisine. Instead, they have all the opportunity of becoming active participants in co-creating meals through technologies like 3D printing and augmented reality. As Mattan (2023) suggested: "Imagine a table where your plate tells the story of your food's origin, while a holographic chef explains how it was made". For Generation Alpha, this is less science fiction and more foundational memory, that is memories and experiences that shape how to see the world (Anderson, 2024).

Children today share the opportunity of designing their own edible shapes, choosing textures and even adjusting nutritional content using food printers. This not only boosts creativity and engagement but fosters early habits of action and curiosity around food. A 13-year-old girl shared:

> *"Oh, I love experimenting with food! It's so fun to build my own snacks. I just made pasta starfish that tasted like cheese and broccoli. Imagine that! It's like turning dinner into an art project. You get to mix textures, flavours, and shapes in ways no one else has tried before. Who knows, maybe one day my wild food creations will end up in an actual restaurant!"* (Female, born 2012)

Generation Alpha is embracing the fun and creativity of digital food customisation, turning virtual kitchens into playgrounds for innovation. They are adventurous and pleasure-seeking foodies (Witherington, 2025). With interactive platforms allowing them to design their own meals and drinks, they are co-creating their food online in ways previous generations never imagined. Whether crafting a pasta starfish that tastes like cheese and broccoli or experimenting with AI-generated recipes, they see food not just as sustenance but as an extension of their creativity. This shift in food engagement highlights a future where personalisation, technology and playfulness merge to redefine how younger generations interact with what they eat. Such playful interaction reinforces positive food behaviours while aligning with parental desires for healthier eating.

Global Flavours, Local Roots

Much like Gen Z, Generation Alpha has a palate shaped by global influences. Sushi rolls, dumplings and tacos are not novelties, they are staples. While Gen Z helped mainstream these foods, Generation Alpha is expected to expand their global reach even further (Moors, 2024). With today's hyperconnectivity, trends spread instantly across continents. A viral food sensation from one part of the world might appear in Sydney school lunchboxes or on dinner tables elsewhere in hours or within days. Generation Alpha's exposure to international cuisines is not just passive, as they actively engage with new flavours, formats, ingredients and cultural fusions, shaping an even broader culinary landscape.

Despite this global enthusiasm, Generation Alpha does not abandon local food culture, they blend both. As authenticity and local sourcing of food are non-negotiable for Gen Z, they are becoming also non-negotiable for Generation Alpha, reflecting the influence of their parents' desire for both global flavours and local integrity. As one 12-year-old remarked:

> *"I like ramen, but I want to know where the eggs and veggies come from. My mum says it's better when it's from around here and it's organic. I think this is a good thing."* (Female, born 2013)

This perspective reflects the rise of a hybrid food identity, influenced by parents globally inspired, yet locally grounded. Generation Alpha builds on Gen Z's global palate, embracing internationally influenced foods like sushi rolls, dumplings and tacos with an added emphasis on participation and creativity. Do-it-yourself (DIY)-style meals, such as build-your-own tacos and pizza kits, appeal to Generation Alpha's love for creativity, autonomy and engagement, making food a hands-on experience rather than just a meal.

These trends align with the growing role of interactive dining, blending education, taste exploration and play to create memorable food experiences. This early exposure helps Generation Alpha form strong connections between food, health and cultural understanding, setting the stage for a new era of culinary innovation.

3D Printing: Where Art Meets Appetite

Another new opportunity is 3D food printing which is not just a spectacle, it is a culinary shift, blending precision, sustainability and personalisation. Food scientists are crafting intricate edible creations, layering flavours and nutrients to create everything from balanced meals to visually stunning desserts (Zoopadm, 2023). For Generation Alpha's parents, this technology presents a compelling mix of health, fun and control, offering a glimpse into a future where meals are interactive and adaptable.

Beyond aesthetics, 3D printing tackles real food system challenges, such as reducing waste by transforming surplus into gourmet creations, while making alternative proteins like algae and insects more accessible in child-friendly formats. A Millennial mother of three Generation Alpha children captured this excitement firsthand:

> *"My son is obsessed with printing dinosaur-shaped snacks. He tried them during the holiday activities at the Powerhouse Museum in Sydney, and now he just cares about the fact that he can make them himself. They were made from veggies and cricket flour, but he doesn't even ask what's in them. He just calls them 'power bites' and thinks the 3D printer is the coolest part."* (Female, born 1987)

This innovation is not just limited to home kitchens, other different settings such as school canteens and healthcare facilities could soon explore 3D food printers for tailored, nutritionally optimised meals, turning food into both medicine and art. These advancements open possibilities for addressing dietary deficiencies, allergies and even picky eating through customisation and creative engagement.

In the world of food technology, 3D-printed meals are pushing the boundaries and redefining what is possible. Chefs and food scientists now design intricate edible structures, layer by layer, merging art, technology and nutrition into one seamless experience. For Gen Z and Generation Alpha families, this is not just about innovation — it is about making food interactive, customisable and fun.

Importantly, 3D food printing offers a range of benefits that go beyond novelty, making it a valuable tool in nutrition, creativity and sustainability. One of its most promising advantages is customisation, allowing meals to be tailored to specific textures, nutrients and dietary needs, ensuring optimal health and personalisation. It also fosters engagement, particularly for children, by enabling them to design their own food shapes, mixtures, flavours and colours, making the eating experience both interactive and enjoyable. Additionally, sustainability plays a key role, as this technology incorporates alternative proteins such as insects and algae while also utilising surplus ingredients to reduce food waste, addressing critical global challenges in resource efficiency. According to Zoopadm (2023), customisable meals created with 3D food printers encourage interactive food experiences while promoting balanced nutrition, demonstrating how this innovation has the potential to reshape the future of food.

Beyond simply addressing sustainability, 3D-printed food could empower families to take greater control over dietary choices, making it especially impactful for healthcare settings or young, selective eaters (Zoopadm, 2023). For Generation Alpha, 3D printing of food is likely to become normalised in the same way as its use for creating prototypes and architectural models or printing automobile parts and medical implants (Murray, 2023).

The Augmented Reality Revolution

The transformation of augmented reality is set to revolutionise dining, offering a digital twist that will impact all generations. However, Generation Alpha's expectations are anticipated to be uniquely high and multifaceted. This generation want tech-enhanced meals that entertain, food that tells stories, and menus that align with both their health and sustainability values. As augmented reality advances, it promises to turn everyday meals into multi-sensory, theatrical experiences. Imagine interactive menus where children build their meals like puzzles, tabletop games that keep families engaged while they wait for the meal, and immersive augmented reality environments that bring themes to life (Ferrandez, 2021). While Gen Z will adapt to these innovations, it is Generation Alpha who will truly embrace and expect them as an essential part of their dining experience.

Restaurants are already pivoting to meet the expectations of a tech-native generation. Augmented reality-enhanced menus now allow diners to preview meals in 3D, while virtual wine lists and live holographic cooking tutorials are turning dining into education and entertainment (Ferrandez, 2021; Mattan, 2023). These experiences particularly resonate with Generation Alpha, who are visual and tactile learners. As one Gen Z sibling observed:

> *"My little brother doesn't read menus... he taps on the holograms on the screen to pick his food. It's like gaming, but you eat it."* (Male, born 2006)

Virtual chef demonstrations offer another layer that could be bringing storytelling and transparency to the dining table. Watching how a dish is prepared in real time, complete with cultural or nutritional backstories, can deepen Generational Alpha's understanding and appreciation for food.

Moreover, augmented reality is transforming food literacy, making it more accessible, engaging and visually immersive for Generation Alpha, regardless of their background. By integrating educational elements directly into the dining experience, Generational Alpha can interact with visually rich culinary knowledge in ways that feel fun, intuitive and are non-intimidating. For brands and policymakers, the challenge lies in ensuring innovation aligns with integrity. Generational Alpha will not just demand transparency in ingredients, they will expect clear sourcing, ethical supply chains, trustworthy algorithms and company values that reflect their own. For both Gen Z and Generation Alpha, a product is not valuable simply because it is trendy. It must also carry meaning. Gen Z, in particular, prioritises transparency and reliability, a trustworthy option that fits seamlessly into their routines. Young people are seeking foods that balance back-to-basics nutrition, convenience and modern efficiency, while still embracing nostalgic, traditional foods that offer comfort and familiarity (Moors, 2024).

Storytelling on the Plate

Dining spaces are no longer just physical places to eat — they are transforming into platforms for immersive storytelling. Using augmented reality, meals are now paired with digital narratives, ethical sourcing details, ingredient origins and even historical or cultural anecdotes (Ferrandez, 2021; Mattan, 2023). This shift elevates food from a mere product to an emotional journey. As we embrace this future-forward approach, it is clear that today's dining experiences, once considered science fiction, are becoming reality. Just as Gen Z has begun reshaping food expectations, Generation Alpha will further redefine what dining means.

Similarly to Gen Z, Generation Alpha is growing with heightened awareness of the environment. Earth-conscious dining is not an afterthought. It is part of the experience. Restaurants and food brands are using storytelling and technology to teach sustainability through taste. Earth-conscious treats, crafted from seasonal and locally sourced ingredients, introduce young eaters to the importance of regional food systems, helping them build a connection with where their food comes from (Rojsatapong, 2025). Complementing this, AR-driven ingredient origin stories bring food journeys to life, offering immersive explanations of how ingredients travel from farm to fork. Together, these elements not only enhance knowledge and awareness but also create emotional investment in sustainable eating habits, a key factor in shaping long-term consumer behaviour and fostering a deeper appreciation for mindful food choices.

Hyper-Personalisation beyond Vanilla and Chocolate

For Gen Z and Generation Alpha personalisation in food extends far beyond simple flavour preferences. While previous generations viewed choice as selecting between basic options like vanilla or chocolate, today's emerging generation of consumers expect meals tailored to their individual biology and emotional state of mind. For Generation

Alpha, this could mean adaptive nutrition powered by AI, with meals customised according to their microbiome, metabolic profile or even mood (Moors, 2024).

In fact, AI-driven wearables and mobile apps are already advancing these possibilities further and it may not be long before snacks "know" whether you have met your individual protein needs for the day. However, this level of personalisation must also resonate emotionally and ethically, not just scientifically (Marinova and Bogueva, 2022). True and meaningful personalisation demands emotional connection and ethical transparency. Brands must embrace authenticity, purpose and hyper-personalisation to engage deeply with consumers while maintaining ethical transparency and integrity (Kiernan, 2025). Even highly technical foods, such as 3D-printed or lab-grown meals, must taste authentic, feel satisfying and align with the values of sustainability and wellness. This emphasis on ethical sourcing and emotional connection is apparent in how younger consumers evaluate food innovations. As one Generation Alpha representative put it:

> *"I tried 3D-printed nuggets at the Powerhouse Museum once — they were pretty cool! But honestly, I'd be even more into them if I knew they used pea protein from farms near Sydney or other ingredients I could trace back to where they came from. There's something about knowing where your food comes from that just makes it feel more real."* (Male, born 2011)

The above quote reveals how local sourcing and emotional relevance significantly shape Generation Alpha's acceptance of innovations. While Generation Alpha embraces food technology innovation, they are not willing to sacrifice the essence of food. As one Gen Z consumer observed:

> *"Kids aren't robots... they still want food that tastes like home, even if it was printed. I can't imagine they will be ignoring the real essence of food being healthy and nutritious."* (Female, born 2003)

Moving forward, culinary innovation must strike a balance. It must blend novelty with nutrition, fun with function and technology with transparency (see Figure 16.1).

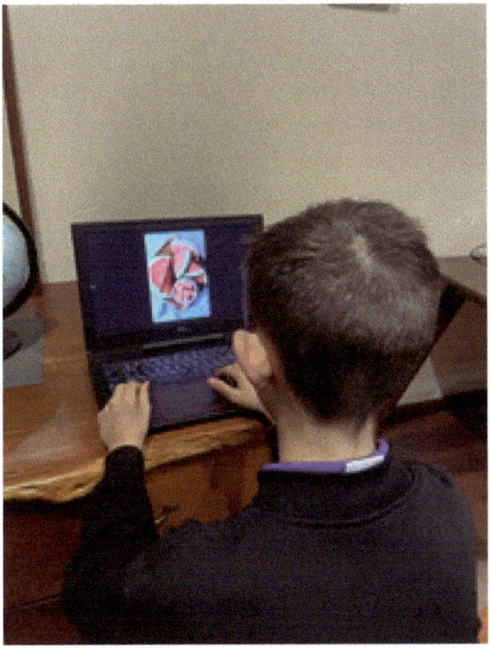

Figure 16.1. Looking for his favourite food.

Gen Z's Perspective on Their Younger Siblings

Now stepping into adulthood, Gen Z is observing Generation Alpha with a mix of admiration and concern. Their reflections offer valuable insight into generational shifts in food culture:

> *"They're so good with tech, it's scary. But I also worry they'll miss out on the slow, messy parts of cooking that we grew up with."* (Female, born 2002)

This reflection captures Gen Z's admiration for Generation Alpha's tech fluency, paired with concern about what might be lost. The "slow, messy parts of cooking" represent hands-on, emotional and communal experiences that Gen Z deeply values. While they respect Generation Alpha's effortless digital adaptability, they worry that automation and convenience could replace meaningful food traditions. This sentiment also reflects a nostalgia for the tangible, imperfect rituals of food preparation, underscoring a broader concern — a fear that as food becomes more digitised through smart kitchens, delivery boots or AI-curated meals, Generation Alpha may lose the joy of hands-on cooking and shared meals. This concern is echoed in other reflections from Gen Z. While they recognise Generation Alpha's ability to shape food culture, they also worry about over-reliance on technology in everyday food decisions:

> *"I think they'll be way more empowered than we were. Like, they won't just follow food trends... they'll help build them. But they also need to learn how to disconnect sometimes. We learned this the hard way, by dealing with mental health struggles, anxiety and distancing ourselves from social media and tech advances"* (Male, born 2001)

The above opinion highlights the generational shift where Generation Alpha is not just consuming trends but actively creating them. They are immersed in a digital food ecosystem, where AI, augmented reality and food technology influence what they eat but also how they engage with food. However, this rapid evolution also prompts concerns about balance, the need to step away from digital convenience and embrace real-world, hands-on food experiences:

> *"They're growing up with lab-grown meat and AI-based menus, which is probably cool. But I hope they also learn how to make a sandwich without asking Siri."* (Female, born 2005)

Such a comment further reinforces the fear of lost practical skills. Generation Alpha's food future is high-tech and hyper-personalised, but will they retain the ability to engage with food in a simple, instinctive way? As food becomes increasingly automated, there is a risk that basic food literacy, traditional preparation skills and cultural cooking rituals may fade. Ultimately, these perspectives from Gen Z reveal a nuanced generational transition, one driven by technological advancement and customisation, but shadowed by concerns about detachment from foundational food experiences.

Producing the Food We Need

In the future, Generation Alpha will also have to find employment in the food production sector. It is highly unlikely that they will look like the traditional pictures of farmers we know. Drones and precision agriculture techniques based on large arrays of data for observing, measuring, analysing and responding to the needs of individual fields, crops and farm animals will be the norm (Monteiro *et al.*, 2021). Many of them will be wearing lab coats engaged in making food in industrialised environments disconnected from the land and the soil. Hopefully, the problem of loss and waste which currently account for more than 30% of all food produced globally −13.2% loss between harvest and retail and another 19% waste in retail, households and catering (United Nations, 2024), will be significantly reduced and even eliminated. In Australia, the amount of land used to grow food that is ultimately wasted is 25 million hectares or larger than the size of the state of Victoria (Department of Climate Change, Energy, the Environment and Water, 2025). The new technologies used for food production will need to also reduce any associated greenhouse gas emissions, water use, prevent pollution, stop deforestation and natural habitat destruction. Food packaging will similarly have to change, particularly avoiding materials, such as PFAS, which have been proven to be toxic for human health and persistent in the natural environment (Fenton *et al.*, 2021).

Food security is another aspect that Generation Alpha will have to embrace in a different way to Gen Z and other previous generations, it relates to the provision not only of any food or calories but options that are nutritionally healthy and sustainable. This will include many traditional plant-based foods as well as fungi and seaweed. Dietary recommendations across the world are being regularly updated to reflect the challenges we face in ensuring nutritional security, but also to avert dangerous climate change and restore the health of the natural world (Halevy and Trewern, 2023). Examples of such dietary approaches are the Mediterranean diet (Willett *et al.*, 1995; Guasch-Ferré and Willett, 2021), planetary diet (Willett *et al.*, 2019) and Livewell diet (Halevy and Trewern, 2023). In response to the COVID-19 pandemic, the war in Ukraine and other geopolitical tensions, a 2025 study (Stehl *et al.*, 2025) analysed the current food self-sufficiency. It shows that out of 186 countries around the world, only Guyana in South America is self-sufficient (Stehl *et al.*, 2025) in producing all categories of food — fruits, vegetables, legumes, nuts and seeds, starchy staples, fish, dairy and meat, the country's population needs according to the net-zero nature-positive Livewell diet (Halevy and Trewern, 2023). If the current polycrisis persists, self-sufficiency and resilience will become a major consideration for Generation Alpha. Equipped with the latest technologies and knowledge, Generation Alpha will have to find the balance between what is good for their own health and the well-being of the planet; luckily there are many co-benefits that they will have to discover and embrace.

Conclusion: Passing the Plate to a New Food Era

John Wesley's words: "What one generation tolerates, the next generation will embrace", perfectly capture the evolving dynamics between Gen Z and Generation Alpha. Just as Gen Z rejected certain Millennial traits, Generation Alpha is shaping up as rebels in the making, pushing back against Gen Z's conservatism (Eastlake, 2025). The generational progression highlights how values shift and expand over time, redefining not just preferences but fundamental approaches to food, culture and technology.

Moving forward, the challenge lies in ensuring that Generation Alpha's tech-driven food landscape does not erase, but rather enhances the essence of shared meals, hands-on cooking and cultural traditions, things that Gen Z remains deeply passionate about. The handoff from Gen Z to Generation Alpha is not just about continuing trends, it is a transformation in how food integrates into the broader socio-digital framework. Tastes are being transferred, but so are values, tools and expectations.

Generation Alpha will build upon Gen Z's values, but with exponentially more advanced tools, expectations and frameworks. Understanding this generational shift is essential for anyone shaping the future of food, from producers to policymakers, technologists to educators and parents. Generation Alpha is not merely the next wave of consumers; like Gen Z before them, they are architects of a radically new food era, one that will be the subject of future studies.

The future of food is no longer a distant vision. It is unfolding now, driven by Generation Alpha's expectations and the groundwork laid by Gen Z. For those willing to listen, the message is clear — blend technology with tradition, design with purpose and above all, serve with authenticity to do good.

References

Anderson, C. (2024). What is foundational memory? Focus Keeper. https://focuskeeper.co/glossary/what-is-foundational-memory

Bogueva, D. & Marinova, D. (2020). Cultured meat and Australia's Generation Z. *Frontiers in Nutrition*, 7, 148. https://doi.org/10.3389/fnut.2020.00148

Connell, L. (n.d.). Generation Alpha: The new kids of America. What tomorrow's consumers are really thinking. https://www.gwi.com/reports/gen-alpha-us#

Department of Climate Change, Energy, the Environment and Water. (2025). Reducing Australia's food waste. Australian Government. https://www.dcceew.gov.au/environment/protection/waste/food-waste

Eastlake, D. (2025). What's Generation Alpha eating? Food Navigator. https://www.foodnavigator.com/Article/2025/05/14/what-is-generation-alpha-gen-a-eating/

Feger, A. (2023). 3 ways Gen Alpha is different from Gen Z — and what it means for marketers. EMarketer. https://www.emarketer.com/content/3-ways-gen-alpha-different-gen-z-and-what-means-marketers

Fenton, S. E., Ducatman, A, Boobis, A., DeWitt. J. C., Lau, C., Ng, C., Smith, J. S., & Roberts, S. M. (2021). Per- and polyfluoroalkyl substance toxicity and human health review: Current state of knowledge and strategies for informing future research. *Environmental Toxicology and Chemistry*, 40(3), 606–630. https://doi.org/10.1002/etc.4890

Ferrandez, C. (2021). Augmented reality restaurant experiences: 5 examples. Poplar Studio. https://poplar.studio/blog/augmented-reality-restaurant-experiences-5-examples/

Guasch-Ferré, M. & Willett, W. C. (2021). The Mediterranean diet and health: A comprehensive overview. *Journal of Internal Medicine*, 290(3), 549–566. https://doi.org/10.1111/joim.13333

Haidt, J. (2024). *The Anxious Generation: How the Great Rewiring of Childhood is Causing an Epidemic of Mental Illness.* Alen Lane, London.

Halevy, S. & Trewern, J. (2023). Eating for net zero: How diet shift can enable a nature positive net-zero transition in the UK. WWF-UK. https://www.wwf.org.uk/sites/default/files/2023-05/Eating_For_Net_Zero_Full_Report.pdf

Howarth, J. (2025). Generation Alpha: Statistics, data and trends (2025). https://explodingtopics.com/blog/generation-alpha-stats

Hunter, T. (2022). Our post-pandemic food future. Food and Drink Business. https://www.foodanddrinkbusiness.com.au/news/our-post-pandemic-food-future

Iberdrola. (2025). Generation Alpha: Generation Alpha will lead a 100% digital world. https://www.iberdrola.com/talent/alpha-generation

Kiernan, P. (2025). Branding in 2025: Authenticity, purpose, and personalization. Though Lab. https://www.thoughtlab.com/blog/branding-in-2025-authenticity-purpose-and-personal/

King, J. (2024). Guide to Gen Alpha, the generation marketers need to start attracting. Emarketer. https://www.emarketer.com/learningcenter/guides/guide-generation-alpha/

Kumar, N. (2024). Generation Alpha Statistics 2025 (Population & Literacy Data). Demand Sage. https://www.demandsage.com/generation-alpha-stats/

Marinova, D. & Bogueva, D. (2022). *Food in a Planetary Emergency.* Springer, Singapore.

Mattan, M. (2023). Augmented Reality restaurant experiences Examples. Brand XR. https://www.brandxr.io/eat-with-your-eyes-augmented-reality-restaurant-examples-revolutionizing-the-dining-experience

McCrindle. (2025). Understanding Generation Alpha. https://mccrindle.com.au/article/topic/generation-alpha/generation-alpha-defined

Monteiro, A., Santos, S., & Gonçalves, P. (2021). Precision agriculture for crop and livestock farming-brief review. *Animals*, 11(8), 2345. https://doi.org/10.3390/ani11082345

Moors, R. (2024). How Gen Z and Gen Alpha are shaping the future of food. FI Global Insights. https://insights.figlobal.com/foodtech/how-gen-z-and-gen-alpha-are-shaping-the-future-of-food

Murray, C. (2023). Common applications of 3D printing: From aerospace to fashion. Forbes. https://www.forbes.com/sites/conormurray/2023/04/21/common-applications-of-3d-printing-from-aerospace-to-fashion/

Razorfish. (2023). Exploring Generation Alpha. A look into the future. Razorfish Study Unveils Generation Alpha Insights, Preparing Brands for the Next Consumer Cohort. https://www.razorfish.com/articles/news/razorfish-study-unveils-generation-alpha-insights/

Rojsatapong, C. L P. (2025). Food trends for kids in 2025: A new era of young foodies. https://www.khaosodenglish.com/life/food/2025/01/11/food-trends-for-kids-in-2025-a-new-era-of-young-foodies/

Stehl, J., Vonderschmidt, A., Vollmer, S., Alexander, P., & Jaaks, L. M. (2025). Gap between national food production and food-based dietary guidance highlights lack of national self-sufficiency. *Nature Food.* https://doi.org/10.1038/s43016-025-01173-4

Trifonova, V. & Moran, S. (n.d.). Generation Alpha: The real picture. GWI. https://www.gwi.com/reports/gen-alpha#get-pdf

United Nations. (2024). Stop food loss and waste. For the people. For the Planet. https://www.un.org/en/observances/end-food-waste-day

Willett, W. C., Sacks, F., Trichopoulou, A., Drescher, G., Ferro-Luzzi, A., Helsing, E., & Trichopoulos, D. (1995). Mediterranean diet pyramid: A cultural model for healthy eating. *The American Journal of Clinical Nutrition*, 61(6 Suppl), 1402S–1406S. https://doi.org/10.1093/ajcn/61.6.1402S

Willett, W., Rockstrõm, J., Loken, B., Springmann, M., Lang, T., Vermeulen, S., Garnett, T.... Murray, C. J. L. (2019). Food in the Anthropocene: The EAT–Lancet Commission on healthy diets from sustainable food systems. *The Lancet*, 393(10170): 447–492. https://doi.org/10.1016/S0140-6736(18)31788-4

Witherington, A. (2025). Meet Gen Alpha: The next generation of food consumers. Mintel. https://www.mintel.com/insights/food-and-drink/meet-gen-alpha-next-generation-food-consumers

Zoopadm. (2023). 7 futuristic food technology inventions transforming how the world dines. https://www.zoopindia.com/blog/7-futuristic-food-technology-inventions-now

Chapter 17

Conclusion: The Legacy of Gen Z Shaping the Future

"I am not young enough to know everything."
— Oscar Wilde

Confidence has always been part of young people's nature and is justified simply by the fact that the future belongs to them. However, Generation Z brings something more to the table — not just confidence, but consciousness. In this book, we tried to link the future of food with the confidence of the current young people, namely Gen Z. This generation is different from the previous as it also has easy access to knowledge and holds values that span beyond food as a nourishment or tradition. Empowered by unprecedented access to knowledge and digital tools, Gen Z's approach to food is layered, value-driven and deeply intentional. For them food is far more than sustenance or tradition. It is as much an identity as it is activism and a way to change and build a world that is more sustainable, fairer, transparent and resilient.

Although Oscar Wilde's words are to a certain extent sarcastic, they speak to a timeless truth that youth dare to believe in solutions, even amid uncertainty. Young people have the confidence and self-assurance that they can find answers to all problems life throws at them. Many of these problems are the result from the actions of previous generations — climate change, biodiversity loss, the rise of non-communicable diseases, pollution with pervasive microplastics and forever chemicals, decreased soil fertility, oceanic collapse and coral bleaching, loss of native habitats, food inequality, wars . . . and the list goes on. They did not create these crises, but they have inherited the task of navigating through them and cleaning them up. Will Gen Z be able to find all the solutions needed for a world in a polycrisis?

In this book we only focussed on food as a lens for understanding young people's attitudes and behaviours but Gen Z's presence and influence extend far beyond the dinner table and is already apparent in all fields of life. We tried to explore how Gen Z responds with agency, creativity and a healthy dose of scepticism to the challenges and burdens passed down to them. Gen Z is not simply reacting to the world, it is reshaping it. Guided by values of sustainability, inclusivity and well-being, they are leaving behind old and outdated perceptions and ways of doing things. What distinguishes Gen Z is not just its concern, but its commitment. It questions legacy systems and actively reimagines them.

In the realm of food, this generation refuses to accept the status quo. It is not content with simply inheriting a broken food system — it wants to rebuild it, from farm to fork, in ways that reflect planetary health and human well-being and dignity. At the same time, Gen Z draws meaning from the past, reviving forgotten practices, reinterpreting traditions and restoring cultural continuity. Importantly, Gen Z does not just preserve the past, it reinvents it, making it meaningful and relevant for those who care deeply about other people and the planet.

Food in this sense, is a perfect example and a powerful symbol of synthesis between the old and the new, the analogue and the digital, the inherited and the invented. Across the various chapters of this book we examined this synthesis through multiple lenses and aspects of food — from a transition to plant-rich diets and decreased alcohol consumption to the frontier of food innovations that aim to change the way food is produced, including cultured meat and seafood, precision fermentation, plant-based milks, the use of AI-driven food technologies, alternative proteins sourced from fungi, algae and edible insects, to the way nourishment is delivered to individuals through the evolution of personalised nutrition. We have also explored how Gen Z engages with packaging, transparency, labelling and food safety standards expecting sustainability not as an exception, but as the norm. Gen Z brings a new perspective to all of these aspects.

Despite its enthusiasm for innovation, Gen Z does not offer blind approval. Without being drastically different, young people are optimistic, but hesitant to give the tick of approval and cautiously sceptical towards the new technological solutions. Their optimism is also cautious and their support is conditional. They demand accountability and want to know: Who made this food? How was it produced? What impact does it have? Their values centre on transparency, trust and traceability, a moral compass for future food systems. They want to see sustainability mainstreamed in the way food systems operate. It is difficult to judge negatively their attitude as they are the ones who will have to handle the mistakes and consequences of the past. In addition to being consumers, many of them will also contribute to shaping the global and local food systems. They will be farmers, scientists, policymakers, designers, inventors, researchers, nutritionists, entrepreneurs and educators. It would be a mistake to dismiss their hesitations. After all, they are the ones who will live the longest with the consequences of today's decisions. Moreover, they will sit at every node of the food system and reshape it with the insights and ethics they bring with their generation.

Now that we are towards the end of this book, we acknowledge that it only partially explored the food systems as well as it is good to reflect about what we canvassed and what we did not cover. Our focus was on the future in terms of the emerging large group of consumers — Gen Z, and the new foods. Many of these new foods are manifestation of different approaches to the existing food systems which have failed to deliver sustainable and equitable solutions (Marinova and Bogueva, 2022). They are the product of human creativity and engineering approach to resolving food-related challenges but are also surrounded with many unknowns, the need for further research and understanding the constraints experienced by Gen Z and other consumers. Let us briefly delve in what we did not cover and essentially took for granted — the role of established plant-based foods that have long offered sustainable choices, as well as the issue about global mental health crises and emotional well-being and their impact on Gen Z's food preferences, and the growing prevalence of non-communicable diseases which similarly influence attitudes towards food. These aspects will increasingly shape how food is perceived, consumed and produced.

Rediscovering the Power of Plants

As we close this exploration into Gen Z and the future of food, it is fitting to turn our attention to something more ancient, more elemental — the enduring power of traditional plant-based foods. While much of this book explored food innovations, technological advances and the evolving preferences of young consumers, we must also recognise the quiet revolution taking place in the everyday choices many Gen Z make and deserves a lot of attention. It is the resurgence of basic whole-plant foods that have nourished humanity for millennia. Vegetables, fruits, grains, nuts, legumes, seeds and roots — these staples form the foundation of sustainable, health-promoting diets. They all have significantly lower environmental footprints (see Table 17.1 for greenhouse gas emissions associated with selected foods) compared to animal-derived food products — beef, lamb, pork, poultry, cheese, eggs, fish and shellfish (e.g. molluscs or crustaceans), and are beneficial for human health (Marinova and Bogueva, 2022). Unlike their animal-based counterparts, plant foods require less water, land and energy, and produce fewer greenhouse gas emissions. Also, beyond the environmental metrics, they offer remarkable nutritional value. They are rich in essential vitamins, minerals, fibre, phytonutrients; many of these foods also provide complex carbohydrates, healthy fats and crucially protein. They are valuable contributors to a balanced human diet.

In fact, legumes offer a twofold benefit: they nourish the body and regenerate the soil. Legumes, including soybeans, have the benefits of enriching the soil content with nitrogen allowing healthy growth for other plants, particularly when planted on a rotation basis.

Table 17.1. Greenhouse gas emissions (in CO_2-equivalent) associated with selected foods.

Food item	kg of CO_2e per kg of food	kg of CO_2e per 100 g of protein	kg of CO_2e per 1000 of kilocalories
Beef	70.6	35.5	25.9
Breads and pastas	1.6	1.3	0.6
Cheese	23.9	10.8	6.2
Eggs	4.7	4.2	3.2
Fish	13.6	6.0	7.6
Fruit	0.9	10.4	1.5
Lamb	39.7	19.9	12.5
Legumes	2.0	0.9	0.5
Milk	3.2	9.5	5.2
Nuts	0.4	0.3	0.1
Pork	12.3	7.6	5.2
Poultry	9.9	5.7	5.3
Rice and grains	3.6	4.8	0.9
Shellfish	26.9	18.2	26.1
Tofu	3.2	2.0	1.2
Vegetables	0.7	6.8	3.3

Source: United Nations (n.d.).

Their natural ability to fix atmospheric nitrogen reduces the need for synthetic fertilisers and enhances the fertility of agricultural ecosystems. This makes them not only a nutritional powerhouse but also a regenerative ally for sustainable agriculture, something Gen Z increasingly values.

Proteins which represent chains of essential, nonessential and conditional amino acids, have for long been a particular concern in dietary intake as the human body cannot store them and needs them on a regular basis (Cleveland Clinic, 2021). The eleven nonessential amino acids can be made by the body but the nine essential amino acids can only be supplied by food. Some nonessential amino acids are classified as conditional because they are needed only in times of stress or illness (MedlinePlus, 2023). Proteins are essential for human survival but they can equally be supplied by animal and plant sources (MedlinePlus, 2023).

Indeed, one of the enduring myths still influencing global food systems is the assumption that quality proteins must come from animal sources. This belief has shaped cultural narratives and marketing strategies for decades. However, science tells a different story. Foods containing all essential proteins are called complete proteins. In addition to animal-sourced foods, the plant-based legumes, including lentils, beans, chickpeas and soybeans, along with grains such as quinoa and buckwheat are not only protein-rich but also contain essential amino acids. Soy, quinoa and buckwheat in particular are complete proteins in their own right. Various combinations of plant foods, e.g. legumes with rice, also achieve the same effect and result in the intake of complete proteins, enabling diverse populations to meet protein needs without resorting to high-impact animal-based foods. Hence, it is wrong to equate proteins only with animal-sourced foods.

Gen Z is not naïve about health trends. Many of them, particularly athletes, gym enthusiasts and wellness-focused youth, understand the fine line between protein needs and protein excess. Animal-based proteins are easier to absorb; however, the excessive consumption of proteins — consuming "too much of the good thing", has proven over time to have poor health impacts and negative health outcomes (Tharrey *et al.*, 2018). Relying on plant-based foods is a much smarter way to help the body make the proteins it needs.

Young people, including Gen Z athletes, gym goers and those who try to build muscles, are fully aware of the risks from having too much proteins. As one of them explains, his diet consists of: "lots and lots of good food . . . so you can just constantly have protein synthesising in your body" (LaVine, 2025, para. 9, 14). The science is clear and with a well-established view that diets with reduced meat consumption lead to longer and healthier lifespan and lower risks of non-communicable diseases, including heart disease, diabetes and certain cancers (Singh *et al.*, 2003). It is no wonder that Gen Z, already confronting a world where mental health, body image and lifestyle diseases intersect, is leaning into food choices that support both physical and emotional well-being.

Food as Medicine and Gen Z's Reclaiming of Preventive Nutrition

Following our discussion on the benefits of traditional plant-based foods, it is important to consider how Gen Z is actively reframing the purpose of food, not only as sustenance but as medicine. While there is no evidence that Hippocrates ever said the famous words he is being credited for: "let thy food be thy medicine and thy medicine be thy food" (Witkamp and van Norren, 2018), this sentiment resonates strongly with Gen Z and underpins many of its dietary philosophies and attitudes towards food. A possible reason for this is because this generation, shaped by uncertainty, is experiencing mental health problems from a young age, including eco-anxiety (Bardales Salguero *et al.*, 2024), and finds empowerment in the idea that food can foster resilience and healing.

With Gen Z we are witnessing a generational shift towards nutritional literacy — an increased awareness about the link between lifestyle and wellness and a resurrection of disease prevention and management through dietary choices and better nutrition, coupled with a desire for control over one's well-being in a world often perceived as chaotic. In the United States, young people report being more health-conscious than the previous generations, viewing healthy eating as a priority and a central lifestyle value rather than a trend (Texas Health, 2023). An example of this is the consumption of functional foods, a term which describes food products that deliver some health benefits. Across the world Gen Z is turning to functional foods as part of its healthy eating strategy. This is not confined to Western contexts. One in three young consumers in South and Southeast Asia is embracing functional foods, designed not just to nourish, but to enhance mood, immunity and energy (Innova Market Insights, 2024).

Gen Z's changing attitudes towards healthy eating are also manifested towards resenting smoking, drug and alcohol use and are combined with the increasing importance of physical fitness (Texas Health, 2023). This is further linked with sustainability and social equity values and facilitated by the use of technology and social media. A global study across 27 countries shows that young people are more likely than older generations to feel ashamed about lifestyles which are unhealthy and not environmentally friendly (Texas Health, 2023). These sentiments are particularly pronounced in their food choices.

Technology and social media have accelerated these patterns. Wellness influencers, recipe-sharing platforms, AI-guided nutrition apps and viral health challenges, all contribute to a digital ecosystem where food is a tool for both self-care and social expression. Beyond technology, this food-as-medicine movement is deeply personal. It is about healing, individually and collectively.

A Generation of Contrasts: Health Trends and Public Health Warnings

Gen Z is not a monolith. Not all young people are equally tuned up to healthy eating. While many lead the charge toward plant-based eating, mindful consumption and reduced meat intake, others are grappling with the early onset of chronic diseases. In many western countries, including the USA, UK and Australia, a rising proportion of young people are experiencing the onset of non-communicable diseases, such as diabetes type 2, chronic respiratory disease and cardiovascular disease, at much younger age than ever before. Alarmingly, the greatest burden of disease among people aged 10–24 years stems from non-communicable diseases (Wang *et al.*, 2023). In a study of Australian adolescents, 45.6% reported at least one non-communicable disease or developmental condition, such as attention-deficit/hyperactivity disorder (ADHD), autism or irritable bowel syndrome, often associated with higher levels of consumption of sugar-sweetened beverages, ultra-processed foods or alcohol, alongside sedentary screen-heavy lifestyles (Osman *et al.*, 2025). These trends compound the mental health burden of Gen Z and highlight stark inequalities in access to nutritional education and affordable healthy food. This also leads to depression, anxiety and psychological distress emphasising the need for proactive public health policies endorsing the importance of food as preventative medicine.

The fight against non-communicable diseases which are prevalent in older people starts with children and adolescents, and food is a major aspect of this. Gen Z has the advantage of easy access to information through social media which can be a tool to make changes towards better and healthier food choices. Despite the contrast, this generational moment offers a rare opportunity: to harness Gen Z's digital savviness and values-driven mindset for impactful public health interventions. Policy must align with Gen Z's lived reality — supporting food environments that promote well-being, empower self-management and destigmatise nutrition as a form of care. The movement toward treating food as a form of preventative medicine must be nurtured, legislated and democratised.

Looking ahead: From Gen Z to Generation Alpha and beyond

If Gen Z is rewriting the food narrative with ethics, innovation and balance, then the generations that follow — Alpha, Beta and Gamma, are poised to take that foundation and build something radically new. We discussed this important perspective in one of the previous chapters of the book. In the trajectory of life on this planet, human generations come and go, and they all leave their mark.

Generation Alpha is already growing up in a food environment shaped by Gen Z. These children are immersed in school curriculums that include climate science, mental health literacy and digital nutrition tools. They are surrounded by plant-based lunchboxes, carbon-labelled menus, AI-powered kitchen gadgets (Miller, 2024; Witherington, 2025) and a broader social discourse about equity, justice and planetary well-being. Generation Alpha is growing up in a social environment influenced by Gen Z's values and attitudes toward food and sustainability. These children are not just passive recipients of change, they are active participants in a world where food is both a personal and planetary decision.

Generation Alpha will not simply mimic Gen Z. If Gen Z is cautious and values-driven, Generation Alpha may perhaps be more boldly experimental, seamlessly blending technology and tradition in how they cook, eat and engage with food. It may become more comfortable with lab-grown meat, edible packaging and biotech-enhanced nutrition, such as vitamin-fortified crops and AI-personalised meal plans (Mridul, 2025; Hefferon, 2022), signalling a shift from novelty to norm.

Looking further ahead, if Gen Z is defined by adaptation, then Generation Beta, born from 2025 onward, might define themselves through integration, viewing climate-positive eating and synthetic biology in food not as innovations, but as expectations and as a norm. Their food culture could be shaped by hyper-personalised nutrition, smart appliances and immersive food experiences that merge digital and physical worlds.

This generational flow from Z to Alpha to Beta suggests a continuum, not a rupture. Gen Z's legacy would not rest solely in their market impact or policy influence, but by the values they have passed on with ethical, sustainable and curious mindsets. These values serve as a generational toolkit, not a fixed rulebook, but a flexible and dynamic set of principles and technologies that empower future generations to face and navigate complexity with creativity and confidence.

The Legacy of Gen Z: A Foundation, Not a Finish Line

As time goes by, each generation becomes older and more mature. As of 2025, the oldest members of Gen Z are turning 30 and are now truly embedded in all fabrics of society. Increasingly, the entire cohort is becoming part of the global consumers of food on their own volition. They are shaping the food system not just as consumers but as creators and leaders. What will their legacy be? What will Gen Z leave to younger people and the world? Will this legacy be much more positive compared to their predecessors? How many and which aspects of the current environmental, social, geopolitical, demographic and food security polycrisis are they going to overcome?

Gen Z is leaving behind more than trends. It is embedding social values such as curiosity in exploring new ingredients and cultural cuisines, compassion in demanding food justice and ethical supply chains, courage in resisting greenwashing and making bold choices, even when inconvenient. Through this, a foundation is laid for a food system that is less extractive and ecosystems-draining and more generative.

Most importantly, we must recognise that Gen Z's legacy does not end with them. Their legacy is a platform, not a pinnacle. The generations that follow will inherit their challenges and their wisdom. Generation Alpha, which is now entering its formative years and eventually Beta and Gamma, will not start from scratch. They will grow up immersed in digital ecosystems, where augmented reality menus, AI-generated recipes, climate-positive foods, personalised microbiome-based nutrition and immersive food experiences may be commonplace. Their worldview will be scaffolded by Gen Z's example: critical yet hopeful, tech-savvy yet grounded in ethical responsibility. They will grow up believing that food should not just feed, but heal, unite and restore.

Gen Z's real gift may not be in solving every crisis, but in laying a foundation of awareness, urgency and care. If they can pass on their values — curiosity, compassion and courage, then future generations may not have to start from scratch, but from a place of shared vision and collective momentum. As with Gen Z, future generations will one day inherit the legacy of the decisions made today.

Conclusion: Leaving the Table with Hunger for Change

In the end, the story of food is the story of us — our values, our culture, our fears and our hopes. This book has sought to capture one pivotal chapter of that story through the lens of Gen Z. What we have found is a generation actively reshaping the food systems, not perfectly, but with intention. From embracing traditional plant-based staples to leveraging AI for smart nutrition, from treating food as medicine to driving a cultural shift toward sustainability, Gen Z has demonstrated that food choices are not just personal, they are political, cultural and planetary.

The conversations we have had with all Gen Z representatives make us believe that there is cautiously optimistic and promising future for the new technological advances and solutions in the way food is produced and delivered to the consumer. There is also great value in rediscovering what nature has long provided to us — real plants, fungi and algae, as well as in reframing and treating food as a major core aspect of human health, identity and planetary care.

Put simply, working on this book made us believe that Gen Z is young enough to know everything about food. It is also wise enough to ask what comes next.

References

Bardales Salguero, R., Bogueva, D., & Marinova, D. (2024). Australia's university Generation Z and its concerns about climate change, *Sustainable Earth Reviews*, 7, 8. https://doi.org/10.1186/s42055-024-00075-w

Cleveland Clinic. (2021). Amino Acids. https://my.clevelandclinic.org/health/articles/22243-amino-acids

Hefferon, K. L. (2022). Next-gen biotech crops for human nutrition. In: Ramesh, S. V., Praveen, S. (eds) *Conceptualizing Plant-Based Nutrition*. Springer, Singapore. https://doi.org/10.1007/978-981-19-4590-8_11

Innova Market Insights. (2024). Functional Beverages Trends in South & Southeast Asia. https://www.innovamarketinsights.com/trends/functional-beverages-trends-in-asia/

LaVine, C. (2025). The rise of the protein trend in the Gen Z gym culture. The Daily Universe. https://universe.byu.edu/metro/the-rise-of-the-protein-trend-in-the-gen-z-gym-culture

Marinova, D. & Bogueva, D. (2022). *Food in a Planetary Emergency*. Springer, Singapore.

MedlinePlus. (2023). Protein in diet. https://medlineplus.gov/ency/article/002467.htm

Miller, N. (2024). 7 Best AI-powered kitchen gadgets that will change how you cook forever. TechGadgets.ai. https://techgadgets.ai/ai-powered-kitchen-gadgets

Mridul, A. (2025). Gen Alpha embraces alternative proteins — but will kids actually eat them? Green Queen. https://www.greenqueen.com.hk/gen-alpha-food-trends-vegan-lab-grown-plant-based-meat/

Osman, B., Sunderland, M., Devine, E.K., Thornton, L., Jacka, F., & Teesson, M. (2025). Prevalence of noncommunicable diseases and developmental conditions in 5014 Australian adolescents, and their correlations with diet, other lifestyle behaviours and mental health. *Australian and New Zealand Journal of Public Health*, 49(2), 100225. https://doi.org/10.1016/j.anzjph.2025.100225

Singh, P. N., Sabaté, J., & Fraser, G. E. (2003). Does low meat consumption increase life expectancy in humans? *The American Journal of Clinical Nutrition*, 78(3), 526S–532S. https://doi.org/10.1093/ajcn/78.3.526S

Texas Health. (2023). Study Shows Younger Generations Are More Health-Conscious Than Previous Generations. https://www.texashealth.org/areyouawellbeing/Health-and-Well-Being/Study-Shows-Younger-Generations-Are-More-Health-Conscious-Than-Previous-Generations

Tharrey, M., Mariotti, F., Mashchak, A., Barbillon, P., Delattre, M., & Fraser, G. E. (2018). Patterns of plant and animal protein intake are strongly associated with cardiovascular mortality: The Adventist Health Study-2 cohort. *International Journal of Epidemiology*, 47(5), 1603–1612. https://doiorg/10.1093/ije/dyy030

United Nations. (n.d.). Food and Climate Change: Healthy diets for a healthier planet. https://www.un.org/en/climatechange/science/climate-issues/food

Wang, H., Song, Y., Ma, J., Ma, S., Shen, L., Huang, Y., ... Zou, Z. (2023). Burden of non-communicable diseases among adolescents and young adults aged 10–24 years in the South-East Asia and Western Pacific regions, 1990–2019: A systematic analysis for the Global Burden of Disease Study 2019. *The Lancet Child & Adolescent Health*, 7(9), 621–635. https://doi.org/10.1016/S2352-4642(23)00148-7

Witherington, A. (2025). Meet Gen Alpha: The next generation of food consumers. Mintel, https://www.mintel.com/insights/food-and-drink/meet-gen-alpha-next-generation-food-consumers/

Witkamp, R. F. & van Norren, K. (2018). Let thy food be thy medicine... when possible. *European Journal of Pharmacology*, 836, 102–114. https://doi.org/10.1016/j.ejphar.2018.06.026

Index

D

E

F

I

J

K

L

M

N

O

P

Q

R

S

www.ingramcontent.com/pod-product-compliance
Lightning Source LLC
LaVergne TN
LVHW051057210926
848486LV00002B/138